FUNDAMENTALS OF Astrodynamics

SECOND EDITION

ROGER R. BATE
DONALD D. MUELLER
JERRY E. WHITE
WILLIAM W. SAYLOR

DOVER PUBLICATIONS
GARDEN CITY, NEW YORK

Bibliographical Note

This Dover edition, first published in 2020, is a revised and updated second edition of the work originally published by Dover in 1971.

Library of Congress Cataloging-in-Publication Data

Names: Bate, Roger R., 1923- author. | Mueller, Donald D., 1933- author. | White, Jerry E., 1937- author. | Saylor, William W., author.
Title: Fundamentals of astrodynamics / Roger R. Bate, Donald D. Mueller, Jerry E. White; William W. Saylor.
Description: Second edition. | Garden City, New York : Dover Publications 2020. | Includes index. | Summary: "Developed at the U.S. Air Force Academy, this teaching text is widely known and used throughout the astrodynamics and aerospace engineering communities. Completely revised and updated, this second edition takes into account new developments of the past four decades, especially regarding information technology. Central emphasis is placed on the use of the universal variable formulation, although classical methods are also discussed. The development of the basic two-body and n-body equations of motion serves as a foundation for all that follows. Subsequent topics include orbit determination and the classical orbital elements, coordinate transformations, and differential correction. The Kepler and Gauss problems are treated in detail, and two-body mechanics are applied to the ballistic missile problem. Perturbations, integration schemes and error, and analytic formulations of several common perturbations are introduced. Example problems and exercises appear throughout the text, along with photographs, diagrams, and drawings. Four helpful appendixes conclude the book"— Provided by publisher.
Identifiers: LCCN 2019045205 | ISBN 9780486497044 (paperback) | ISBN 0486497046 (paperback)
Subjects: LCSH: Orbital mechanics. | Astrodynamics.
Classification: LCC TL1050 .B33 2020 | DDC 629.4/11—dc23
LC record available at https://lccn.loc.gov/2019045205

Manufactured in the United States of America
49704607 2023
www.doverpublications.com

This textbook is dedicated to the members of the United States Air Force who have died in combat, are missing in action, or are prisoners of war.

Preface

The first edition of Fundamentals of Astrodynamics was published in 1971 and has been a popular textbook and reference for many in the aerospace field. Of course, many of those same professionals have been responsible for an incredible number of successes in technologies, programs, missions, computational tools, and educational textbooks and courses.

Updating the book, lovingly referred to as "BMW" in honor of the original author's last names, has been approached with care to maintain the same tenor while making content changes to be consistent with industry practices that have evolved during the Space Age. This second edition is still viewed as a primary introduction to the study of astrodynamics. Numerous excellent textbooks have been published to extend the study of astrodynamics and many advanced topics in orbital mechanics are well covered in those books. The science and engineering of satellites and rockets, and all of their subsystems and components, have expanded significantly and there are also many textbooks that provide excellent coverage of those topics.

The modern use of units (km, kg, and s) has been included to reflect common industry practices. There have been some changes to the content of the book where a newer approach is more consistent with the use of computers rather than slide rules. The use of canonical units made sense in the original edition, but that has been minimized here and appears when their use helps with understanding some concepts. Two-line elements and the NORAD SGP4 orbit prediction methods are also introduced.

Chapter 1 develops the foundation for the rest of the book in the development of the basic two-body and n-body equations of motion. Chapter 2 treats orbit determination from various types of observations. It also introduces the classical orbital elements, coordinate transformations, the non-spherical earth, and differential correction. Chapter 3 then develops orbital transfer maneuvers such as the Hohmann transfer. Chapters 4 and 5 introduce time of flight with emphasis on the universal variable solution. These chapters treat the classic Kepler and Gauss problems in detail. Chapter 6 discusses the application of two-body mechanics to the ballistic missile problem, including launch error analysis in targeting on a rotating earth. Chapters 7 and 8 are further specialized applications to lunar and interplanetary flight. Chapter 9 is a brief introduction to perturbation analysis emphasizing special perturbations. That also includes a discussion of integration schemes and errors and analytic formulations of several common perturbations.

Since the first edition already discussed orbit transfer and rendezvous calculations, a chapter has been added to include the relative motion equations for two satellites in close proximity based on the Clohessey-Wiltshire-Hills formulations.

Perhaps the biggest change in astrodynamics since the first edition has been the extensive use of computers, programming tools, and commercially available software to perform calculations—and most importantly—enable visualization of orbits and satellite operations. Several of the sample problems and end-of-chapter exercises have been updated. The original Appendix D projects in the first edition have been replaced with a project in Appendix C focused on developing a computer solution to the Gauss targeting problem using universal variables.

Contributions to the ideas and methods of this text have been made by many present and former members of the Department of Astronautics. Also special thanks is given to the many instructors in the Air Force's Advanced Orbital Mechanics course who have developed tools, techniques, and productive approaches for teaching the material.

Special thanks are owed to several individuals who labored mightily behind the scenes to bring this project to fruition. The original material existed only in paper form and all of the text, equations, and figures have been recreated with appropriate modifications, corrections, and extensions. Mary Saylor devoted countless hours to typing the initial draft and Anita Shute and Linda Gentsch provided expert guidance and extensive editing support to bring the book to completion. And many thanks are owed for a significant effort by former Schriever Chair Maarten Meerman who provided a thorough technical review. Errors in the first printing of the 2nd edition have been corrected in this second printing.

United States Air Force Academy — W.W.S
Colorado
October 2020

Table of Contents

About the Authors

Roger R. Bate (1923–2009) graduated from the United States Military Academy at West Point in 1947, and received a degree in physics as a Rhodes Scholar from Magdelan College, Oxford. He served with the Army Corps of Engineers during the Korean War, and in 1959 was assigned to the U.S. Air Force Academy, and was recommissioned in the Air Force. He earned a Ph.D. in control systems from Stanford University in 1966, and was the first permanent professor in the Astronautics Dept. at the Air Force Academy. While there he co-authored Fundamentals of Astrodynamics, and later became Vice Dean of the Academy. After retiring from the Air Force in 1973, with the rank of Brigadier General (USAF Ret.) he joined Texas Instruments where he became Chief Computer Scientist, and later was a consultant with the Software Engineering Institute at Carnegie Mellon University. He passed away on March 18, 2009.

Donald D. Mueller was born in St. Louis in 1933. He received a degree in mechanical engineering with a minor in rocket propulsion from the University of Illinois and a commission in the Air Force through ROTC. In the Air Force he flew the F-86 and F-100, and received a Masters in Astronautical Engineering from the Air Force Institute of Technology. He was assigned to the zero gravity research program at the Aerospace Medical Research Laboratories where he wrote several technical reports on the effects of zero gravity and also designed and conducted weightless familiarization training for the Gemini and Apollo astronauts. Two of his early published reports concerned relative motion during orbital rendezvous and the possible dangers associated with the use of tetherlines in space. He joined the Astronautics Department at the Air Force Academy in 1964 and became one of the authors of Fundamentals of Astrodynamics. He retired from the Air Force in 1977 as Director of Fighter Requirements at Tactical Air Command Headquarters. Following his retirement from the Air Force he became head of the Physics Dept. at Colorado Technical University, from which he is now also retired.

Jerry E. White received a B.S. in Electrical Engineering from the University of Washington in 1959, a Masters Degree in Astronautics from the Air Force Institute of Technology in 1964, and a Ph.D. in Astronautics from Purdue University in 1970. He was on active duty in the Air Force for more than thirteen years following his graduation from the University of Washington. During that time he fulfilled numerous assignments including as a space program mission controller at Cape Kennedy, and as an associate professor at the Air Force

Academy during which he co-authored Fundamentals of Astrodynamics. He resigned from active service in 1973, and then spent 24 years in the Air Force Reserves where he served in various capacities, including with what is now Air Force Space Command, the National Security Agency, and, lastly, as assistant to the commander of Air Force Materiel Command, at Wright-Patterson AFB, Ohio. He was also a former Vice Chairman of Education for the Air Force Association. He retired from the Air Force Reserves in 1997 as a Major General.

William W. Saylor, who joined the authorship team of Fundamentals of Astrodynamics for this second edition, graduated from the United States Military Academy at West Point in 1972 with a degree in Nuclear Engineering. He served in the Army in Germany as a combat engineer officer and received a Master's Degree in Nuclear Engineering and Magnetic Fusion Technology from the Massachusetts Institute of Technology. He worked as a nuclear engineer in the power industry and then spent twelve years at Los Alamos National Laboratory on various energy and defense programs. From 2005 to 2011 he was the General Bernard Schriever Chair in the Department of Astronautics at the United States Air Force Academy. Since leaving the Air Force Academy in 2013, he has been providing consulting services for technology advances in satellite communications and imaging radar satellites, while developing complex systems concepts in the energy and transportation areas.

Chapter 1

Two-Body Orbital Mechanics

> On Christmas Day, 1642, the year Galileo died, there was born in the Manor House of Woolsthorpe-by-Colsterworth a male infant so tiny that his mother told him in later years, he might have been put into a quart mug, and so frail that he had to wear a bolster around his neck to support his head. This unfortunate creature was entered in the parish register as 'Isaac sonne of Isaac and Hanna Newton.' There is no record that the wise men honored the occasion, yet this child was to alter the thought and habit of the world.
>
> -James R. Newman[1]

1.1 Historical Background and Basic Laws

If Christmas Day 1642 ushered in the age of reason it was only because two men, Tycho Brahe and Johann Kepler, who chanced to meet only 18 months before the former's death, laid the groundwork for Newton's greatest discoveries some 50 years later.

It would be difficult to imagine a greater contrast between two men working in the same field of science than existed between Tycho Brahe and Kepler.

Tycho, the noble and aristocratic Dane, was exceptional in mechanical ingenuity and meticulous in the collection and recording of accurate data on the positions of the planets. He was utterly devoid of the gift of theoretical speculation and mathematical power.

Kepler, the poor and sickly mathematician, unfitted by nature for accurate observations, was gifted with the patience and innate mathematical perception needed to unlock the secrets hidden in Tycho's data.[2]

1.1.1 Kepler's Laws

Since the time of Aristotle, who taught that circular motion was the only perfect and natural motion and that the heavenly bodies, therefore, necessarily moved in circles, the planets were assumed to revolve in circular paths or combinations of smaller circles moving on larger ones. But, now that Kepler had the accurate observations of Tycho to, refer to he found immense difficulty in reconciling any such theory with the observed facts. From 1601 until 1606 he tried fitting various geometrical curves to Tycho's data on the positions of Mars. Finally, after struggling for almost a year to remove a discrepancy of only 8 minutes of arc (which a less honest man might have chosen to ignore!), Kepler hit upon the ellipse as a possible solution. It fit. The orbit was found and in 1609 Kepler published his first two laws of planetary motion. The third law followed in 1619.[3]

These laws, which mark an epoch in the history of mathematical science, are as follows:

Kepler's Laws

First Law – The orbit of each planet is an ellipse, with the Sun at a focus.

Second Law – The line joining the planet to the Sun sweeps out equal areas in equal times.

Third Law – The square of the period of a planet is proportional to the cube of its mean distance from the Sun. The semi-major axis of the ellipse is also called the mean distance from the Sun.

Still, Kepler's laws are a description, not an explanation, of planetary motion. It remained for the genius of Isaac Newton to unravel the mystery of "why?"

In 1665 Newton was a student at the University of Cambridge when an outbreak of the plague forced the university to close down for 2 years. Those 2 years were to be the most creative period in Newton's life. The 23-year-old genius conceived the law of gravitation and the laws of motion and developed the fundamental concepts of differential calculus during the long vacation of 1666, but owing to small discrepancies in his explanation of the Moon's motion he tossed his papers aside. The world was not to learn of his momentous discoveries until some 20 years later!

To Edmond Halley, discoverer of Halley's Comet, is due the credit for bringing Newton's discoveries before the world. One day in 1685 Halley and two of his contemporaries, Christopher Wren and Robert Hooke, were discussing the theory of Descartes, which explained the motion of the planets by means of whirlpools and eddies that swept the planets around the Sun. Dissatisfied with this explanation, they speculated whether a force "similar to magnetism" and falling off inversely with the square of distance, might not require the planets to move in precisely elliptical paths. Hooke thought that this would be easy to prove whereupon Wren offered Hooke 40 shillings if he could produce the proof within 2 weeks. The 2 weeks passed and nothing more was heard from Hooke. Several months later Halley was visiting Newton at Cambridge and, without mentioning the bet, casually posed the question, "If the Sun pulled the planets with a force inversely proportional to the square of their distances, in what paths ought they to go?" To Halley's utter and complete astonishment Newton replied without hesitation, "Why, in ellipses, of course. I have already calculated it and have the proof among my papers somewhere. Give me a few days and I shall find it for you." Newton was referring to the work he had done some 20 years earlier and only in this casual way was his greatest discovery made known to the world!

Halley, when he recovered from his shock, advised his reticent friend to develop completely and to publish his explanation of planetary motion. The result took 2 years in preparation and appeared in 1687 as *The Mathematical Principles of Natural Philosophy* or, more simply, the *Principia*, undoubtedly one of the supreme achievements of the mind.[4]

1.1.2 Newton's Laws of Motion

In Book 1 of the Principia Newton introduces his three laws of motion:

Newton's Laws

First Law – Every body continues in its state of rest or of uniform motion in a straight line unless it is compelled to change that state by forces impressed upon it.

Second Law – The time-rate change of momentum is proportional to the force impressed and is in the same direction as that force.

Third Law – To every action there is always opposed an equal reaction.

The second law can be expressed mathematically as follows:

$$\sum \mathbf{F} = m\ddot{\mathbf{r}} \tag{1-1}$$

where $\Sigma\mathbf{F}$ is the vector sum of all the forces acting on the mass and is the vector acceleration of the mass measured relative to an inertial reference frame shown as XYZ in Figure 1-1. Note that Equation (1-1) applies only for a system of constant mass.

1.1.3 Newton's Law of Universal Gravitation

Besides enunciating his three laws of motion in the *Principia,* Newton formulated the law of gravity by stating that any two bodies attract one another with a force proportional to the product of their masses and inversely proportional to the square of the distance between them. We can express this law mathematically in vector notation as

$$\mathbf{F}_g = -\frac{GMm}{r^2}\frac{\mathbf{r}}{r} \tag{1-2}$$

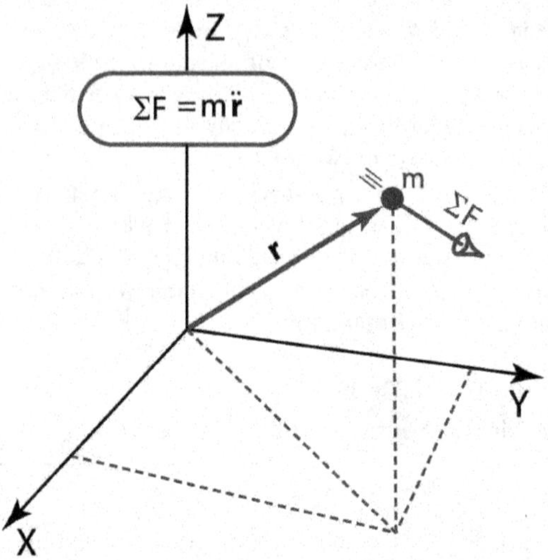

Figure 1-1 Newton's Law of Motion.

where $\mathbf{F_g}$ is the force on mass m due to mass M and **r** is the vector from M to m. The universal gravitational constant, G, has the approximate value of 6.673 $\times 10^{-20}$ km^3/kg-s^2.

In the next sections we will apply Equation (1-3) to Equation (1-1) and develop the equation of motion for planets and satellites.

We will begin with the general n-body problem and then specialize to the problem of two bodies. Figure 1-2 shows Newton's law of gravity.

1.2 THE n-BODY PROBLEM

In this section we shall examine in some detail the motion of a body (e.g., an Earth satellite, a lunar or interplanetary probe or a planet). At any given time in its journey, the body is being acted upon by several masses and may even be experiencing other forces such as drag, thrust and solar radiation pressure.

1.2.1 A System of Bodies

For this examination we shall assume a "system" of n-bodies (m_1, m_2, m_3,... m_n) one of which is the body whose motion we wish to study—call it the ith body, m_i. The vector sum of all gravitational forces and other external forces acting on m_i will be used to determine the equation of motion. To determine the gravitational forces we shall apply Newton's law of universal gravitation. In addition, the ith body may be a rocket expelling mass (i.e. propellants) to produce thrust; the motion may be in an atmosphere where drag effects are present; solar radiation may impart some pressure on the body; etc. All of these

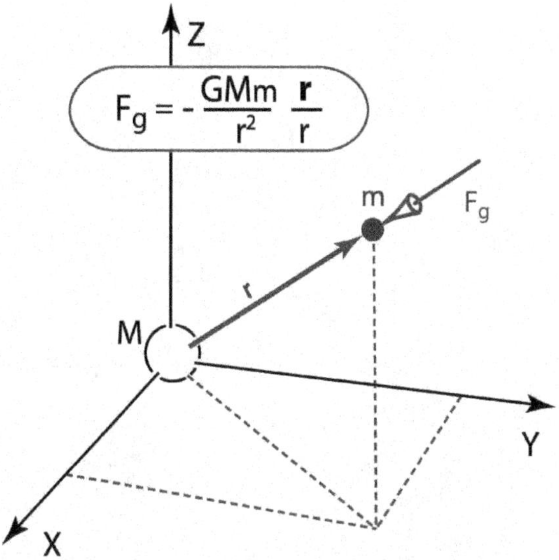

Figure 1-2 Newton's Law of Gravity.

effects must be considered in the general equation of motion. An important force, not yet mentioned, is due to the nonspherical shape of the planets. Earth is flattened at the poles and bulged at the equator; the Moon is elliptical about the poles and around the equator. Newton's law of universal gravitation can be applied to a body using a single term, but only if that body is spherical and if its mass is evenly distributed in spherical shells. Thus, variations are introduced to the gravitational forces due primarily to the shape of the bodies. The magnitude of this force for a near-Earth satellite is on the order of 10^{-3} g. Although small, this force is responsible for several important effects not predictable from the studies of Kepler and Newton. These effects, regression of the line-of-nodes and rotation of the line-of-apsides, are discussed in Chapter 3.

The first step in our analysis will be to choose a "suitable" coordinate system in which to express the motion. This is not a simple task since any coordinate system we choose has a fair degree of uncertainty as to its inertial qualities. Without losing generality let us assume a suitable coordinate system (**X**, **Y**, **Z**) in which the positions of the n masses are known: $\mathbf{r}_1$, $\mathbf{r}_2$,...,$\mathbf{r}_n$. This system is illustrated in Figure 1-3.

By applying Newton's law of universal gravitation, the force $\mathbf{F}_{g_n}$ exerted on m_i by m_n is

$$\mathbf{F}_{g_n} = -\frac{Gm_i\,m_n}{r_{ni}^3}(\mathbf{r}_{ni}) \qquad (1\text{-}3)$$

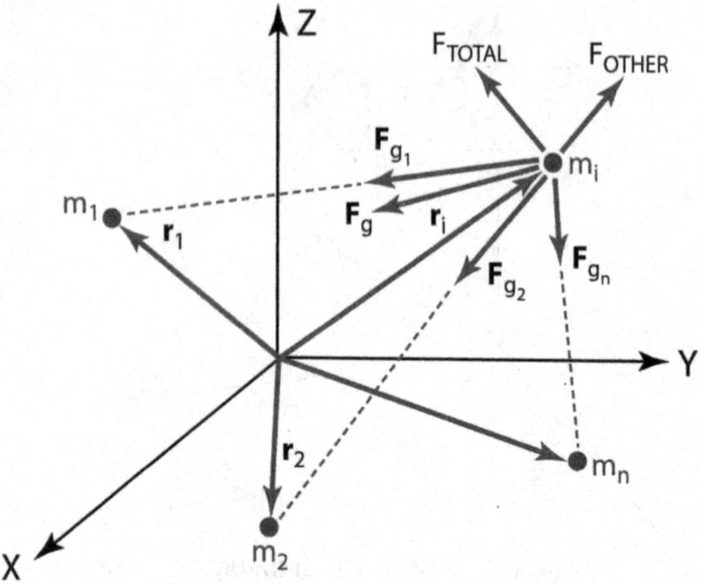

Figure 1-3 The n-Body Problem.

where

$$\mathbf{r}_{ni} = \mathbf{r}_i - \mathbf{r}_n \tag{1-4}$$

The vector sum, $\mathbf{F_g}$, of all such gravitational forces acting on the ith body may be written as

$$\mathbf{F}_g = -\frac{Gm_i m_1}{r_{1i}^3}(\mathbf{r}_{1i}) - \frac{Gm_i m_2}{r_{2i}^3}(\mathbf{r}_{2i}) - \ldots \frac{Gm_i m_n}{r_{ni}^3}(\mathbf{r}_{ni}) \tag{1-5}$$

Obviously, Equation (1-5) does not contain the term

$$-\frac{Gm_i m_i}{r_{ii}^3}(\mathbf{r}_{ii}) \tag{1-6}$$

since the body cannot exert a force itself. We must simplify this equation by using the summation notation so that

$$\boxed{\mathbf{F}_g = -Gm_i \sum_{\substack{j=1 \\ j\neq i}}^{n} \frac{m_j}{r_{ji}^3}(\mathbf{r}_{ji})} \tag{1-7}$$

The other external force, $\mathbf{F}_{OTHER}$, illustrated in Figure 1-3, is composed of drag, thrust, solar radiation pressure, perturbations due to nonspherical shapes, etc. The combined force acting on the ith body we will call $\mathbf{F}_{TOTAL}$, where

$$\mathbf{F}_{TOTAL} = \mathbf{F}_g + \mathbf{F}_{OTHER} \tag{1-8}$$

We are now ready to apply Newton's second law of motion. Thus,

$$\frac{d}{dt}(m_i \mathbf{v}_i) = \mathbf{F}_{TOTAL} \tag{1-9}$$

The time derivative may be expanded to

$$m_i \frac{d\mathbf{v}_i}{dt} + \mathbf{v}_i \frac{dm_i}{dt} = \mathbf{F}_{TOTAL} \tag{1-10}$$

It was previously mentioned that the body may be expelling some mass to produce thrust in which case the second term of Equation (1-10) would not be zero. Certain relativistic effects would also give rise to changes in the mass m_i as a function of time. In other words, it is not always true—especially in space dynamics—that $\mathbf{F} = m\mathbf{a}$. Dividing through by the mass m_i gives the most general equation of motion for the ith body:

$$\boxed{\ddot{\mathbf{r}}_i = \frac{\mathbf{F}_{TOTAL}}{m_i} - \dot{\mathbf{r}}_i \frac{\dot{m}_i}{m_i}} \tag{1-11}$$

where

$\ddot{\mathbf{r}}_i$ is the vector acceleration of the ith body relative to the **X, Y, Z** coordinate system,

m_i is the mass of the ith body, and

$\mathbf{F}_{TOTAL}$ is the vector sum of all gravitational forces given by

$$\mathbf{F}_g = -Gm_i \sum_{\substack{j=1 \\ j \neq i}}^{n} \frac{m_j}{r_{ji}^3}(\mathbf{r}_{ji})$$

and all other external forces

$$F_{OTHER} = F_{DRAG} + F_{THRUST} + F_{SOLAR\ PRESSURE} + F_{PERTURB} + \text{etc., where}$$

$\dot{\mathbf{r}}_i$ is the velocity vector of the ith body relative to the **X, Y, Z** coordinate system and

$\dot{m}_i$ is the same time rate of change of mass of the ith body (due to expelling mass or relativistic effects).

Equation (1-11) is a second-order, nonlinear, vector, differential equation of motion that has defied solution in its present form. It is here therefore that we depart from the realities of nature to make some simplifying assumptions.

Assume that the mass of the ith body remains constant (i.e., unpowered flight: $\dot{m}_i = 0$). Also, assume that drag and other external forces are not present. The only remaining forces then are gravitational. Equation (1-11) reduces to

$$\ddot{\mathbf{r}}_i = -G\sum_{\substack{j=1 \\ j\neq i}}^{n}\frac{m_j}{r_{ji}^3}\left(\mathbf{r}_{ji}\right) \tag{1-12}$$

Let us assume also that m_2 is an Earth satellite and that m_1 is Earth. The remaining masses $m_3, m_4, \ldots, m_n$ may be the Moon, Sun and planets. Then writing Equation (1-12) for i = 1, we get

$$\ddot{\mathbf{r}}_i = -G\sum_{j=2}^{n}\frac{m_j}{r_{j1}^3}\left(\mathbf{r}_{j1}\right) \tag{1-13}$$

And for i = 2, Equation (1-13) becomes

$$\ddot{\mathbf{r}}_2 = -G\sum_{\substack{j=1 \\ j\neq 2}}^{n}\frac{m_j}{r_{j2}^3}\left(\mathbf{r}_{j2}\right) \tag{1-14}$$

From Equation (1-4) we see that

$$\mathbf{r}_{12} = \mathbf{r}_2 - \mathbf{r}_1 \tag{1-15}$$

so that

$$\ddot{\mathbf{r}}_{12} = \ddot{\mathbf{r}}_2 - \ddot{\mathbf{r}}_1 \tag{1-16}$$

Substituting Equations (1-13) and (1-14) into Equation (1-16) gives,

$$\ddot{\mathbf{r}}_{12} = -G\sum_{\substack{j=1 \\ j\neq 2}}^{n}\frac{m_j}{r_{j2}^3}(\mathbf{r}_{j2}) + G\sum_{j=2}^{n}\frac{m_j}{r_{j1}^3}(\mathbf{r}_{j1}) \tag{1-17}$$

or expanding we have

$$\ddot{\mathbf{r}}_{12} = -\left[\frac{Gm_1}{r_{12}^3}(\mathbf{r}_{12}) + G\sum_{j=3}^{n}\frac{m_j}{r_{j2}^3}(\mathbf{r}_{j2})\right] - \left[-\frac{Gm_2}{r_{21}^3}(\mathbf{r}_{21}) - G\sum_{j=3}^{n}\frac{m_j}{r_{j1}^3}(\mathbf{r}_{j1})\right] \tag{1-18}$$

Since $\mathbf{r}_{12} = -\mathbf{r}_{21}$ we may combine the first terms in each bracket. Hence,

$$\ddot{\mathbf{r}}_{12} = -\frac{G(m_1 + m_2)}{r_{12}^3}(\mathbf{r}_{12}) - \sum_{j=3}^{n} Gm_j\left(\frac{\mathbf{r}_{j2}}{r_{j2}^3} - \frac{\mathbf{r}_{j1}}{r_{j1}^3}\right) \tag{1-19}$$

The reason for writing the equation in this form will become clear when we recall that we are studying the motion of a near-Earth satellite where m_2 is the mass of the satellite and m_1 is the mass of Earth. Then $\ddot{\mathbf{r}}_{12}$ is the acceleration of the satellite relative to Earth. The effect of the last term of Equation (1-19) is to account for the perturbing effects of the Moon, Sun and planets on a near-Earth satellite.

To further simplify this equation it is necessary to determine the magnitude of the perturbing effects compared to the force between Earth and the satellite. Table 1-1 lists the approximate relative accelerations (not the perturbative accelerations) for a satellite in a 500 km orbit about Earth. Notice also that the effect of the nonspherical Earth (oblateness) is included for comparison.

Table 1-1: Comparison of Relative Acceleration (in g) for a 500 km Earth Satellite.

Central Body	Acceleration in g on 500 km Earth Satellite
Earth	0.86
Sun	6.0×10^{-4}
Mercury	2.6×10^{-10}
Venus	1.9×10^{-8}
Mars	7.1×10^{-10}
Jupiter	3.2×10^{-8}
Saturn	2.3×10^{-9}
Uranus	8.0×10^{-11}
Neptune	3.6×10^{-11}
Moon	3.3×10^{-6}
Earth oblateness	1.0×10^{-3}

1.3 THE TWO-BODY PROBLEM

Now that we have a general expression for the relative motion of two bodies perturbed by other bodies it would be a simple matter to reduce it to an equation for only two bodies. However, to further clarify the derivation of the equation of relative motion, some of the work of the previous section will be repeated considering just two bodies.

1.3.1 Simplifying Assumptions

There are two assumptions we will make with regard to our model:

1. The bodies are spherically symmetric. This enables us to treat the bodies as though their masses were concentrated at their centers.

2. There are no external nor internal forces acting on the system other than the gravitational forces that act along the line joining the centers of the two bodies.

1.3.2 The Equation of Relative Motion

Before we may apply Newton's second law to determine the equation of relative motion of these two bodies, we must find an inertial (unaccelerated and non-rotating) reference frame for the purpose of measuring the motion or the lack of it. Newton described this inertial reference frame by saying that it was fixed in absolute space, which "in its own nature, without relation to anything external, remains always similar and immovable."[5] However, he failed to indicate how one found this frame that was absolutely at rest. For the time being, let us carry on with our investigation of the relative motion by assuming that we have found such an inertial reference frame and then later return to a discussion of the consequences of the fact that in reality all we can ever find is an "almost" inertial reference frame.

Consider the system of two bodies of masses M and m illustrated in Figure 1-4. Let (**X**′, **Y**′, **Z**′) be an inertial set of rectangular Cartesian coordinates. Let (**X**, **Y**, **Z**) be a set of non-rotating coordinates parallel to (**X**′, **Y**′, **Z**′) and having an origin coincident with the body of mass M. The position vectors of the bodies M and m with respect to the set (**X**′, **Y**′, **Z**′) are $\mathbf{r}_M$ and $\mathbf{r}_m$, respectively. Note that we have defined

$$\mathbf{r} = \mathbf{r}_m - \mathbf{r}_M \qquad (1\text{-}20)$$

Now we can apply Newton's laws in the inertial frame (**X**′, **Y**′, **Z**′) and obtain

$$m\ddot{\mathbf{r}}_m = -\frac{GMm}{r^2}\frac{\mathbf{r}}{r} \qquad (1\text{-}21)$$

and

$$M\ddot{\mathbf{r}}_M = \frac{GMm}{r^2}\frac{\mathbf{r}}{r} \qquad (1\text{-}22)$$

The above equations may be written as

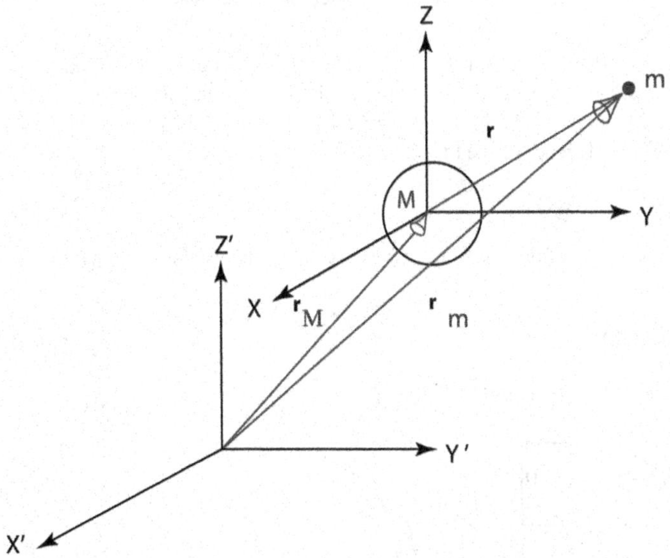

Figure 1-4 Relative Motion of Two Bodies.

$$\ddot{\mathbf{r}}_m = -\frac{GM}{r^3}\mathbf{r} \tag{1-23}$$

and

$$\ddot{\mathbf{r}}_M = \frac{Gm}{r^3}\mathbf{r} \tag{1-24}$$

Subtracting Equation (1-24) from Equation (1-23) we have

$$\ddot{\mathbf{r}} = -\frac{G(M+m)}{r^3}\mathbf{r} \tag{1-25}$$

Equation (1-25) is the vector differential equation of the relative motion for the two-body problem. Note that this is the same as Equation (1-19) without perturbing effects and with $\mathbf{r}_{12}$ replaced by $\mathbf{r}$.

Note that since the coordinate set (**X**, **Y**, **Z**) is non-rotating with respect to the coordinate set (**X′**, **Y′**, **Z′**), the magnitudes and directions of $\mathbf{r}$ and $\ddot{\mathbf{r}}$ as measured in the set (**X**, **Y**, **Z**) will be equal respectively to their magnitudes and directions as measured in the inertial set (**X′**, **Y′**, **Z′**) Thus having postulated the existence of an inertial reference frame in order to derive Equation (1-25), we may now discard it and measure the relative position, velocity and acceleration in a non-rotating, non-inertial coordinate system such as the set (**X**, **Y**, **Z**) with its origin in the central body.

Since our efforts in this text will be devoted to studying the motion of artificial satellites, ballistic missiles, or space probes orbiting about some planet or the Sun, the mass of the orbiting body, m, will be much less than that of the central body, M. Hence we see that

$$G(M + m) \cong GM \tag{1-26}$$

It is convenient to define a parameter, μ (mu), called the gravitational parameter as

$$\mu \equiv GM \tag{1-27}$$

Then Equation (1-25) becomes

$$\boxed{\ddot{\mathbf{r}} + \frac{\mu}{r^3}\mathbf{r} = 0} \tag{1-28}$$

Equation (1-28) is the two-body equation of motion that we will use for the remainder of the text. Remember that the results obtained from Equation (1-28) will be only as accurate as the assumptions (1) and (2) and the assumption that M » m. If the ratio of M to m is less than 1×10^6, then G(M+m) must be used in place of μ to retain sufficient accuracy for most applications. μ will have a different value for each major attracting body. Values for Earth and the Sun are listed in the appendix and values for other planets are included in Chapter 8.

1.4 CONSTANTS OF MOTION

Before attempting to solve the equation of motion to obtain the trajectory of a satellite we shall derive some useful information about the nature of orbital motion. If you think about the model we have created, namely a small mass moving in a gravitational field whose force is always directed toward the center of a larger mass, you would probably arrive intuitively at the conclusions we will shortly confirm by rigorous mathematical proofs. From your previous knowledge of physics and mechanics you know that a gravitational field is "conservative." That is, an object moving under the influence of gravity alone does not lose or gain mechanical energy but only exchanges one form of energy, "kinetic," for another form called "potential energy." You also know that it takes a tangential component of force to change the angular momentum of a system in rotational motion about some center of rotation. Since the gravitational force is always directed radially toward the center of the large mass we would expect that the angular momentum of the satellite about the center of our reference frame (the large mass) does not change. In the next two sections we will prove these statements.

1.4.1 Conservation of Mechanical Energy

The energy constant of motion can be derived as follows:

1. Dot multiply Equation (1-28) by $\dot{\mathbf{r}}$ to get

$$\dot{\mathbf{r}} \bullet \ddot{\mathbf{r}} + \dot{\mathbf{r}} \bullet \frac{\mu}{r^3}\mathbf{r} = 0$$

2. Since in general $\mathbf{a} \bullet \dot{\mathbf{a}} = a\dot{a}$, $\mathbf{v} = \dot{\mathbf{r}}$ and $\dot{\mathbf{v}} = \ddot{\mathbf{r}}$, then

$$\mathbf{v} \bullet \dot{\mathbf{v}} + \frac{\mu}{r^3}\mathbf{r} \bullet \dot{\mathbf{r}} = 0 \text{, so}$$

$$v\dot{v} + \frac{\mu}{r^3} r\dot{r} = 0$$

3. Noticing that $\frac{d}{dt}\left(\frac{v^2}{2}\right) = v\dot{v}$ and $\frac{d}{dt}\left(-\frac{\mu}{r}\right) = \frac{\mu}{r^2}\dot{r}$, we have

$$\frac{d}{dt}\left(\frac{v^2}{2}\right) + \frac{d}{dt}\left(-\frac{\mu}{r}\right) = 0 \text{ or } \frac{d}{dt}\left(\frac{v^2}{2} - \frac{\mu}{r}\right) = 0$$

4. To make step 3 perfectly general we should say that

$$\frac{d}{dt}\left(\frac{v^2}{2} + c - \frac{\mu}{r}\right) = 0$$

where c can be any arbitrary constant since the time derivative of any constant is zero.

5. If the time rate of change of an expression is zero, that expression must be a constant which we will call E. Therefore,

$$E = \frac{v^2}{2} + \left(c - \frac{\mu}{r}\right) = \text{a constant called “specific mechanical energy”}$$

The first term of E is obviously the kinetic energy per unit mass of the satellite. To convince yourself that the second term is the potential energy per unit mass you need only equate it with the work done in moving a satellite from one point in space to another against the force of gravity. But what about the arbitrary constant, c, which appears in the potential energy term? The value of this constant will depend on the zero reference of potential energy. In other words, at what distance r do you want to say the potential energy is zero? This is obviously arbitrary. In your elementary physics courses it was convenient to choose ground level or the surface of Earth as the zero datum for potential energy, in which case the object lying at the bottom of a deep well was found to have a negative potential energy. If we wish to retain the surface of the large mass, e.g., Earth, as our zero reference we would choose $c = \mu/r_\oplus$, where $r_\oplus$ is the radius of Earth. This would be perfectly legitimate but, since c is arbitrary, why not set it equal to zero? Setting c equal to zero is equivalent to choosing our zero reference for potential energy at infinity. The price we pay for this simplification is that the potential energy of a satellite (now simply $-\mu/r$) will always be negative.

We conclude, therefore, that the *specific mechanical energy, E,* of a satellite which is the sum of its kinetic energy per unit mass and its potential energy per unit mass, remains constant along its orbit, neither increasing nor decreasing as a result of its motion. The expression for E is

$$\boxed{E = \frac{v^2}{2} - \frac{\mu}{r}} \tag{1-29}$$

1.4.2 Conservation of Angular Momentum

The angular momentum constant of motion is obtained as follows:

1. Cross-multiply Equation (1-28) by $\mathbf{r}$ to get

$$\mathbf{r} \times \ddot{\mathbf{r}} + \mathbf{r} \times \frac{\mu}{r^3}\mathbf{r} = \mathbf{0}.$$

2. Since in general $\mathbf{a} \times \mathbf{a} = 0$, the second term vanishes and

$$\mathbf{r} \times \ddot{\mathbf{r}} = 0$$

3. Noticing that $\frac{d}{dt}(\mathbf{r} \times \dot{\mathbf{r}}) = \dot{\mathbf{r}} \times \dot{\mathbf{r}} + \mathbf{r} \times \ddot{\mathbf{r}}$ the equation becomes

$$\frac{d}{dt}\left(\mathbf{r} \times \dot{\mathbf{r}}\right) = 0 \quad \text{or} \quad \frac{d}{dt}\left(\mathbf{r} \times \mathbf{v}\right) = 0$$

The expression $\mathbf{r} \times \mathbf{v}$, which must be a constant of motion, is simply the vector $\mathbf{h}$, called *specific angular momentum.* Therefore, we have shown that the specific angular momentum, $\mathbf{h}$, of a satellite remains constant along its orbit and that the expression for $\mathbf{h}$ is

$$\boxed{\mathbf{h} = \mathbf{r} \times \mathbf{v}} \tag{1-30}$$

Since $\mathbf{h}$ is the vector cross product of $\mathbf{r}$ and $\mathbf{v}$ it must always be perpendicular to the plane containing $\mathbf{r}$ and $\mathbf{v}$. But $\mathbf{h}$ is a constant vector so $\mathbf{r}$ and $\mathbf{v}$ must always remain in the same plane. Therefore, we conclude that the satellite's motion must be confined to a plane that is fixed in space. We shall refer to this as the orbital plane.

By looking at the vectors **r** and **v** in the orbital plane and the angle between them (see Figure 1-5) we can derive another useful expression for the magnitude of the vector **h**.

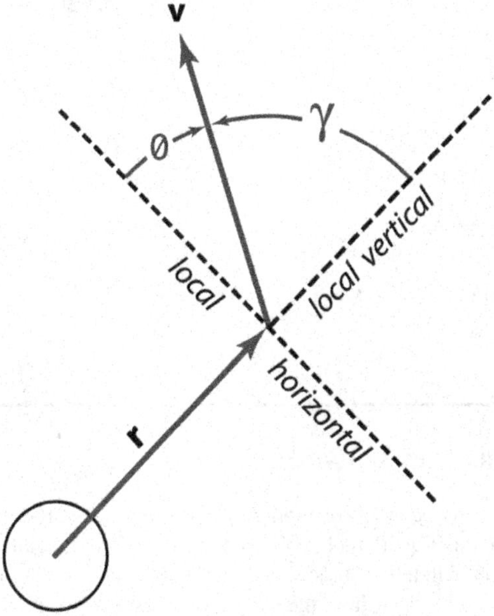

Figure 1-5 Flight-path Angle, ϕ.

No matter where a satellite is located in space it is always possible to define "up and down" and "horizontal." "Up" simply means away from the center of Earth and "down" means toward the center of Earth. So the local vertical at the location of the satellite coincides with the direction of the vector **r**. The local horizontal plane must then be perpendicular to the local vertical. We can now define the direction of the velocity vector, **v**, by specifying the angle it makes with the local vertical as γ (gamma), the zenith angle. The angle between the velocity vector and the local horizontal plane is called ϕ (phi), the flight-path \elevation angle or simply "flight-path angle." From the definition of the cross product the magnitude of **h** is

$$h = rv \sin\gamma \tag{1-31}$$

We will find it more convenient, however, to express **h** in terms of the flight-path angle, ϕ. Since γ and ϕ obviously complementary angles,

$$h = rv \cos\phi \tag{1-32}$$

The sign of ϕ will be the same as the sign of $\mathbf{r} \cdot \mathbf{v}$.

Example Problem. In an inertial coordinate system, the position and velocity vectors of satellite are, respectively, $(1.2756\mathbf{I} + 1.9135\mathbf{J} + 3.1891\mathbf{K})\ 10^4$ km and $(7.9053\mathbf{I} + 15.8106\mathbf{J})$ km/s, where **I**, **J** and **K** are unit vectors. Determine the specific mechanical energy, E, and the specific angular momentum, **h**. Also find the flight-path angle, ϕ.

$r = 3.9318 \times 10^4$ km, $v = 17.678$ km/s

$$E = \frac{v^2}{2} - \frac{\mu}{r} = \underline{\underline{146.0963\ \text{km}^2/\text{s}^2}}$$

$\mathbf{h} = \mathbf{r} \times \mathbf{v} = (-5.0422\mathbf{I} + 2.5211\mathbf{J} + 0.5042\mathbf{K})\ 10^5\ \text{km}^2/\text{s}$

$h = \underline{\underline{5.6598 \times 10^5\ \text{km}^2/\text{s}}}$

$h = rv\cos\phi,\ \cos\phi = \dfrac{h}{rv} = 0.8143$

$\mathbf{r} \cdot \mathbf{v} > 0$, therefore:

$\phi = \cos^{-1} 0.8143 = \underline{\underline{35.42°}}$

1.5 THE TRAJECTORY EQUATION

Earlier we wrote the equation of motion for a small mass orbiting a large central body. While this Equation (1-23) is simple, its complete solution is not. A partial solution that will tell us the size and shape of the orbit is easy to obtain. The more difficult question of how the satellite moves around this orbit as a function of time will be postponed to Chapter 4.

1.5.1 Integration of the Equation of Motion

You recall that the equation of motion for the two-body problem is

$$\ddot{\mathbf{r}} = -\frac{\mu}{r^3}\mathbf{r} \tag{1-33}$$

Crossing this equation into **h** leads to a form that can be integrated:

$$\ddot{\mathbf{r}} \times \mathbf{h} = \frac{\mu}{r^3}(\mathbf{h} \times \mathbf{r}) \tag{1-34}$$

The left side of Equation (1-34) is clearly $d/dt(\dot{\mathbf{r}} \times \mathbf{h})$—try it and see. Looking for the right side to also be the time rate of change of some vector quality, we see that

$$\frac{\mu}{r^3}(\mathbf{h} \times \mathbf{r}) = \frac{\mu}{r^3}(\mathbf{r} \times \mathbf{v}) \times \mathbf{r} = \frac{\mu}{r^3}\left[\mathbf{v}(\mathbf{r}\bullet\mathbf{r}) - \mathbf{r}(\mathbf{r}\bullet\mathbf{v})\right] = \frac{\mu}{r}\mathbf{v} - \frac{\mu\dot{r}}{r^2}\mathbf{r} \tag{1-35}$$

since $\mathbf{r} \bullet \dot{\mathbf{r}} = r\dot{r}$. Note that μ times that derivative of the unit vector is also

$$\mu\frac{d}{dt}\left(\frac{\mathbf{r}}{r}\right) = \frac{\mu}{r}\mathbf{v} - \frac{\mu\dot{r}}{r^2}\mathbf{r} \tag{1-36}$$

We can rewrite Equation (1-34) as

$$\frac{d}{dt}\left(\dot{\mathbf{r}} \times \mathbf{h}\right) = \mu\frac{d}{dt}\left(\frac{\mathbf{r}}{r}\right) \tag{1-37}$$

Integrating both sides gives

$$\dot{\mathbf{r}} \times \mathbf{h} = \mu\frac{\mathbf{r}}{r} + \mathbf{B} \tag{1-38}$$

where $\mathbf{B}$ is the vector constant of integration. If we now dot multiply this equation by $\mathbf{r}$ we get a scalar equation

$$\mathbf{r}\bullet\dot{\mathbf{r}} \times \mathbf{h} = \mathbf{r}\bullet\mu\frac{\mathbf{r}}{r} + \mathbf{r}\bullet\mathbf{B} \tag{1-39}$$

Since, in general, $\mathbf{a} \bullet \mathbf{b} \times \mathbf{c} = \mathbf{a} \times \mathbf{b} \bullet \mathbf{c}$ and $\mathbf{a} \bullet \mathbf{a} = a^2$,

$$h^2 = \mu r + rB\cos\nu \tag{1-40}$$

where ν (nu) is the angle between the constant vector $\mathbf{B}$ and the radius vector $\mathbf{r}$. Solving for $\mathbf{r}$, we obtain

$$\boxed{r = \frac{h^2 / \mu}{1 + (B / \mu)\cos\nu}} \tag{1-41}$$

1.5.2 The Polar Equation of a Conic Section

Equation (1-41) is the trajectory equation expressed in polar coordinates where the polar angle, ν, is measured from the fixed vector $\mathbf{B}$ to $\mathbf{r}$. To determine what kind of a curve it represents we need only compare it to the general equation of a conic section written in polar coordinates with the origin located at a focus and where the polar angle, ν, also referred to as the *true anomaly*, is the angle between $\mathbf{r}$ and the point on the conic nearest the focus:

$$\boxed{r = \frac{p}{1 + e\cos\nu}} \tag{1-42}$$

In this equation, which is mathematically identical in form to the trajectory equation, p is a geometrical constant of the conic called the "parameter" or "semi-latus rectum," as indicated by the label "p" in Figure 1-6. The constant e is called the "eccentricity" and it determines the type of conic section in polar coordinates represented by Equation (1-42).

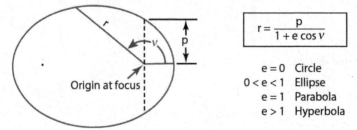

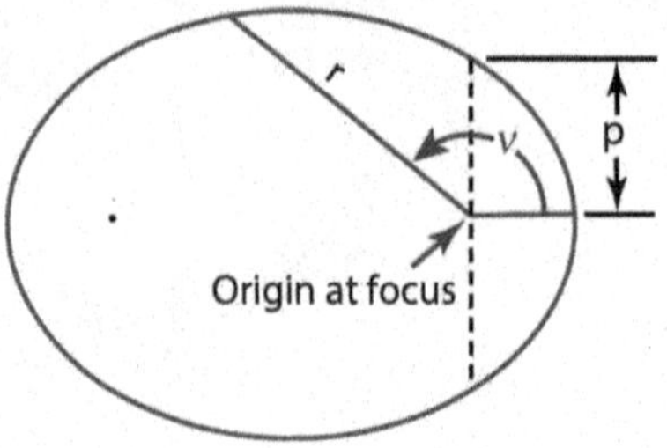

$$r = \frac{p}{1 + e \cos v}$$

$e = 0$	Circle
$0 < e < 1$	Ellipse
$e = 1$	Parabola
$e > 1$	Hyperbola

Figure 1-6 General Equation of any Conic Section in Polar Coordinates.

The similarity in form between the trajectory Equation (1-41) and the equation of a conic section (1-42) not only verifies Kepler's first law but allows us to extend the law to include orbital motion along any conic section path, not just an ellipse.

We can summarize our knowledge concerning orbital motion up to this point as follows:

1. The family of curves called "conic sections" (circle, ellipse, parabola, and hyperbola) represents *the only possible paths* for an orbiting object in the two-body problem.

2. The focus of the conic orbit *must* be located at the center of the central body.

3. The mechanical energy of a satellite (which is the sum of kinetic and potential energy) does not change as the satellite moves along its conic orbit. There is, however, an exchange of energy between the two forms, potential and kinetic, which means that the satellite must slow down as it gains altitude (as r increases) and speed up as r decreases in such a manner that E *remains constant.*

4. The orbital motion takes place in a plane that is *fixed in inertial space*.

5. The specific angular momentum of a satellite about the central attracting body *remains constant.* As **r** and v change along the orbit, the flight-path angle, ϕ, must change so as to keep **h** constant. (See Figure 1-6 and Equation (1-32).)

1.5.3 Geometrical Properties Common to All Conic Sections

Although Figure 1-6 illustrates an ellipse, the ellipse is only one of the family of curves called conic sections. Before discussing the factors

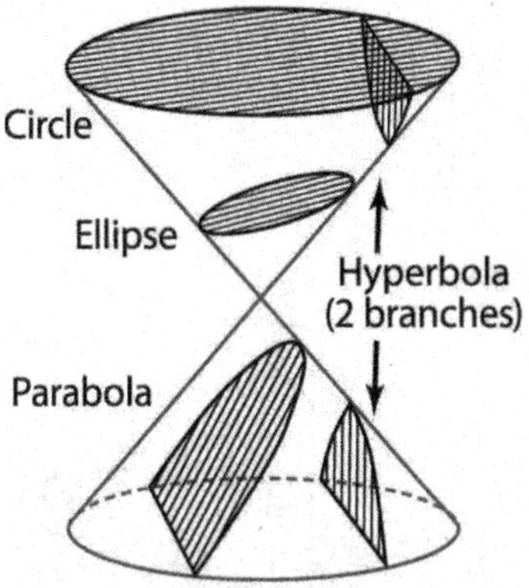

Figure 1-7 The Conic Sections.

that determine which of these conic curves a satellite will follow, we need to know a few facts about conic sections in general.

The conic sections have been known and studied for centuries. Many of their most interesting properties were discovered by the early Greeks. The name derives from the fact that a conic section may be defined as the curve of intersection of a plane and a right circular cone. Figure 1-7 illustrates this definition. If the plane cuts across one nappe (half-cone), the section is an *ellipse*. A *circle* is just a special case of the ellipse where the plane is parallel to the base of the cone. If, in addition to cutting just one nappe of the cone, the plane is parallel to a line in the surface of the cone, the section is a *parabola*. If the plane cuts both nappes, the section is a *hyperbola* having two branches. There are also *degenerate conics* consisting of one or two straight lines, or a single point; these are produced by planes passing through the apex of the cone.

There is an alternate definition of a conic that is mathematically equivalent to the geometrical definition above:

> A conic is a circle or the locus of a point that moves so that the ratio of its absolute distance from a given point (a focus) to its absolute distance from a given line (a directrix) is a positive constant e (the eccentricity).

While the directrix has no physical significance as far as orbits are concerned, the focus and eccentricity are indispensable concepts in the understanding of orbital motion. Figure 1-8 illustrates certain other geometrical dimensions and relationships that are common to all conic sections.

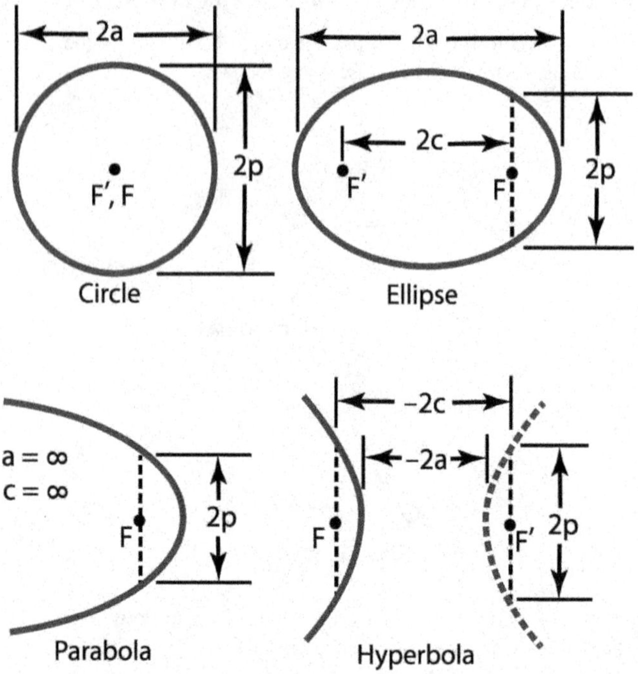

Figure 1-8 Geometrical Dimensions Common to all Conic Sections.

Because of their symmetry, all conic sections have two foci, F and F'. The prime focus, F, marks the location of the central attracting body in an orbit. The secondary or vacant focus, F', has little significance in orbital mechanics. In the parabola, which represents the borderline case between the open and closed orbits, the secondary focus is assumed to lie an infinite distance to the left of F. The width of each curve *at the focus* is a positive dimension called the latus rectum and is labeled 2p in Figure 1-8. The length of the chord passing through the foci is called the major axis of the conic and is labeled 2a. The dimension a is called the semi-major axis. Note that for the circle 2a is simply the diameter, for the parabola 2a is infinite and for the hyperbola 2a is taken as negative. The distance between the foci is given the symbol 2c. For the circle the foci are considered coincident and 2c is zero, for the parabola 2c is infinite and for the hyperbola 2c is taken as a negative. It follows directly from the definition of a conic section given above that, for any conic *except a parabola*,

$$\boxed{e = \frac{c}{a}} \tag{1-43}$$

and

$$\boxed{p = a\left(1 - e^2\right)} \tag{1-44}$$

The extreme endpoints of the major axis of an orbit are referred to as "apses." The point nearest the prime focus is called "periapsis" (meaning the "near apse") and the point farthest from the prime focus is called "apoapsis" (meaning the "far apse"). Depending on what is the central attracting body in an orbital situation these points may also be called "perigee" or "apogee," "perihelion" or "aphelion," "periselenium" or "aposelenium," etc. Notice that for the circle these points are not uniquely defined and for the open curves (parabolas and hyperbolas) the apoapsis has no physical meaning.

The distance from the prime focus to either periapsis or apoapsis (where it exists) can be expressed by simply inserting $\nu = 0°$ or $\nu = 180°$ in the general polar equation of a conic section (Equation (1-42)). Thus, *for any conic*,

$$r_{min} = r_{periapsis} = \frac{p}{1 + e\cos 0°} \tag{1-45}$$

Combining this with Equation (1-45) gives

$$\boxed{r_p = \frac{p}{1 + e} = a(1 - e)} \tag{1-46}$$

Similarly,

$$r_{max} = r_{apoapsis} = \frac{p}{1 + e\cos 180°}$$

and

$$\boxed{r_a = \frac{p}{1 - e} = a(1+e)} \tag{1-47}$$

1.5.4 The Eccentricity Vector

In the derivation of Equation (1-41), the trajectory equation, we encountered the vector constant of integration, **B**, which points toward periapsis. By comparing Equations (1-41) and (1-42) we conclude that $B = \mu\mathbf{e}$. Quite obviously, since **e** is also a constant vector pointing toward periapsis,

$$\mathbf{e} = \mathbf{B}/\mu \tag{1-48}$$

By integrating the two-body equation of motion we obtained the following result:

$$\dot{\mathbf{r}} \times \mathbf{h} = \mu\frac{\mathbf{r}}{r} + \mathbf{B} \tag{1-49}$$

Solving for **B** and noting that

$$\mathbf{B} = \mathbf{v} \times \mathbf{h} - \mu \frac{\mathbf{r}}{r}$$

gives

$$\boxed{\mathbf{e} = \frac{\mathbf{v} \times \mathbf{h}}{\mu} - \frac{\mathbf{r}}{r}} \tag{1-50}$$

We can eliminate **h** from this expression by substituting $\mathbf{h} = \mathbf{r} \times \mathbf{v}$, so

$$\mu\mathbf{e} = \mathbf{v} \times (\mathbf{r} \times \mathbf{v}) - \mu \frac{\mathbf{r}}{r} \tag{1-51}$$

Expanding the vector triple product, we get

$$\mu\mathbf{e} = (\mathbf{v} \bullet \mathbf{v})\mathbf{r} - (\mathbf{r} \bullet \mathbf{v})\mathbf{v} - \mu \frac{\mathbf{r}}{r}$$

Noting that $(\mathbf{v} \cdot \mathbf{v}) = v^2$ and collecting terms gives

$$\boxed{\mu\mathbf{e} = \left(v^2 - \frac{\mu}{r}\right)\mathbf{r} - (\mathbf{r} \bullet \mathbf{v})\mathbf{v}} \tag{1-52}$$

The eccentricity vector will be used in orbit determination in Chapter 2.

1.6 RELATING *E* AND *h* TO THE GEOMETRY OF AN ORBIT

By comparing Equation (1-41) and Equation (1-42) we see immediately that the parameter or semi-latus rectum, p, of the orbit depends only on the specific angular momentum, h, of the satellite. By inspection, *for any orbit*,

$$\boxed{p = h^2 / \mu} \tag{1-53}$$

In order to see intuitively why an increase in h should result in a larger value for p consider the following argument:

Suppose that a cannon were set up on the top of a high mountain whose summit extends above the sensible atmosphere (so that we may neglect atmospheric drag). If the muzzle of the cannon is aimed horizontally and the cannon is fired, Equation (1-32) tells us that $h = rv$ since the flight-path angle, ϕ, is zero. Therefore, progressively increasing the muzzle velocity, v, is equivalent to increasing h, Figure 1-9 shows the family of curves that represents the trajectory or orbit of the cannonball as the angular momentum of the "cannonball satellite" is progressively increased. Notice that each trajectory is a conic section with the focus located at the center of Earth and that as h is increased the parameter p of the orbit also increases just as Equation (1-51) predicts.

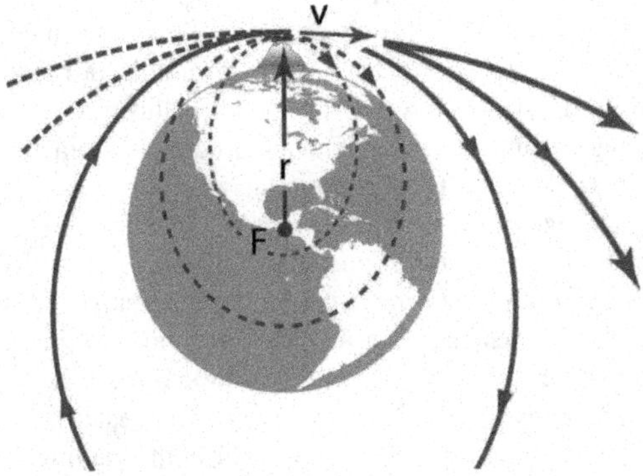

Figure 1-9 "Cannonball Satellite."

As a by-product of this example we note that at periapsis or apoapsis of any conic orbit the velocity vector (which is always tangent to the orbit) is directed horizontally and the flight-path angle, ϕ, is zero. We can then write, as a corollary to Equation (1-32),

$$\boxed{h = r_p v_p = r_a v_a} \tag{1-54}$$

If we write the energy Equation (1-29) for the periapsis point and substitute from Equation (1-54) we obtain

$$\mathcal{E} = \frac{v^2}{2} - \frac{\mu}{r} = \frac{h^2}{2r_p^2} - \frac{\mu}{r_p} \tag{1-55}$$

But from Equation (1-46)

$$r_p = a(1 - e) \tag{1-56}$$

and from Equation (1-44) and Equation (1-53)

$$h^2 = \mu a(1 - e^2) \tag{1-57}$$

Therefore

$$\mathcal{E} = \frac{\mu a(1 - e^2)}{2a^2(1 - e)^2} - \frac{\mu}{a(1 - e)} \tag{1-58}$$

which reduces to

$$\boxed{\mathcal{E} = -\frac{\mu}{2a}} \tag{1-59}$$

This simple relationship, which is valid *for all conic orbits*, tell us that the semi-major axis, a, of an orbit depends only on the specific mechanical energy, E of the satellite (which in turn depends only on the magnitude of **r** and **v** at any point along the orbit). Figure 1-9 serves as well to illustrate the intuitive explanation of this fundamental relationship since progressively increasing the muzzle velocity of the cannon also progressively increases E.

Many students find Equation (1-59) misleading. You should study it together with Figure 1-8, which shows that, for the circle and ellipse, a is positive, while, for the parabola, a is infinite and, for the hyperbola, a is negative. This implies that the specific mechanical energy of a satellite in a closed orbit (circle or ellipse) is negative, while E for a satellite in a parabolic orbit is zero and on a hyperbolic orbit the energy is positive. Thus the energy of a satellite (negative, zero, or positive) alone determines the type of conic orbit the satellite is in.

Since h alone determines p and since E alone determines a, the two together determine e (which specifies the exact shape of conic orbit). This can be shown as follows:

$$p = a(1-e^2); \text{ therefore, } e = \sqrt{1-\frac{p}{a}}$$

$$p = h^2/\mu \text{ and } a = -\mu/2E$$

So, for any *conic orbit*,

$$e = \sqrt{1+\frac{2Eh^2}{\mu^2}} \qquad (1\text{-}60)$$

Notice again that if E is negative, e is positive and less than 1 (an ellipse, or a circle if $e = 0$); if E is zero, e is exactly 1 (a parabola); if E is positive, e is greater than 1 (a hyperbola). But what if h is zero regardless of what value E has? The eccentricity will be exactly 1 but the orbit will not be a parabola! Rather, the orbit will be a degenerate conic (a point or straight line). The student should be aware of this pitfall. Namely, all parabolas have an eccentricity of 1 but an orbit whose eccentricity is 1 need not be a parabola—it could be a degenerate conic.

Example Problem. For a given satellite, $E = -20.0 \text{ km}^2/\text{s}^2$ and $e = 0.2$. Determine its specific angular momentum, semi-latus rectum and semi-major axis.

$$a = -\frac{\mu}{2E} = \underline{\underline{9.96503 \times 10^3 \text{ km}}}$$

$$p = a(1 - e^2) = \underline{\underline{9.566429 \times 10^3 \text{km}}}$$

$$h = \sqrt{p\mu} = \underline{\underline{6.175103 \times 10^4 \text{km}^2/\text{s}}}$$

Example Problem. A radar tracing station tells us that a certain decaying weather satellite has e = 0.1 and perigee altitude = 400 km. Determine its altitude at apogee, specific mechanical energy and specific angular momentum.

$$r_p = r_\oplus + 400 = 6{,}778.14 \text{ km}$$

$$p = r_p\,(1 + e) = 7{,}455.96 \text{ km}$$

$$r_a = \frac{p}{1 - e} = 8{,}284.40 \text{ km}$$

$$\text{altitude at apogee} = r_a - r_\oplus = \underline{\underline{1{,}906.25 \text{ km}}}$$

$$h = \sqrt{p\mu} = \underline{\underline{545{,}200 \text{ km}^2/\text{s}^2}}$$

$$2a = r_a + r_p = 15{,}060 \text{ km}$$

$$E = -\frac{\mu}{2a} = \underline{\underline{-26.4631 \text{ km}^2/\text{s}^2}}$$

1.7 THE ELLIPTICAL ORBIT

The orbits of all the planets in the solar system as well as the orbits of all Earth satellites are ellipses. Since an ellipse is a closed curve, an object in an elliptical orbit travels the same path over and over. The time for the satellite to go once around its orbit is called the period. We will first look at some geometrical results that apply only to the ellipse and then derive an expression for the period of an elliptical orbit.

1.7.1 Geometry of the Ellipse

An ellipse can be constructed using two pins and a loop of thread. The method is illustrated in Figure 1-10. Each pin marks the location of a focus and, since the length of the thread is a constant, the sum of the distances from any point on an ellipse to each focus ($r + r'$) is a constant. When the pencil is at either the periapsis, r_p, or the apoapsis, r_a, it is easy to see that, specifically

$$r + r' = 2a \tag{1-61}$$

By inspection, the radius of periapsis and the radius of apoapsis are related to the major axis of an ellipse as

$$r_p + r_a = 2a \tag{1-62}$$

Also by inspection the distance between the foci is

$$r_a - r_p = 2c \tag{1-63}$$

Since, in general, e is defined as c/a, Equations (1-62) and (1-63) combine to yield

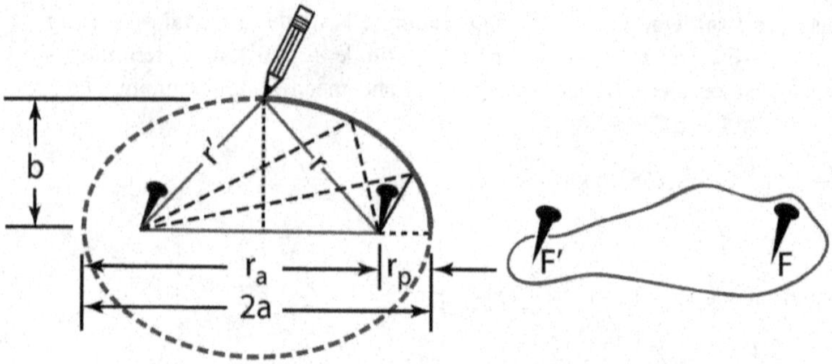

Figure 1-10 Simple Way to Construct an Ellipse.

$$e = \frac{r_a - r_p}{r_a + r_p} \tag{1-64}$$

The width of an ellipse at the center is called the minor axis, 2b. At the end of the minor axis, r and r′ are equal, as illustrated at the right of Figure 1-10. Since $r + r' = 2a$, r and r′ must both be equal to a at this point. Dropping a perpendicular line (dotted line in Figure 1-10) to the major axis, we can form a right triangle from which we conclude that

$$a^2 + b^2 = c^2 \tag{1-65}$$

1.7.2 Period of an Elliptical Orbit

If you refer to Figure 1-11, you will see that the horizontal component of velocity of a satellite is simply $v \cos\phi$, which can also be expressed as $r\dot{\nu}$. Using Equation (1-32) we can express the specific angular momentum of the satellite as

$$h = \frac{r^2 d\nu}{dt} \tag{1-66}$$

which, when rearranged, becomes

$$dt = \frac{r^2}{h} d\nu \tag{1-67}$$

But from elementary calculus we know that the differential element of area, dA, swept out by the radius vector as it moves through an angle $d\nu$ is given by the expression

$$dA = \frac{1}{2} r^2 d\nu \tag{1-68}$$

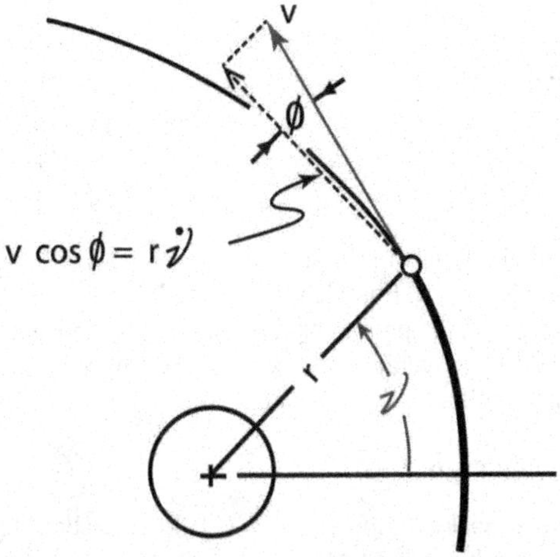

Figure 1-11 Horizontal Component of v.

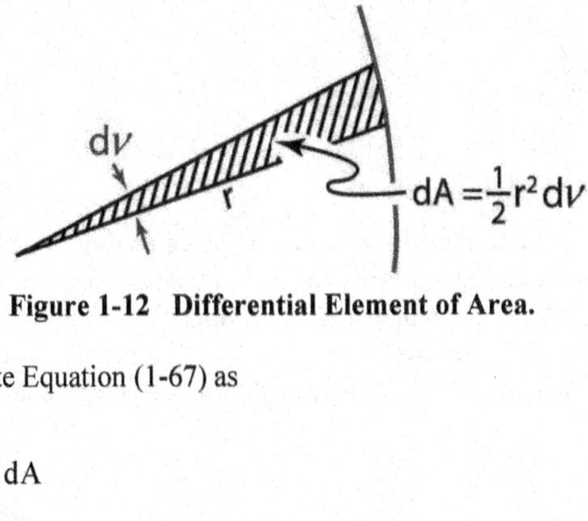

Figure 1-12 Differential Element of Area.

So we can rewrite Equation (1-67) as

$$dt = \frac{2}{h} dA \tag{1-69}$$

Equation (1-69) proves Kepler's second law that "equal areas are swept out by the radius vector in equal time intervals" since h is a constant for an orbit.

During one orbital period the radius vector sweeps out the entire area of the ellipse. Integrating Equation (1-69) for one period gives us

$$\mathbb{P} = \frac{2\pi ab}{h} \tag{1-70}$$

where πab is the total area of an ellipse and $\mathbb{P}$ is the period. From Equations (1-65), (1-43) and (1-44)

$$b = \sqrt{a^2 - c^2} = \sqrt{a^2(1 - e^2)} = \sqrt{ap} \tag{1-71}$$

and, since $h = \sqrt{\mu p}$,

$$\mathbb{P} = \frac{2\pi}{\sqrt{\mu}} a^{\frac{3}{2}} \tag{1-72}$$

Thus, the period of an elliptical orbit around a given body depends *only on the size of the semi-major axis*, a. Equation (1-72), incidentally, proves Kepler's third law that "the square of the period is proportional to the cube of the mean distance" since a, being the average of a periapsis and apoapsis radii, is the "mean distance" of the satellite from the prime focus.

1.8 THE CIRCULAR ORBIT

The circle is just a special case of an ellipse so all the relationships we just derived for the elliptical orbit including the period are also valid for the circular orbit. Of course, the semi-major axis of a circular orbit is just its radius, so Equation (1-72) is simply

$$\mathbb{P} = \frac{2\pi}{\sqrt{\mu}} r_{cs}^{\frac{3}{2}} \tag{1-73}$$

1.8.1 Circular Satellite Speed

The speed necessary to place a satellite in a circular orbit is called circular speed. Naturally, the satellite must be launched in the horizontal direction at the desired altitude to achieve a circular orbit. The latter condition is called circular velocity and implies both the correct speed and direction. We can calculate the speed required for a circular orbit of radius r_{cs} from the energy equation:

$$E = \frac{v^2}{2} - \frac{\mu}{r} = -\frac{\mu}{2a} \tag{1-74}$$

If we remember that $r_{cs} = a$, we obtain

$$\frac{v_{cs}^2}{2} - \frac{\mu}{r_{cs}} = -\frac{\mu}{2r_{cs}} \tag{1-75}$$

which reduces to

$$\boxed{v_{cs} = \sqrt{\frac{\mu}{r_{cs}}}} \tag{1-76}$$

Notice that the greater the radius of the circular orbit the less speed is required to keep the satellite in this orbit. For a low altitude Earth orbit, circular speed is about 7.6 km/s while the speed required to keep the Moon in its orbit around Earth is only about 1.022 km/s.

1.9 THE PARABOLIC ORBIT

The parabolic orbit is rarely found in nature although the orbits of some comets approximate a parabola. The parabola is interesting because it represents the borderline case between open and closed orbits. An object traveling a parabolic path is on a one-way trip to infinity and will never retrace the same path again.

1.9.1 Geometry of the Parabola

There are only a few geometrical properties peculiar to the parabola that you should know. One is that the two arms of a parabola become more and more parallel as one extends them further and further to the left of the focus in Figure 1-13. Another is that, since the eccentricity of parabola is exactly 1, the periapsis radius is just

$$r_p = \frac{p}{2} \tag{1-77}$$

which follows from Equation (1-46). Of course, there is no apoapsis for a parabola and it may be thought of as an "infinitely long ellipse."

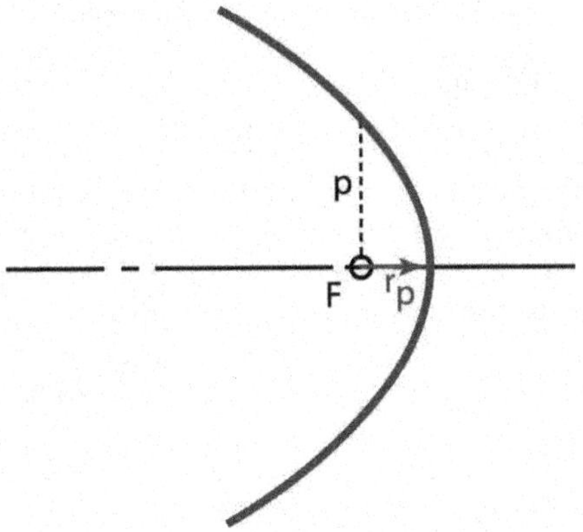

Figure 1-13 Geometry of the Parabola.

1.9.2 Escape Speed

Even though the gravitational field of the Sun or a planet theoretically extends to infinity, its strength decreases so rapidly with distance that only a finite amount of kinetic energy is needed to overcome the effects of gravity and allow an object to coast to an infinite distance without "falling back." The speed that is just sufficient to do this is called the escape speed. A space probe that is given escape speed in any direction will travel on a parabolic escape trajectory. Theoretically, as its distance from the central body approaches infinity, its speed approaches zero. We can calculate the speed necessary to escape by writing the energy equation for two points along the escape trajectory: first at a general point a distance, r, from the center where the "local escape speed" is v_{esc}, and then at infinity where the speed will be zero:

$$E = \frac{v_{esc}^2}{2} - \frac{\mu}{r} = \cancelto{0}{\frac{v_\infty^2}{2}} - \cancelto{0}{\frac{\mu}{r_\infty}} = 0$$

from which we get

$$\boxed{v_{esc} = \sqrt{\frac{2\mu}{r}}} \qquad (1\text{-}78)$$

Since the specific mechanical energy, E, must be zero if the probe is to have zero speed at infinity and since $E = -\mu/2a$, the semi-major axis a, of the escape trajectory must be infinite, which confirms that it is a parabola. Note that $v_{esc} = \sqrt{2}v_{cs}$ for the same altitude.

As you would expect, the farther away you are from the central body (larger value of r) the less speed it takes to escape the remainder of the gravitational field. The escape speed from the surface of Earth is about 11.179 km/s while from a point 7,000 km above the surface it is only 7.719 km/s.

Example Problem. In Figure 1-14, a space probe is to be launched on an escape trajectory from a circular parking orbit that is at an altitude of 200 km above Earth. Calculate the minimum escape speed required to escape from the parking orbit altitude. (Ignore the gravitational forces of the Sun and other planets.) Sketch the escape trajectory and the circular parking orbit.

a. Escape Speed:

Earth's gravitational parameter is

$$\mu = 398{,}600.4418 \text{ km}^3/\text{s}^2$$

Radius of a circular orbit is

$$r = r_{Earth} + \text{Altitude Circular Orbit}$$
$$= 6{,}578.14 \text{ km}$$

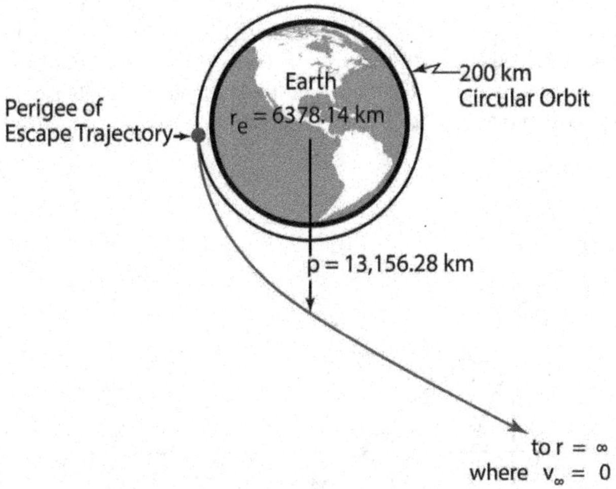

Figure 1-14 Escape Trajectory for Example Problem.

From Equation (1-79)

$$v_{esc} = \sqrt{\frac{2\mu}{r}} = 11.0086 \text{ km/s}$$

b. Sketch of escape trajectory and circular parking orbit:

From the definition of escape speed the energy constant is zero on the escape trajectory, which is therefore parabolic. The parameter p is determined by Equation (1-46), if we remember that the eccentricity is 1.0 for the escape trajectory:

$$p = r_p (1 + e)$$
$$= 6{,}578.14 \times 2 = 13{,}156.28 \text{ km}$$

1.10 THE HYPERBOLIC ORBIT

Meteors which strike Earth and interplanetary probes sent from Earth travel hyperbolic paths relative to Earth. A hyperbolic orbit is necessary if we want the probe to have some speed left over after it escapes Earth's gravitational field. The hyperbola is an unusual and interesting conic section because it has two branches. Its geometry is worth a few moments of study.

1.10.1 Geometry of the Hyperbola

The arms of a hyperbola are asymptotic to two intersecting straight lines (the asymptotes). If we consider the left-hand focus, F, as the prime focus (where the center of our gravitational body is located), then only the left branch of the hyperbola represents the possible orbit. If, instead, we assume a force of repulsion between our satellite and the body located at F (such as the force between two like-charged electrical particles), then the right-hand branch represents the orbit. The parameters a, b and c are labeled in Figure 1-15. Obviously,

$$c^2 = a^2 + b^2 \tag{1-79}$$

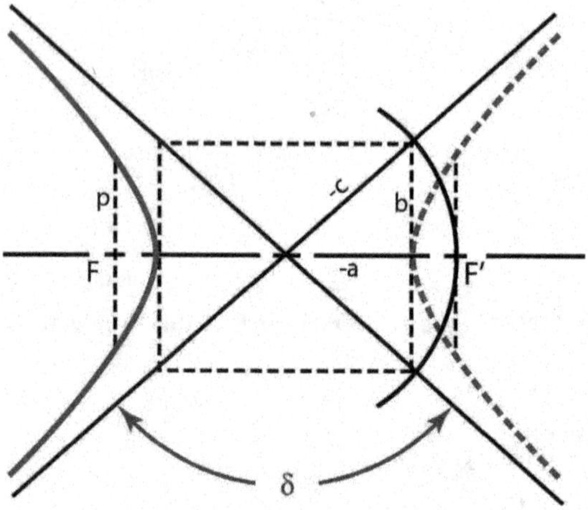

Figure 1-15 Geometry of the Hyperbola.

for the hyperbola. The angle between the asymptotes, which represents the angle through which the path of a space probe is turned by an encounter with a planet, is labeled δ (delta) in Figure 1-15. The turning angle, δ, is related to the geometry of the hyperbola as follows:

$$\sin\frac{\delta}{2} = \frac{a}{c} \tag{1-80}$$

But since $e = c/a$ Equation (1-80) becomes

$$\boxed{\sin\frac{\delta}{2} = \frac{1}{e}} \tag{1-81}$$

The greater the eccentricity of the hyperbola, the smaller will be the turning angle, δ.

1.10.2 Hyperbolic Excess Speed

If you give a space probe exactly escape speed, it will just barely escape the gravitational field, which means that its speed will be approaching zero as its distance from the force center approaches infinity. If, on the other hand, we give our probe more than escape speed at a point near Earth we would expect the speed at a great distance from Earth to be approaching some finite constant value. This residual speed that the probe would have left over even at infinity is called "hyperbolic excess speed." We can calculate this speed from the energy equation written for two points on the hyperbolic escape trajectory—a point near Earth called the "burnout point" and a point an infinite distance from Earth where the speed will be the hyperbolic excess speed, v_∞.

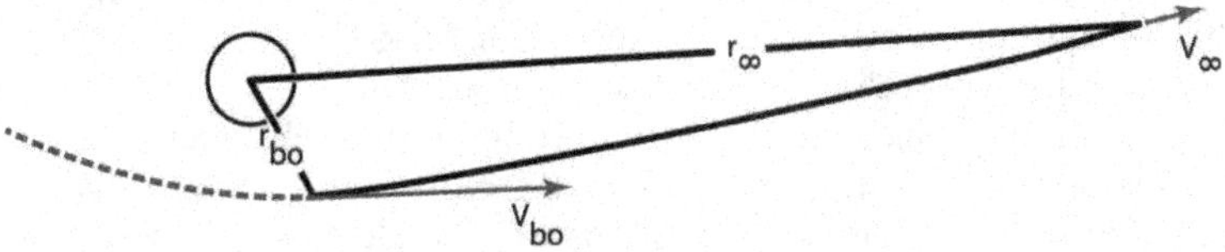

Figure 1-16 Hyperbolic Excess Speed.

Since specific mechanical energy does not change along an orbit, we may equate (E) at the burnout point and (E) at infinity:

$$E = \frac{v_{bo}^2}{2} - \frac{\mu}{r_{bo}} = \frac{v_\infty^2}{2} - \cancelto{0}{\frac{\mu}{r_\infty}} \tag{1-82}$$

from which we conclude that

$$v_\infty^2 = v_{bo}^2 - \frac{2\mu}{r_{bo}} = v_{bo}^2 - v_{esc}^2 \tag{1-83}$$

Note that if v_∞ is zero (as it is on a parabolic trajectory) v_{bo} becomes simply the escape speed.

1.10.3 Sphere of Influence

It is, of course, absurd to talk about a space probe "reaching infinity" and in this sense it is meaningless to talk about escaping a gravitational field completely. It is a fact, however, that once a space probe is a great distance (say a million km) from Earth, for all practical purposes it has escaped. In other words, it has already slowed down to very nearly its hyperbolic excess speed. It is convenient to define a sphere around every gravitational body and say that when a probe crosses the edge of this "sphere of influence" it has escaped. Although it is difficult to get even two people to agree on exactly where the sphere of influence should be drawn, the concept is convenient and is widely used, especially in lunar and interplanetary trajectories.

1.11 CANONICAL UNITS

Astronomers have made great progress in determining the precise distance and mass of objects in space. Such fundamental qualities as the mean distance from Earth to the Sun, the mass and mean distance of the Moon and the mass of the Sun are known with some accuracy. However, for the purposes of understanding the relationships among key variables it is useful to assume the mass of the Sun to be "1 mass unit" and the mean distance from Earth to the Sun be our unit of distance, which is called one "astronomical unit." All other masses and distances can then be given in terms of these assumed units even though we do not know precisely the absolute value of the Sun's mass and distance in kilograms or kilometers. Astronomers call this normalized system of units "canonical units."

While the use of canonical units has been decreased in the second edition, the following sections have been retained as examples of how canonical units can be useful in simplifying concepts and calculations for educational purposes. Several example problems and chapter exercises also retain the use of canonical units where they help illuminate the underlying concepts.

1.11.1 The Reference Orbit

We will use a system of units based on a hypothetical circular reference orbit. In a two-body problem where the Sun is the central body the reference orbit will be a circular orbit whose radius is one astronomical unit (AU). For other problems where Earth, the Moon, or some other planet is the central body the reference orbit will be a minimum altitude circular orbit just grazing the surface of the planet.

We will define our distance unit (DU) to be the radius of the reference body. If we now define our time unit (TU) such that the speed of a satellite in the hypothetical reference orbit is 1 DU/TU, then the value of the gravitational parameter, μ, will turn out to be 1 DU^3/TU^2.

Unless it is perfectly clear which reference orbit the units in your problem are based on, you will have to indicate this by means of a subscript on the symbol DU and TU. This is most easily done by annexing as a subscript the astronomer's symbol for the Sun, Earth, or other planet. The most commonly used symbols are:

☉ The Sun	♂ Mars
☾ The Moon	♃ Jupiter
☿ Mercury	♄ Saturn
♀ Venus	♅ Uranus
⊕ Earth	♆ Neptune

The concept of the reference orbit is illustrated in Figure 1-17.

Values for the commonly used astrodynamic constants and their relationship to canonical units are listed in the appendices.

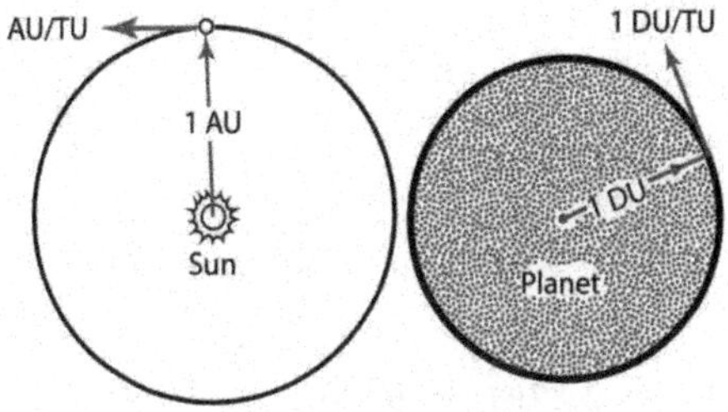

Figure 1-17 Reference Circular Orbits.

Example Problem. A space object is sighted in an altitude of 3,189.069 km above Earth traveling at 7.90537 km/s and a flight path angle of 0° at the time of sighting. Using canonical units determine *E*, h, p, e, r_a, and r_p.

Convert altitude and speed to Earth canonical units:

$$\text{Alt} = 3{,}189.069\ \text{km}/6{,}378.136\ \text{km} = 0.5\ \text{DU}_\oplus$$

The gravitational parameter and Earth radius are

$$\mu_\oplus = 3.986012\times10^5\ \text{km}^3/\text{s}^2 = 1\,\text{DU}_\oplus^3/\text{TU}_\oplus^2$$

$$r_\oplus = 1\,\text{DU}_\oplus$$

The radius of the object from the center of Earth is:

$$r = r_\oplus + \text{Alt} = 1.5\ \text{DU}_\oplus$$

Find *E* from Equation (1-29):

$$E = \frac{v^2}{2} - \frac{\mu_\oplus}{r} = -0.167\ \text{DU}_\oplus^2/\text{TU}_\oplus^2 = -10.417\ \text{km}^2/\text{s}^2$$

Find h from Equation (1-32):

$$h = rv\cos\phi = 1.50\,\text{DU}_\oplus^2/\text{TU}_\oplus = 7.56324\times10^4\ \text{km}^2/\text{s}$$

Find p from Equation (1-53):

$$p = \frac{h^2}{\mu_\oplus} = 2.25\,\text{DU}_\oplus = 1.43508\times10^4\ \text{km}$$

Find e from Equation (1-60):

$$e = \sqrt{1 + \frac{2Eh^2}{\mu^2}} = 0.5$$

Find r_a from Equation (1-47):

$$r_a = \frac{p}{1 - e} = 4.5\,DU_\oplus = 2.87017 \times 10^4 \text{ km}$$

Find r_p from Equation (1-46):

$$r_p = \frac{p}{1 + e} = 1.5\,DU_\oplus = 9.56722 \times 10^3 \text{ km}$$

Exercises

1.1 The position and velocity of a satellite at a given instant are described by

$\mathbf{r} = 2\mathbf{I} + 2\mathbf{J} + 2\mathbf{K}$ (Distance Units)

$\mathbf{v} = -0.4\mathbf{I} + 0.2\mathbf{J} + 0.4\mathbf{K}$ (Distance Units per Time Units)

where **IJK** is a non-rotating geocentric coordinate system. Find the specific angular momentum and total specific mechanical energy of the satellite.

(Answer: $\mathbf{h} = 0.4\mathbf{I} - 1.6\mathbf{J} + 1.2\mathbf{K}$ DU^2/TU
$E = -0.1087$ DU^2/TU^2)

1.2 For a certain satellite the observed velocity and radius at $\nu = 90°$ are observed to be 13.716 km/s and 7,408 km, respectively. Find the eccentricity of the orbit.

(Answer: $e = 1.581$)

1.3 An Earth satellite is observed to have a height of perigee of 185.2 km and a height of apogee of 1,111.2 km.

Find the period of the orbit.

1.4 Why, in general, is a completely determined closed solution of the n-body problem an impossibility if $n \geq 3$?

1.5 For a certain Earth satellite it is known that the semi-major axis, a, is 9,144 km. The orbit eccentricity is 0.2.

a. Find its perigee and apogee distances from the center of Earth.

b. Find the specific energy of the trajectory.

c. Find the semi-latus rectum or parameter (p) of the orbit.

d. Find the length of the position vector at a true anomaly of 135°.

(Answer: r = 10,223 km)

1.6 Find the equation for the velocity of a satellite as a function of total specific mechanical energy and distance from the center of Earth.

1.7 Prove that $r_{apoapsis} = a(1 + e)$.

1.8 Identify each of the following trajectories as either circular, elliptical, hyperbolic or parabolic:

a. $r = 3$ DU
$v = 1.5$ DU/TU

b. $r_{perigee} = 1.5$ DU
$p = 3$ DU

c. $\mathcal{E} = -1/3\ DU^2/TU^2$
$p = 1.5$ DU

d. $\mathbf{r} = \mathbf{J} + 0.2\mathbf{K}$
$v = 0.9\mathbf{I} + 0.123\mathbf{K}$

e. $\mathbf{r} = 1.01\mathbf{K}$
$\mathbf{v} = \mathbf{I} + 1.4\mathbf{K}$

1.9 A space vehicle enters the sensible atmosphere of Earth (100 km) with a velocity of 7.62 km/s at a flight-path angle of –60°. What was its velocity and flight-path angle at an altitude of 200 km during descent?

(Answer: v = 7.503 km/s, $\phi = -59.991$ deg)

1.10 Show that two-body motion is confined to a plane fixed in space.

1.11 A sounding rocket is fired vertically. It achieves a burnout speed of 3.4 km/s at an altitude of 30 km. Determine the maximum altitude attained. (Neglect atmospheric drag.)

1.12 Given that e = c/a, derive values for e for circles, ellipses and hyperbolas.

1.13 Show by means of differential calculus that the position vector is an extremum (maximum or minimum) at the apses of the orbit.

1.14 Given the equation $r = \dfrac{p}{1 + e\cos\nu}$ plot at least four points, and sketch and identify the locus and label the major dimensions for the following conic sections:

a. $p = 2, \quad e = 0$

b. $p = 6, \quad e = 0.2$

c. $p = 6, \quad e = 0.6$

d. $p = 3, \quad e = 1$

e. $p = 2, \quad e = 2$

(Hint: Polar graph paper would be of help here!)

1.15 Starting with:

$$m_k \ddot{\mathbf{r}}_k = \sum_{\substack{j=1 \\ j \neq k}}^{m} \frac{G m_j m_k}{r_{jk}^2} \frac{\mathbf{r}_j - \mathbf{r}_k}{r_{jk}}$$

where r_j is the vector from the origin of an inertial frame to any j^{th} body and r_{jk} is the scalar distance between j^{th} and k^{th} bodies ($r_{jk} \equiv r_{kj}$)

a. Show that

$$\sum_{k=1}^{m} m_k \ddot{\mathbf{r}}_k = 0$$

b. Using the definition of a system's mass center show:

$$\mathbf{r}_c = \mathbf{a}t + \mathbf{b}$$

where $\mathbf{r}_c$ is the vector from the inertial origin to the system mass center and **a** and **b** are constant vectors.

c. What is the significance of the equation derived in part b above?

1.16 A satellite is injected into an elliptical orbit with a semi-major axis equal to 25,513 km. When it is precisely at the end of the semi-minor axis it receives an impulsive velocity change just sufficient to place it into an escape trajectory. What was the magnitude of the velocity change?

(Answer: $\Delta \mathbf{v} = 1.637$ km/s)

1.17 Show that the speed of a satellite on an elliptical orbit at either end of the *minor* axis is the same as the local circular satellite speed at that point.

1.18 Show that when an object is located at the intersection of the semi-minor axis of an elliptical orbit the eccentricity of the orbit can be expressed as $e = -\cos\nu$.

1.19 * **SSN** (Space Surveillance Network) detects an unidentified object with the following parameters:

altitude = 0.5 DU

speed = $\sqrt{2/3}$ DU/TU

flight-path angle = 30°

Is it possible that this object is a space probe intended to escape Earth, an Earth satellite or a ballistic missile?

1.20 * Prove that the flight-path angle is equal to 45° when $\nu = 90°$ on all parabolic trajectories.

1.21 * Given two spherically symmetric bodies of considerable mass, assume that the only force that acts is a repulsive force, proportional to the product of the masses and inversely proportional to the cube of the distance between the masses that acts along the line connecting the centers of the bodies. Assume that Newton's second law holds ($\Sigma F = ma$) and derive a differential equation of motion for these bodies.

1.22 * A space vehicle destined for Mars was first launched into a 200 km circular parking orbit.

a. What was the speed of the vehicle at injection into its parking orbit?

The vehicle coasted into orbit for a period of time to allow system checks to be made and then was restarted to increase its velocity to 11.386 km/s which placed it on an interplanetary trajectory toward Mars.

b. Find $\mathbf{e}$, $\mathbf{h}$ and $\mathcal{E}$ relative to Earth for the escape orbit. What kind of orbit is it?

c. Compare the velocity at 2,000,000 km from Earth with the hyperbolic excess velocity, v_∞. Why are the two so nearly alike?

List of References

1. Newman, James R. "Commentary on Isaac Newton," in *The World of Mathematics*. Vol. 1. New York, NY, Simon and Schuster, 1956.

2. Lodge, Sir Oliver. "Johann Kepler," in *The World of Mathematics*. Vol. 1. New York, NY, Simon and Schuster, 1956.

3. Koestler, Arthur. *The Watershed*. Garden City, NY, Anchor Books, Doubleday & Company, Inc., 1960.

4. Turnbull, Herbert Westren. *The Great Mathematicians*. 4th ed. London, Methuen & Co., Ltd., 1951.

5. Newton, Sir Isaac. *Principia*. Motte's translation revised by Cajori. Vol. 1. Berkeley and Los Angeles, University of California Press, 1962.

Chapter 2

Orbit Determination from Observations

Finally, as all our observations, on account of the imperfection of the instruments and of the senses, are only approximations to the truth, an orbit based only on the six absolutely necessary data may be still liable to considerable errors. In order to diminish these as much as possible, and thus to reach the greatest precision attainable, no other method will be given except to accumulate the greatest number of the most perfect observations and to adjust the elements, not so as to satisfy this or that set of observations with absolute exactness, but so as to agree with all in the best possible manner.

-Carl Friedrich Gauss[1]

2.1 HISTORICAL BACKGROUND

The first method of finding the orbit of a body from three observations was devised by Newton and is given in the *Principia*. The most publicized result of applying Newton's method of orbit determination belongs to our old friend Sir Edmond Halley.

In 1705, shortly after the publication of the *Principia* (1687), Halley set to work calculating the orbits of 24 comets that had been sighted between 1337 and 1698, using the method that Newton had described. Newton's arguments, presented in a condensed form, were so difficult to understand that of all his contemporaries only Halley was able to master his technique and apply it to the calculation of the orbits of those comets for which there was a sufficient number of observations.

In a report on his findings published in 1705 he states: "Many considerations incline me to believe the comet of 1531 observed by Apianus to have been the same as that described by Kepler and Longomontanus in 1607 and which I again observed when it returned in 1682. All the elements agree; . . . whence I would venture confidently to predict its return, namely in the year 1758."[2]

In a later edition of the same work published in 1752, Halley confidently identified his comet with those that had appeared in 1305, 1380 and 1456 and remarked, "wherefore, if according to what we have already said it should return again about the year 1758, candid posterity will not refuse to acknowledge that this was first discovered by an Englishman."[2]

No one could be certain he was right, much less that it might not succumb to a churlish whim and bump into Earth. The realization that stability of Earth's orbit depends "on a nice balance between the velocity with which Earth is falling toward the Sun and its tangential velocity at right angles to that fall" did not unduly disturb the handful of mathematicians and astronomers who understood what that equilibrium meant. But there were many others, of weaker faith in arithmetic and geometry, who would have preferred a less precarious arrangement.[3]

On Christmas day, 1758, Halley's Comet reappeared, just as he said it would. Recent investigations have revealed, in ancient Chinese chronicles, an entire series of earlier appearances of Halley's Comet—the earliest in 467 BC. The comet has appeared since in 1835, 1910 and 1986; its next visit is expected in 2061.

Newton's method of determining a parabolic orbit from observations depended on a graphical construction that, by successive approximations, led to the elements. The first completely analytical method for solving the same problem was given by Euler in 1744 in his *Theory of the Motion of Planets and Comets*. To Euler belongs the discovery of the equation connecting two radius vectors and the subtended chord of the parabola with the interval of time during which the comet describes the corresponding arc.

Lambert, in his works of 1761 to 1771, gave a generalized formulation of the theorem of Euler for the case of elliptical and hyperbolic orbits. But Lambert was a geometrician at heart and was not inclined to analytical development of his method. Nevertheless, he had an unusual grasp of the physics of the problem and actually anticipated many of the ideas that were ultimately carried out by his successors in better and more convenient ways.

Lagrange, who at age 16 was made Professor of Mathematics at Turin, published three memoirs on the theory of orbits, two in 1778 and one in 1783. It is interesting to note that the mathematical spark was kindled in the young Lagrange by reading a memoir of Halley. From the very first his writings were elegance itself; he would set to mathematics all the problems that his friends brought him, much as Schubert would set to music any stray rhyme that took his fancy. As one would expect, Lagrange brought to the incomplete theories of Euler and Lambert generality, precision, and mathematical elegance.

In 1780 Laplace published an entirely new method of orbit determination. The basic ideas of his method are given later in this chapter and are still of fundamental importance.

The theory of orbit determination was really brought to fruition through the efforts of the brilliant young German mathematician, Gauss, whose work led to the rediscovery of the asteroid Ceres in 1801 after it had become "lost." Gauss also invented the "method of least squares" to deal with the problem of fitting the best possible orbit to a large number of observations.

Today, with the availability of radar, laser systems and space-based position reference systems, referred to as global navigation satellite systems (GNSS), the problem of orbit determination is much simpler. Before showing you how it is done, however, we must digress a moment to explain how the size, shape and orientation of an orbit in three-dimensional space can be described by six quantities called "orbital elements."

2.2 COORDINATE SYSTEMS

Our first requirement for describing an orbit is a suitable inertial reference frame. In the case of orbits around the Sun such as planets, asteroids, comets and some deep-space probes describe, the heliocentric-ecliptic coordinate system is convenient. For satellites of Earth, we will want to use the geocentric-equatorial system. In order to describe these coordinate systems we will give the position of the origin, the orientation of the fundamental plane (i.e. the **X**–**Y** plane), the principal direction (i.e. the direction of the **X**-axis), and the direction of the **Z**-axis. Since the **Z**-axis must be perpendicular to the fundamental plane it is only necessary to specify which direction is positive. The **Y**-axis is always chosen so as to form a right-handed set of coordinate axes.

2.2.1 The Heliocentric-Ecliptic Coordinate System

As the name implies, the heliocentric-ecliptic system has its origin at the center of the Sun. The $\mathbf{X}_\varepsilon$–$\mathbf{Y}_\varepsilon$ or fundamental plane coincides with the "ecliptic," which is the plane of Earth's revolution around the Sun. The line of intersection of the ecliptic plane and Earth's equatorial plane defines the direction of the $\mathbf{X}_\varepsilon$-axis as shown in Figure 2-1. On the first day of spring a line joining the center of Earth and the center of the Sun points in the direction of the positive $\mathbf{X}_\varepsilon$-axis. This is called the vernal equinox direction and is given the symbol ♈ by astronomers because it is used to point in the direction of the constellation Aries (the ram). As you know, Earth wobbles slightly and its axis of rotation shifts in direction slowly over the centuries. This effect is known as precession and causes the line of intersection of Earth's equator and ecliptic to shift slowly. As a result the heliocentric-ecliptic system is not really an inertial reference frame and where extreme precision is required it would be necessary to specify that $\mathbf{XYZ}_\varepsilon$ coordinates of an object were based on the vernal equinox direction of a particular year or "epoch."[4]

2.2.2 The Geocentric-Equatorial Coordinate System

The geocentric-equatorial system has its origin at Earth's center. The fundamental plane is the equator and the positive **X**-axis points in the vernal equinox direction. The **Z**-axis points in the direction of the north pole. It is important to keep in mind when looking at Figure 2-2 that the **XYZ** system is not fixed to Earth and turning with it; rather, the geocentric-equatorial frame is non-rotating with respect to the stars (except for precession of the equinoxes) and Earth turns relative to it.

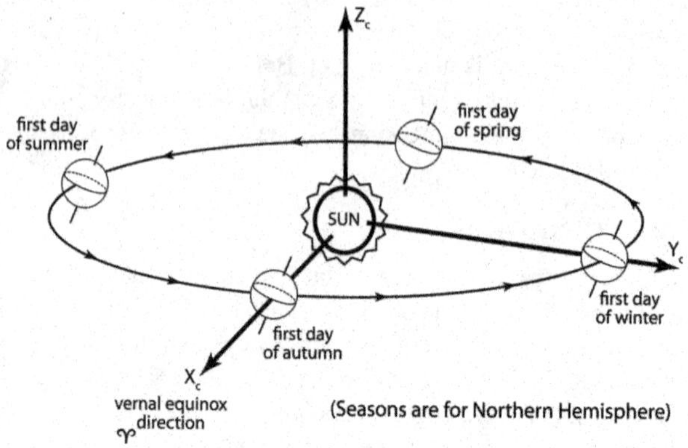

Figure 2-1 Heliocentric-Ecliptic Coordinate System.

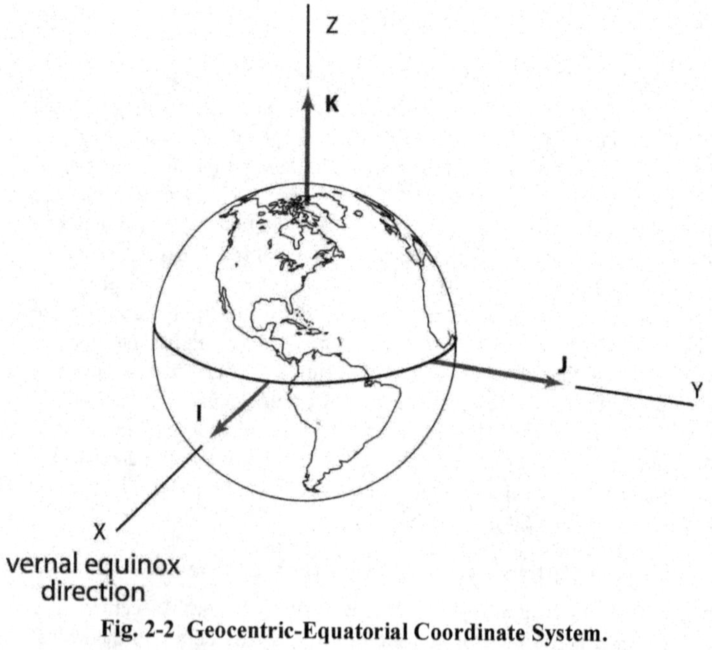

Fig. 2-2 Geocentric-Equatorial Coordinate System.

Unit vectors, **I, J** and **K**, shown in Figure 2-2, lie along the **X**, **Y** and **Z** axes, respectively, and will be useful in describing vectors in the geocentric-equatorial system.

2.2.3 The Right Ascension-Declination System

A coordinate system closely related to the geocentric-equatorial frame is the right ascension-declination system. The fundamental plane is the "celestial equator"—the extension of Earth's equatorial plane to a fictitious sphere of infinite radius called the "celestial sphere." The position of an object projected against the celestial sphere is described by two angles called right ascension and declination.

As shown in Figure 2-3, the right ascension, α, is measured eastward in the plane of the celestial equator from the vernal equinox direction. The declination, δ, is measured northward from the celestial equator to the line of sight.

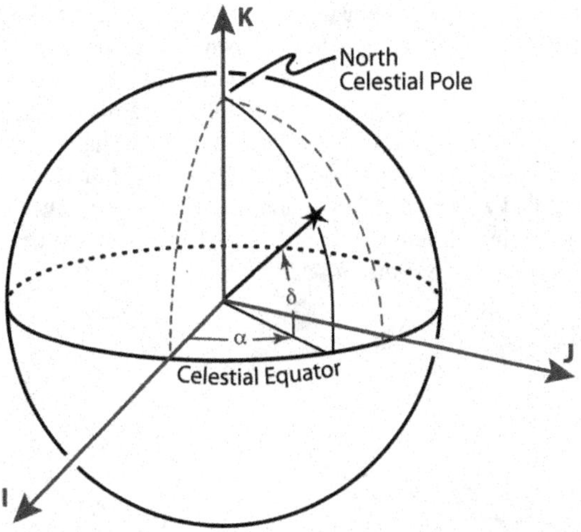

Figure 2-3 Right Ascension-Declination Coordinate System.

It may be implemented in spherical or rectangular coordinates, both defined by an origin at the center of Earth. The primary direction is toward the vernal equinox and a standard right-handed convention is used. The origin at the center of Earth means the coordinates are geocentric, that is, as seen from the center of Earth as if it were transparent. The fundamental plane and the primary direction mean that the coordinate system, while aligned with Earth's equator and pole, does not rotate with Earth but remains relatively fixed against the background stars. A right-handed convention means that coordinates are positive toward the north and toward the east in the fundamental plane.

Astronomers use the right ascension-declination system to catalog star positions accurately. Because of the enormous distances to the stars, their coordinates remain essentially unchanged even when viewed from opposite sides of Earth's orbit around the Sun.

While star positions are known accurately to fractions of an arc second, the optical systems that track satellites against the background of stars are limited in accuracy by atmospheric turbulence, lighting conditions and the finite number of pixels within the field of view (FOV) of the ground-based telescopes used in the tracking network. Individual measurements of a satellite's position from the tracking network are typically accurate to ~0.002° in topocentric right ascension and declination at the time of observation. Orbit determination from such optical sightings of a satellite will be discussed in a later section.

2.2.4 The Perifocal Coordinate System

One of the most convenient coordinate frames for describing the motion of a satellite is the perifocal coordinate system shown in Figure 2-4. Here the fundamental plane is the plane of the satellite's orbit. The coordinate axes are named $\mathbf{x}_\omega$, $\mathbf{y}_\omega$ and $\mathbf{z}_\omega$ The $\mathbf{x}_\omega$-axis points toward the periapsis; the $\mathbf{y}_\omega$-axis is rotated 90° in the direction of orbital motion and lies in the orbital plane; the $\mathbf{z}_\omega$-axis along h completes the right-handed perifocal system. Unit vectors in the direction of $\mathbf{x}_\omega$, $\mathbf{y}_\omega$ and $\mathbf{z}_\omega$ are called **P**, **Q** and **W**, respectively. Several space-based systems are being developed and deployed that can be capable of providing more accurate observations of satellites.

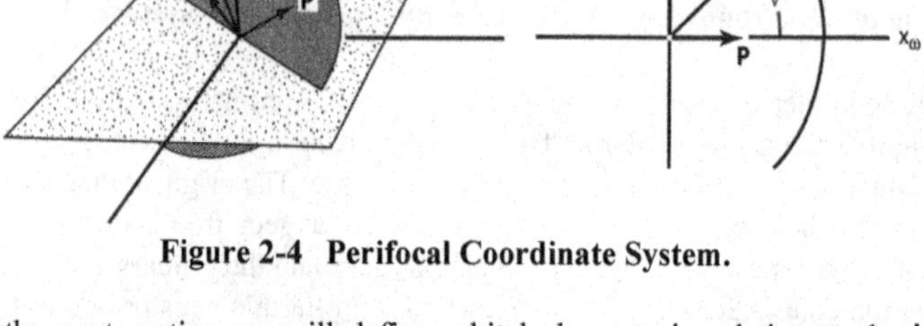

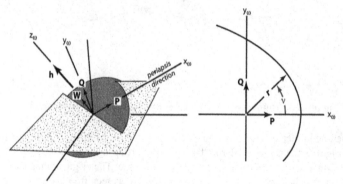

Figure 2-4 Perifocal Coordinate System.

In the next section we will define orbital elements in relation to the **IJK** vectors. When describing a heliocentric orbit, you may use the same definitions applied to $\mathbf{XYZ_e}$ axes. Another system, the topocentric-horizon coordinate system will be introduced in a later section.

2.3 CLASSICAL ORBITAL ELEMENTS

Five independent quantities called "orbital elements" are sufficient to completely describe the size, shape and orientation of an orbit. A sixth element is required to pinpoint the position of the satellite along the orbit at a particular time. The classical set of six orbital elements is defined with the help of Figure 2-5 as follows:

1. a, *semi-major axis*—a constant defining the size of the conic orbit.
2. e, *eccentricity*—a constant defining the shape of the conic orbit.
3. i, *inclination*—the angle between the **K** unit vector and the angular momentum vector **h**.
4. Ω, *longitude of the ascending node*—the angle, in the fundamental plane, between the **I** unit vector and the point where the satellite crosses through the fundamental plane in a northerly direction (ascending node) measured counterclockwise when viewed from the north side of the fundamental plane.
5. ω, *argument of periapsis*—the angle, in the plane of the satellite's orbit, between the ascending node and the periapsis point, measured in the direction of the satellite's motion. Strictly speaking, for a satellite in Earth orbit, the correct term is argument of perigee.
6. T, *time of periapsis passage*—the time when the satellite was at a periapsis. Several other options for the sixth orbital element are discussed below.

The above definitions are valid whether we are describing the orbit of an Earth satellite in the geocentric-equatorial system or the orbit of a planet in the heliocentric-ecliptic system. Only the definition of the unit vectors and the fundamental plane would be different.

It is common when referring to Earth satellites to use the term "argument of perigee" for ω. Similarly, the term "argument of perihelion" is used for Sun-centered orbits. In the remainder of this chapter we shall tacitly assume that we are describing the orbit of an Earth satellite in the geocentric-equatorial system using **IJK** unit vectors.

The list of six orbital elements defined above is by no means exhaustive. Frequently the semi-latus rectum, p, is substituted for a in the above list. Obviously, if you know a and e you can compute p.

Instead of argument of perigee, the following is sometimes used: Π, *longitude of perigee*—the angle from **I** to perigee measured eastward to the ascending node (if it exists) and then in the orbital plane to perigee. This is only strictly accurate when the orbit inclination is zero; otherwise it is an approximation since the two angles summed are in different planes. Refer to Figure 2-12 for the general case. If both Ω and ω are defined, then

$$\boxed{\Pi = \Omega + \omega} \tag{2-1}$$

If there is no perigee (circular orbit), then both ω and Π are undefined.

Any of the following may be substituted for time of periapsis passage and would suffice to locate the satellite at t_0:

ν_0, *true anomaly at epoch*—the angle, in the plane of the satellite's orbit, between perigee and the position of the satellite at a particular time, t_0, called the "epoch."

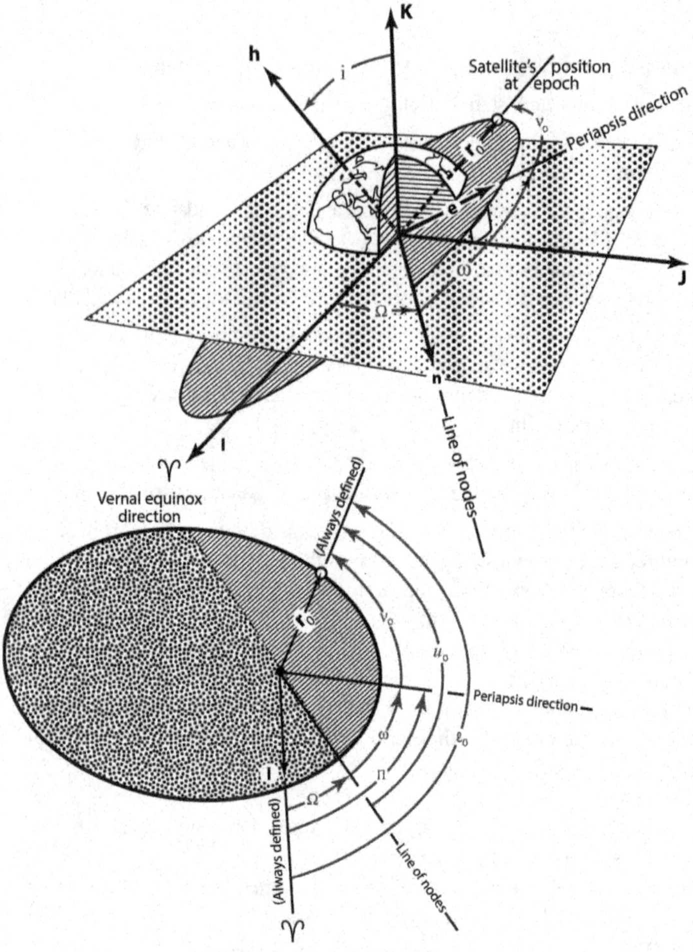

Figure 2-5 Orbital Elements.

u_0, *argument of latitude at epoch*—the angle, in the plane of the orbit, between the ascending node (if it exists) and the radius vector to the satellite at time t_0. If ω and ν_0 are both defined, then

$$\boxed{u_0 = \omega + \nu_0} \tag{2-2}$$

If there is no ascending node (equatorial orbit), then both ω and u_0 are undefined.

ℓ_0, *true longitude at epoch*—the angle between **I** and $\mathbf{r_0}$ (the radius vector to the satellite at t_0) measured eastward to the ascending node (if it exists) and then in the orbital plane $\mathbf{r_0}$. If Ω, ω and ν_0 are all defined, then

$$\boxed{\ell_0 = \Omega + \omega + \nu_0 = \Pi + \nu_0 = \Omega + u_0} \tag{2-3}$$

If there is no ascending node (equatorial orbit) then $\ell_0 = \Pi + \nu_0$. If there is no periapsis (circular orbit), then $\ell_0 = \Omega + \mu_0$. If the orbit is both circular and equatorial, ℓ_0 is simply the true angle from **I** to $\mathbf{r_0}$, both of which are *always* defined. As is the case with the longitude of perigee, the true longitude of epoch equation (Eq. (2-3)) is exact only for circular, equatorial orbits.

Two other terms frequently used to describe orbital motion are "direct" (or "prograde") and "retrograde." *Direct means easterly.* This is the direction in which the Sun, Earth and most of the planets and their moons rotate on their axes and the direction in which all of the planets revolve around the Sun. Retrograde is the opposite of direct. From Figure 2-5 you can see that inclinations between 0° and 90° imply direct orbits and inclinations between 90° and 180° are retrograde.

2.4 DETERMINING THE ORBITAL ELEMENTS FROM r AND v

Let us assume that a radar site on Earth is able to provide us with the vectors **r** and **v** representing the position and velocity of a satellite relative to the geocentric-equatorial reference frame at a particular time t_0. How do we find the six orbital elements that describe the motion of the satellite? The first step is to form the three vectors, **h**, **n** and **e**.

2.4.1 Three Fundamental Vectors—h, n and e

We have already encountered the angular momentum vector, **h**:

$$\boxed{\mathbf{h} = \mathbf{r} \times \mathbf{v}} \tag{2-4}$$

Thus

$$\mathbf{h} = \begin{bmatrix} \mathbf{I} & \mathbf{J} & \mathbf{K} \\ r_I & r_J & r_K \\ v_I & v_J & v_K \end{bmatrix} = h_I\mathbf{I} + h_J\mathbf{J} + h_K\mathbf{K} \tag{2-5}$$

An important thing to remember is that **h** is a vector *perpendicular to the plane of the orbit.*

The node vector, **n**, is defined as

$$\boxed{\mathbf{n} \equiv \mathbf{K} \times \mathbf{h}} \tag{2-6}$$

Thus

$$\mathbf{n} = \begin{bmatrix} \mathbf{I} & \mathbf{J} & \mathbf{K} \\ 0 & 0 & 1 \\ h_I & h_J & h_K \end{bmatrix} = n_I\mathbf{I} + n_J\mathbf{J} + n_K\mathbf{K} = -h_J\mathbf{I} + h_I\mathbf{J} \tag{2-7}$$

From the definition of a vector cross product, **n** must be perpendicular to both **K** and **h.** To be perpendicular to **K**, **n** would have to lie in the equatorial plane. To be perpendicular to **h**, **n** would have to lie in the orbital plane. Therefore, **n** must lie in both the equatorial and orbital planes, thus at their intersection, which is called the "line of nodes." Specifically, **n** *is a vector pointing along the line of nodes in the direction of the ascending node.* The magnitude of **n** is of no consequence to us. We are only interested in its direction.

The vector **e** is obtained from

$$\boxed{\mathbf{e} = \frac{1}{\mu}\left[\left(v^2 - \frac{\mu}{r}\right)\mathbf{r} - (\mathbf{r} \cdot \mathbf{v})\,\mathbf{v}\right]} \tag{2-8}$$

and is derived in Chapter 1 (Equation (1-52)). The vector **e** points *from the center of Earth (focus of the orbit) toward perigee with a magnitude exactly equal to the eccentricity of the orbit.*

All three vectors, **h**, **n** and **e**, are illustrated in Figure 2-5. Study this figure carefully. An understanding of it is essential to what follows.

2.4.2 Solving for the Orbital Elements

Now that we have **h**, **n** and **e** we can proceed rather easily to obtain the orbital elements. The parameter and eccentricity follow directly from **h** and **e** while all the remaining orbital elements are simply angles between two vectors whose components are now known. If we know how to find the angle between two vectors the problem is solved. In general, the cosine of the angle, α. between two vectors **a** and **b** is found by dotting the two vectors together and dividing by the product of their magnitudes. Since

$$\mathbf{a} \cdot \mathbf{b} = ab \cos \alpha$$

then

$$\cos \alpha = \frac{\mathbf{a} \cdot \mathbf{b}}{ab} \quad \text{(see Figure 2-6)} \tag{2-9}$$

Of course, being able to evaluate the cosine of an angle does not mean that you know the angle. You still have to decide whether the angle is smaller or greater than 180°. The answer to this quadrant resolution problem must come from other information in the problem as we shall see.

We can outline the method of finding the orbital elements as follows:

1. $p = h^2/\mu$
2. $e = |\mathbf{e}|$
3. Since the inclination i is the angle between **K** and **h**,

$$\cos i = \frac{h_K}{h} \quad \text{(where inclination can range from 0° to 180°)} \tag{2-10}$$

4. Since Ω is the angle between **I** and **n**,

$$\cos \Omega = \frac{n_I}{n} \quad \text{(where if } n_J > 0 \text{ then } \Omega \text{ is less than 180°)} \tag{2-11}$$

5. Since ω is the angle between **n** and **e**,

$$\cos \omega = \frac{\mathbf{n} \cdot \mathbf{e}}{ne} \quad \text{(where if } e_K > 0 \text{ then } \omega \text{ is less than 180°)} \tag{2-12}$$

6. Since ν_o is the angle between **e** and **r**,

$$\cos \nu_o = \frac{\mathbf{e} \cdot \mathbf{r}}{er} \quad \text{(where if } \mathbf{r} \cdot \mathbf{v} > 0 \text{ then } \nu_o \text{ is less than 180°)} \tag{2-13}$$

7. Since u_o is the angle between **n** and **r**,

$$\cos u_o = \frac{\mathbf{n} \cdot \mathbf{r}}{nr} \quad \text{(where if } r_K > 0 \text{ then } u_o \text{ is less than 180°)} \tag{2-14}$$

8. $\ell_o = \Omega + \omega + \nu_o = \Omega + u_o$ (see Figure 2-12)

All of the quadrant checks in parentheses make physical sense. If they don't make sense to you, look at the geometry of Figure 2-6 and study it until they do. The quadrant check for ν_o is nothing more than a method of determining whether the satellite is between periapsis and apoapsis (where flight-path angle is always positive) or between apoapsis and periapsis (where ϕ is always negative). With this hint see if you can fathom the logic of checking $\mathbf{r} \cdot \mathbf{v}$.

The three orbits illustrated in the following example problem should help you visualize what we have been talking about up to now.

Find the orbital elements of the three orbits in the following illustrations.

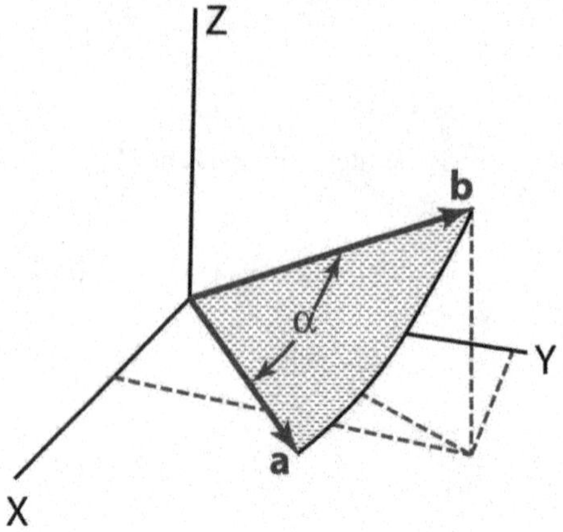

Figure 2-6 Angle Between Vectors.

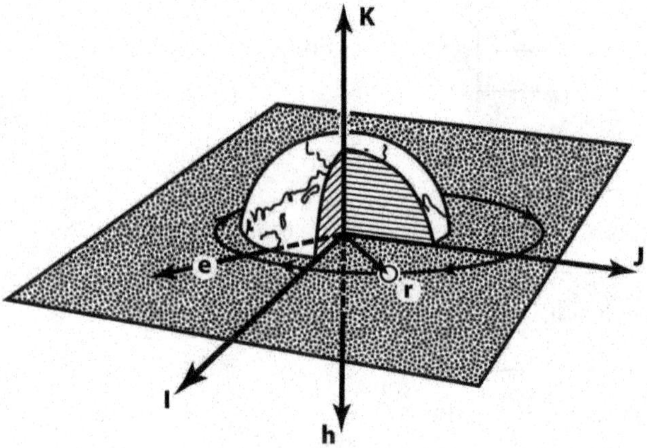

Figure 2-7 Orbit 1.

Orbit 1 (retrograde equatorial)

p = 9,567 km	$\Pi = 45°$
e = 0.2	$\nu_0 = 270°$
$i = 180°$	u_0 = undefined
Ω = undefined	$\ell_0 = 315°$

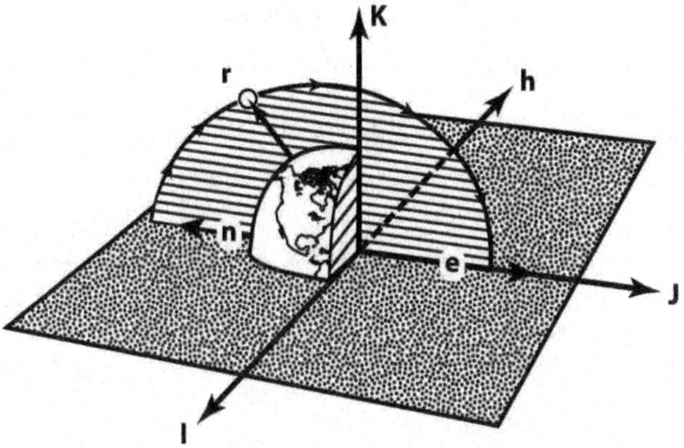

Figure 2-8 Orbit 2.

Orbit 2 (polar)

$p = 9{,}567$ km	$\omega = 180°$
$e = 0.2$	$\nu_0 = 225°$
$i = 90°$	$u_0 = 45°$
$\Omega = 270°$	$\ell_0 = 315°$

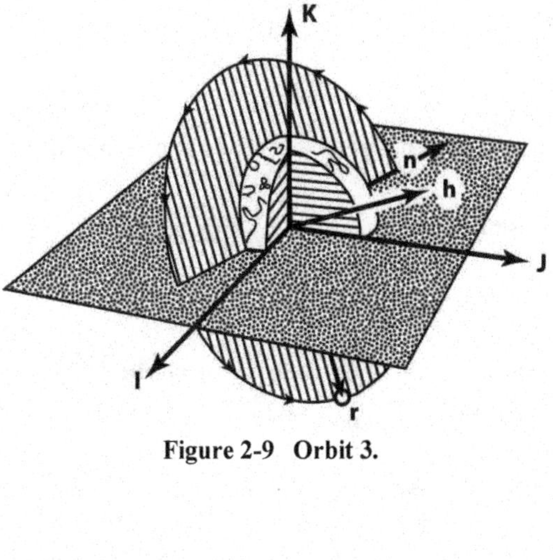

Figure 2-9 Orbit 3.

Orbit 3 (direct circular)

$p = 9{,}567$ km	$\omega =$ undefined
$e = 0$	$\nu_o =$ undefined
$i = 60°$	$u_o = 270°$
$\Omega = 170°$	$\ell_0 = 420°$

Example Problem. A radar tracks a meteoroid and from the tracking data the following inertial position and velocity vectors are found (expressed in the geocentric-equatorial coordinate system):

$\mathbf{r} = 12{,}756.2\mathbf{I}$ km

$\mathbf{v} = 7.90537\mathbf{J}$ km

Determine the six orbital elements for the observed meteoroid.

Find p using Equation (2-4):

$$\mathbf{h} = \mathbf{r} \times \mathbf{v} = 1.008432\mathbf{K} \times 10^5 \text{ km}^2/\text{s}$$

Then from Equation (1-53)

$$p = \frac{h^2}{\mu} = \underline{\underline{25{,}512.60 \text{ km}}}$$

Find eccentricity, e, using Equation (2-8):

$$\mathbf{e} = \frac{1}{\mu}\left[\left(v^2 - \frac{\mu}{r}\right)\mathbf{r} - (\mathbf{r} \cdot \mathbf{v})\mathbf{v}\right] = [1, 0, 0]$$

$$e = |\mathbf{e}| = \underline{\underline{1}}$$

Since $h \neq 0$ and $e = 1$, the path of the meteoroid is parabolic with respect to Earth.

Find inclination, i, using Equation (2-10):

$$i = \cos^{-1}\left(\frac{\mathbf{h} \cdot \mathbf{K}}{h}\right) = \underline{\underline{0°}}$$

Therefore, the meteoroid is traveling in the equatorial plane.

Find the longitude of the ascending node, Ω

From Equation (2-11),

$$\Omega = \cos^{-1}\left(\frac{\mathbf{n} \cdot \mathbf{I}}{n}\right)$$

There is no ascending node because its trajectory does not cross the equatorial plane and, therefore, for this case Ω is undefined.

Find the argument of periapsis, ω.

From Equation (2-12),

$$\omega = \cos^{-1}\left(\frac{\mathbf{n} \cdot \mathbf{e}}{ne}\right)$$

Again, since there is no ascending node, ω is also undefined. In lieu of ω, the longitude of periapsis Π can be determined in this case.

Find longitude of periapsis, Π.

Since the orbital plane is coincident with the equatorial plane ($i = 0°$) Π is measured from the **I**-axis to periapsis, which is colocated with the **e** vector. Therefore, from the definition of the dot product,

$$\Pi = \cos^{-1}\left(\frac{\mathbf{e} \bullet \mathbf{I}}{e}\right) = \underline{\underline{0°}}$$

it is determined that perigee is located along the **I** axis.

Find the true anomaly, ν_0.

From Equation (2-13),

$$\nu_0 = \cos^{-1}\left(\frac{\mathbf{e} \bullet \mathbf{r}}{er}\right) = \underline{\underline{0°}}$$

and it is found that the meteoroid is presently at periapsis.

Find true longitude of the epoch, ℓ_0.

From Equation (2-3),

$$\ell_0 = \Pi + \nu_0 = \underline{\underline{0°}}$$

Example Problem. The following inertial position and velocity vectors are expressed in a geocentric equatorial coordinate system for an observed space object:

$$\mathbf{r} = 8{,}750\mathbf{I} + 5{,}100\mathbf{J} + 0\mathbf{K} \text{ km}$$

$$\mathbf{v} = -3.000\mathbf{I} + 5.200\mathbf{J} + 5.900\mathbf{K} \text{ km/s}$$

Determine the six orbital elements for the observed object.

Find **p**.

First find **h** by Equation (2-4):

$$\mathbf{h} = \mathbf{r} \times \mathbf{v} = 30{,}090\mathbf{I} - 51{,}625\mathbf{J} + 60{,}800\mathbf{K}\ \text{km}^2/\text{s}$$

Then, from Equation (1-53),

$$p = \frac{h^2}{\mu} = \underline{\underline{18,232.0\ \text{km}}}$$

Find **e**.

From Equation (2-8),

$$\mathbf{e} = \frac{1}{\mu}\left[\left(v^2 - \frac{\mu}{r}\right)\mathbf{r} - (\mathbf{r}\cdot\mathbf{v})\mathbf{v}\right] = 0.693360\mathbf{I} + 0.399423\mathbf{J} - 0.03996\mathbf{K}$$

and the magnitude of the **e** vector is

$$\mathbf{e} = \underline{\underline{0.8002}}$$

Find the inclination, i:

From Equation (2-10),

$$i = \cos^{-1}\left(\frac{\mathbf{h}\cdot\mathbf{K}}{h}\right) = \underline{\underline{44.5029°}}$$

Find the longitude of ascending node, Ω.

First, we must determine the ascending node vector **n**.

From Equation (2-6),

$$\mathbf{n} = \mathbf{K} \times \mathbf{h} = \frac{3}{4\sqrt{2}}(\sqrt{3}\mathbf{I} + \mathbf{J}) = 51{,}625\mathbf{I} + 30{,}090\mathbf{J} + 0\mathbf{K}$$

From Equation (2-11),

$$\Omega = \cos^{-1}\left(\frac{\mathbf{n}\cdot\mathbf{I}}{n}\right) = \underline{\underline{30.2361°}}$$

Find the argument for periapsis, ω:

From Equation (2-12),

$$\omega = \cos^{-1}\left(\frac{\mathbf{n}\cdot\mathbf{e}}{ne}\right) = \underline{\underline{-0.41°}}$$

Find the true anomaly, ν_0.

From Equation (2-13),

$$\nu_o = \cos^{-1}\left(\frac{\mathbf{e} \cdot \mathbf{r}}{er}\right) = \underline{\underline{0.41^\circ}}$$

2.5 DETERMINING r AND v FROM THE ORBITAL ELEMENTS

In the last section we saw how to determine a set of classical orbital elements from the vectors **r** and **v** at some epoch. Now we will look at the inverse problem of determining **r** and **v** when the six classical elements are given.

This is both an interesting and a practical exercise since it represents one way of solving a basic problem of astrodynamics—that of updating the position and velocity of a satellite to some future time. Suppose you know $\mathbf{r}_0$ and $\mathbf{v}_0$ at some time t_0. Using the techniques presented in the last section you could determine the elements p, e, i, Ω, ω and ν_0. Of these six elements, the first five are constant (if we accept the assumptions of the restricted two-body problem) and only the true anomaly, ν, changes with time. In Chapter 4 you will learn how to determine the change in true anomaly, Δν, that occurs in a given time, $t - t_0$. This will enable you to construct a new set of orbital elements and the only step remaining is to determine the new **r** and **v** from this "updated" set.

The method, shown below, consists of two steps; first we must express **r** and **v** in perifocal coordinates and then transform **r** and **v** to geocentric-equatorial components.

2.5.1 Expressing r and v in the Perifocal System

Let's assume that we know p, e, i, Ω, ω and ν. We can immediately write an expression for **r** in terms of the perifocal system (see Figure 2-4):

$$\boxed{\mathbf{r} = r\cos\nu\,\mathbf{P} + r\sin\nu\,\mathbf{Q}} \tag{2-15}$$

where the scalar magnitude r can be determined from the polar equation of a conic:

$$r = \frac{p}{1 + e\cos\nu} \tag{2-16}$$

To obtain **v** we only need to differentiate **r** in Equation (2-15), keeping in mind that the perifocal coordinate frame is "inertial" and so the time derivatives are null $(\dot{P} = \dot{Q} = 0)$, and

$$\mathbf{v} = \dot{\mathbf{r}} = (\dot{r}\cos\nu - r\dot{\nu}\sin\nu)\mathbf{P} + (\dot{r}\sin\nu + r\dot{\nu}\cos\nu)\mathbf{Q} \tag{2-17}$$

This expression for **v** can be simplified by recognizing that $h = r^2\dot{\nu}$ (see Figure 1-11), $p = h^2/\mu$ and differentiating Equation (2-16) above to obtain

$$\dot{r} = \sqrt{\frac{\mu}{p}}\ e \sin \nu \tag{2-18}$$

and

$$r\dot{\nu} = \sqrt{\frac{\mu}{p}}\ (1 + e \cos\nu) \tag{2-19}$$

Making these substitutions in the expression for **v** and simplifying yields

$$\boxed{\mathbf{v} = \sqrt{\frac{\mu}{p}}\ [-\sin \nu\, \mathbf{P} + (e + \cos\nu)\, \mathbf{Q}]} \tag{2-20}$$

Example Problem. A space object has the following orbital elements as determined by NORAD's space track system:

$p = 14{,}351$ km $\quad \Omega = 30°$

$e = 0.5 \quad \omega = 0°$

$i = 45° \quad \nu_o = 0°$

Express the **r** and **v** vectors for the space object in the perifocal coordinate system.

Before Equation (2-15) can be applied, the magnitude of the **r** vector must be determined first by Equation (2-16):

$$r = \frac{p}{1 + e\cos\nu} = 9{,}567.2 \text{ km}$$

From Equation (2-15),

$$\mathbf{r} = r\cos \nu\, \mathbf{P} + r\sin \nu\, \mathbf{Q} = 9{,}567.2\mathbf{P} \text{ km}$$

From Equation (2-20),

$$\mathbf{v} = \sqrt{\frac{\mu}{p}}[\sin \nu\mathbf{P} + (e + \cos \nu)\mathbf{Q}] = 7.9054\mathbf{Q} \text{ km/s}$$

2.6 COORDINATE TRANSFORMATIONS

Before we discuss the transformation of **r** and **v** to the geocentric-equatorial frame we will review coordinate transformations in general.

A vector may be expressed in any coordinate frame. It is common in astrodynamics to use rectangular coordinates although occasionally spherical polar coordinates are more convenient. A rectangular coordinate frame is usually defined by specifying its origin, its fundamental (x–y) plane, the direction of the positive z-axis, and the principal (x) direction in the fundamental plane. Three unit vectors are then defined to indicate the directions of the three mutually perpendicular axes. Any other vector can be expressed as a linear combination of these three unit vectors. This collection of unit vectors is commonly referred to as a "basis."

2.6.1 What a Coordinate Transformation Does

A coordinate transformation merely changes the basis of a vector—nothing else. The vector still has the same length and direction after the coordinate transformation and it still represents the same thing. For example, suppose you know the south, east, and zenith components of the "velocity of a satellite relative to the topocentric-horizon frame". (See Section 2.7.2) The phase in quotes describes what the vector represents. The "basis" of the vector is obviously the set of unit vectors pointing south, east and up.

Now suppose you want to express this vector in terms of a different basis, say the **IJK** unit vectors of the geocentric-equatorial frame (perhaps because you want to add another vector to it and this other vector is expressed in **IJK** components). A simple coordinate transformation will do the trick. The vector will still have the same magnitude and direction and will represent the same thing, namely, the "velocity of a satellite relative to the topocentric-horizon frame" even though you now express it in terms of geocentric-equatorial coordinates. In other words, changing the basis of vector does not change its magnitude, direction or what it represents.

A vector has only two properties that can be expressed mathematically—magnitude and direction. Certain vectors, such as position vectors, have a definite starting point, but this point of origin cannot be expressed mathematically and does not change in a coordinate transformation. For example, suppose you know the south, east and zenith components of "the vector from a radar site on the surface of Earth to a satellite." A simple change of basis will enable you to express this vector in terms of **IJK** components:

$$\boldsymbol{\rho} = \rho_I \mathbf{I} + \rho_J \mathbf{J} + \rho_K \mathbf{K}$$

The transformation did not change what the vector represents so it is still the vector "*from* the radar site *to* the satellite." In other words, expressing a vector in coordinates of a particular frame does not imply that the vector has its tail at the origin of that frame.

2.6.2 Change of Basis Using Matrices

Changing from one basis to another can be streamlined by using matrix methods. Suppose we have two coordinate frames XYZ and X′ Y′ Z′ related by a simple rotation through a positive angle α about the Z-axis. Figure 2-10 illustrates the two frames. We will define a positive rotation about any axis by means of the "right-hand rule"—if the thumb of the right hand is extended in the direction of the positive coordinate axis, the fingers curl in the sense of a positive rotation.

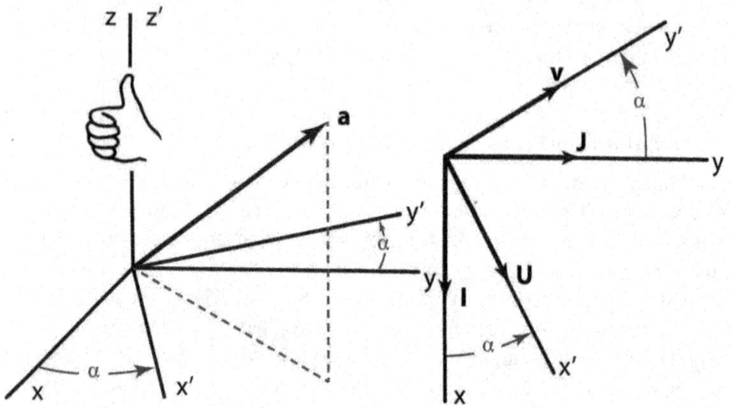

Figure 2-10 Rotation about z-Axis.

Let us imagine three unit vectors **I**, **J** and **K** extending along the X, Y and Z axes, respectively, and another set of unit vectors **U**, **V** and **W** along the X′, Y′ and Z′ axes. Now suppose we have a vector **a** that may be expressed in terms of the **IJK** basis as

$$\mathbf{a} = a_I \mathbf{I} = a_J \mathbf{J} = a_K \mathbf{K} \tag{2-21}$$

or in terms of the **UVW** basis as

$$\mathbf{a} = a_U \mathbf{U} + a_V \mathbf{V} + a_W \mathbf{W} \tag{2-22}$$

From Figure 2-10 we can see that the unit vectors **U**, **V** and **W** are related to **I**, **J** and **K** by the following set of equations:

$$\mathbf{U} = \mathbf{I}\,(\cos\alpha) + \mathbf{J}\,(\sin\alpha) + \mathbf{K}\,(0) \tag{2-23}$$

$$\mathbf{V} = \mathbf{I}\,(-\sin\alpha) + \mathbf{J}\,(\cos\alpha) + \mathbf{K}\,(0) \tag{2-24}$$

$$\mathbf{W} = \mathbf{I}\,(0) + \mathbf{J}\,(0) + \mathbf{K}\,(1) \tag{2-25}$$

Substituting these equations into Equation (2-22) above and equating it with Equation (2-21) yields

$$a_U = a_I\,(\cos\alpha) + a_J\,(\sin\alpha) + a_K\,(0) \tag{2-26}$$

$$a_V = a_I(-\sin\alpha) + a_J(\cos\alpha) + a_K(0) \tag{2-27}$$

$$a_W = a_I(0) + a_J(0) + a_K(1) \tag{2-28}$$

We can express this last set of equations very compactly if we use matrix notations and think of the vector **a** as a triplet of numbers representing a column matrix. We will use subscripts to identify the basis, thus

$$\mathbf{a}_{IJK} = \begin{bmatrix} a_I \\ a_J \\ a_K \end{bmatrix} \text{ and } \mathbf{a}_{UVW} = \begin{bmatrix} a_U \\ a_V \\ a_W \end{bmatrix} \tag{2-29}$$

The coefficients of a_I, a_J and a_K in Equations (2-26), (2-27) and (2-28) should be taken as the elements of a three-by-three "transformation matrix" which we can call $\tilde{\mathbf{A}}$

$$\tilde{\mathbf{A}} = \begin{bmatrix} \cos\alpha & \sin\alpha & 0 \\ -\sin\alpha & \cos\alpha & 0 \\ 0 & 0 & 1 \end{bmatrix} \tag{2-30}$$

The set of Equations (2-26)–(2-28) can then be represented as

$$\mathbf{a}_{UVW} = \tilde{\mathbf{A}}\,\mathbf{a}_{IJK} \tag{2-31}$$

which is really matrix shorthand for

$$\begin{bmatrix} a_U \\ a_V \\ a_W \end{bmatrix} = \begin{bmatrix} \cos\alpha & \sin\alpha & 0 \\ -\sin\alpha & \cos\alpha & 0 \\ 0 & 0 & 1 \end{bmatrix} \begin{bmatrix} a_I \\ a_J \\ a_K \end{bmatrix} \tag{2-32}$$

Applying the rules of matrix multiplication to Equation (2-32) yields the set of Equations (2-26)–(2-28) above.

2.6.3 Summary of Transformation Matrices for Single Rotation of Coordinate Frame

Arguments similar to those above may be used to derive transformation matrices that will represent rotations of the coordinate frame about the X or Y axes. These are summarized below.

Rotation about the X axis. The transformation matrix ($\tilde{\mathbf{A}}$) corresponding to a single rotation of the coordinate frame about the positive X axis through α positive angle α is

$$\tilde{\mathbf{A}} = \begin{bmatrix} 1 & 0 & 0 \\ 0 & \cos\alpha & \sin\alpha \\ 0 & -\sin\alpha & \cos\alpha \end{bmatrix} \tag{2-33}$$

Rotation about the Y axis. The transformation matrix $\tilde{\mathbf{B}}$ corresponding to a single rotation of the coordinate frame about the positive Y axis through a positive angle β is

$$\tilde{\mathbf{B}} = \begin{bmatrix} \cos\beta & 0 & -\sin\beta \\ 0 & 1 & 0 \\ \sin\beta & 0 & \cos\beta \end{bmatrix} \tag{2-34}$$

Rotation about the Z axis. The transformation matrix $\tilde{\mathbf{C}}$ corresponding to a single rotation of the coordinate frame about the positive Z axis through a positive angle γ is

$$\tilde{\mathbf{C}} = \begin{bmatrix} \cos\gamma & \sin\gamma & 0 \\ -\sin\gamma & \cos\gamma & 0 \\ 0 & 0 & 1 \end{bmatrix} \tag{2-35}$$

2.6.4 Successive Rotations About Several Axes

So far we have learned how to use matrices to perform a simple change of basis where the new set of unit vectors is related to the old by a simple rotation about one of the coordinate axes. Let us now look at a more complicated transformation involving more than one rotation.

Suppose we know the **IJK** components of some general vector, **a**, in the geocentric-equatorial frame and we wish to find its **SEZ** components in the topocentric-horizon frame (see Section 2.7.1).

Figure 2-21 shows the angular relationship between two frames. Starting with the **IJK** frame, we can first rotate it through a positive angle θ about the Z (**K**) axis and then rotate it through a positive angle (90° – L) about the Y axis to bring it into angular alignment with the **SEZ** frame.

The three components of **a** after the first rotation can be found from

$$\begin{bmatrix} \cos\theta & \sin\theta & 0 \\ -\sin\theta & \cos\theta & 0 \\ 0 & 0 & 1 \end{bmatrix} \begin{bmatrix} a_I \\ a_J \\ a_K \end{bmatrix} \tag{2-36}$$

The above expression is actually a column matrix and represents the three components of **a** in the intermediate frame. We can now multiply this column matrix by the appropriate matrix corresponding to the second rotation and obtain

$$\begin{bmatrix} a_S \\ a_E \\ a_Z \end{bmatrix} = \begin{bmatrix} \sin L & 0 & -\cos L \\ 0 & 1 & 0 \\ \cos L & 0 & \sin L \end{bmatrix} \begin{bmatrix} \cos\theta & \sin\theta & 0 \\ -\sin\theta & \cos\theta & 0 \\ 0 & 0 & 1 \end{bmatrix} \begin{bmatrix} a_I \\ a_J \\ a_K \end{bmatrix} \tag{2-37}$$

Since matrix multiplication is associative we can multiply the two simple rotation matrices together to form a single transformation matrix and write

$$\begin{bmatrix} a_S \\ a_E \\ a_Z \end{bmatrix} = \begin{bmatrix} \sin L \cos\theta & \sin L \sin\theta & -\cos L \\ -\sin\theta & \cos\theta & 0 \\ \cos L \cos\theta & \cos L \sin\theta & \sin L \end{bmatrix} \begin{bmatrix} a_I \\ a_J \\ a_K \end{bmatrix} \tag{2-38}$$

Keep in mind that the order in which you multiply the two rotation matrices is important since matrix multiplication is not commutative, (i.e., $\tilde{\mathbf{A}}\tilde{\mathbf{B}} \neq \tilde{\mathbf{B}}\tilde{\mathbf{A}}$). Since matrix multiplication can represent rotation of an axis system we can infer from this that the order in which rotations are performed is not irrelevant. For example if you are in an airplane and you rotate in pitch 45° nose-up and then roll 90° right you will be in a different attitude than if you first roll 90° right and then pitch up 45°.

Equation (2-38) represents the transformation from geocentric-equatorial to topocentric coordinates. We can write Equation (2-38) more compactly as

$$\mathbf{a}_{SEZ} = \tilde{\mathbf{D}}\mathbf{a}_{IJK} \tag{2-39}$$

where **D** is the overall transformation matrix.

Now, if we want to perform the inverse transformation (from topocentric to geocentric), we need to find the inverse of matrix $\tilde{\mathbf{D}}$, which we call $\tilde{\mathbf{D}}^{-1}$:

$$\mathbf{a}_{IJK} = \tilde{\mathbf{D}}^{-1}\mathbf{a}_{SEZ} \tag{2-40}$$

In general the inverse of a matrix is difficult to calculate. Fortunately, all transformation matrices between rectangular frames have the unique property that they are orthogonal.

A three-by-three matrix is called "orthogonal" if the rows and the columns are scalar components of mutually perpendicular unit vectors. The inverse of any orthogonal matrix is equal to its transpose. The transpose of the matrix $\tilde{\mathbf{D}}$, denoted by $\tilde{\mathbf{D}}^{\mathbf{T}}$, is the new matrix whose rows are the old columns and whose columns are the old rows (in the same order). Hence

$$\begin{bmatrix} a_I \\ a_J \\ a_K \end{bmatrix} = \begin{bmatrix} \sin L \cos\theta & -\sin\theta & \cos L \cos\theta \\ \sin L \sin\theta & \cos\theta & \cos L \sin\theta \\ -\cos L & 0 & \sin L \end{bmatrix} \begin{bmatrix} a_S \\ a_E \\ a_Z \end{bmatrix} \tag{2-41}$$

The angles through which one frame must be rotated to bring its axes into coincidence with another frame are commonly referred to as "Euler angles." A maximum of three Euler angle rotations is sufficient to bring any two frames into coincidence.

By memorizing the three basic rotation matrices and the rules for matrix multiplication you will be able to perform any desired change of basis no matter how complicated.

2.6.5 Transformation from the Perifocal to the Geocentric-Equatorial Frame

The perifocal coordinate system is related geometrically to the **IJK** frame through the angles Ω, i and ω as shown in Figure 2-11.

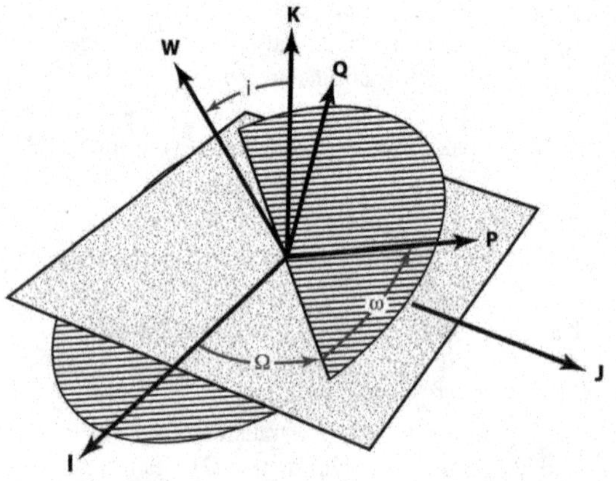

Figure 2-11 Relationship between PQW and IJK.

The transformation coordinates between the **PQW** and **IJK** systems can be accomplished by means of the rotation matrix, $\tilde{\mathbf{R}}$.

Thus if a_I, a_J, a_K and a_P, a_Q, a_W are the components of a vector **a** in each of the two systems, then

$$\begin{bmatrix} a_I \\ a_J \\ a_K \end{bmatrix} = \tilde{\mathbf{R}} \begin{bmatrix} a_P \\ a_Q \\ a_W \end{bmatrix} \tag{2-42}$$

Since we will often know the elements of the orbit, it may be convenient to use those angles in the transformation to the **IJK** frame.

To do this we can use direction cosines, which can be found using the cosine law of spherical trigonometry. For example, from Figure 2-12, we see that the cosine of the angle between **I** and **P** can be calculated from their dot product, **I** • **P**. Recall from vector theory that the dot product simply gives the projection of one vector upon another. **P** can then be projected into the **IJK** frame by simply taking its dot product with **I**, **J** and **K**. Thus

$$\tilde{\mathbf{R}} = \begin{bmatrix} \mathbf{I \cdot P} & \mathbf{I \cdot Q} & \mathbf{I \cdot W} \\ \mathbf{J \cdot P} & \mathbf{J \cdot Q} & \mathbf{J \cdot W} \\ \mathbf{K \cdot P} & \mathbf{K \cdot Q} & \mathbf{K \cdot W} \end{bmatrix} = \begin{bmatrix} R_{11} & R_{12} & R_{13} \\ R_{21} & R_{22} & R_{23} \\ R_{31} & R_{32} & R_{33} \end{bmatrix} \tag{2-43}$$

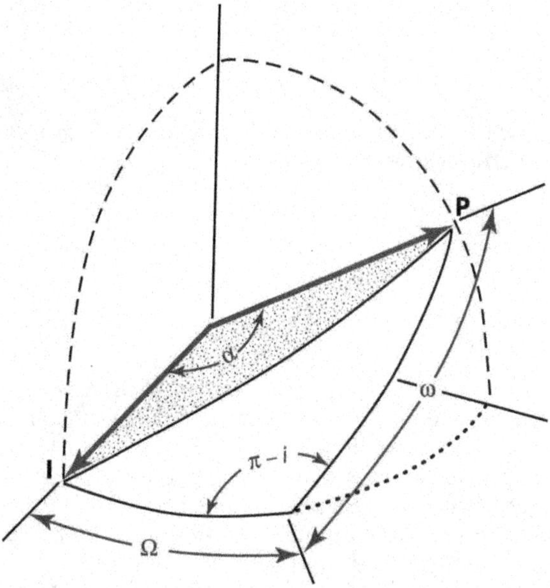

Figure 2-12 Angle between I and P.

In Figure 2-12, the angle between **I** and **P** forms one side of a spherical triangle whose other two sides are Ω and ω. The included angle is $\pi - i$. The law of cosines for spherical triangles is

$$\cos a = \cos b \cos c + \sin b \sin c \cos A$$

where A, B and C are the three angles and a, b and c are the opposite sides. Now

$$R_{11} = \mathbf{I} \cdot \mathbf{P} = \cos\Omega \ \cos\omega + \sin\Omega \ \sin\omega \ \cos(\pi - i)$$

$$= \cos\Omega \ \cos\omega - \sin\Omega \ \sin\omega \ \cos i$$

Similarly, the elements of $\tilde{\mathbf{R}}$ are

$$R_{11} = \cos\Omega\cos\omega - \sin\Omega\sin\omega\cos i$$

$$R_{12} = -\cos\Omega\sin\omega - \sin\Omega\cos\omega\cos i$$

$$R_{13} = \sin\Omega\sin i$$

$$R_{21} = \sin\Omega\cos\omega + \cos\Omega\sin\omega\cos i$$

$$R_{22} = -\sin\Omega\sin\omega + \cos\Omega\cos\omega\cos i$$

$$R_{23} = -\cos\Omega\sin i$$

$$R_{31} = \sin\omega\sin i$$

$$R_{32} = \cos\omega\sin i$$

$$R_{33} = \cos i$$

Having determined the elements of the rotation matrix, it only remains to find **r** and **v** in terms of **IJK** components. Thus

$$\begin{bmatrix} r_I \\ r_J \\ r_K \end{bmatrix} = \tilde{\mathbf{R}} \begin{bmatrix} r_P \\ r_Q \\ r_W \end{bmatrix} \quad \text{and} \quad \begin{bmatrix} v_I \\ v_J \\ v_K \end{bmatrix} = \tilde{\mathbf{R}} \begin{bmatrix} v_P \\ v_Q \\ v_W \end{bmatrix} \tag{2-44}$$

This method is not recommended unless other transformations are not practical. Often it is possible to go to the **IJK** frame using Equations (2-15) and (2-20) with **P** and **Q** known in **IJK** coordinates.

Special precautions must be taken when the orbit is equatorial or circular or both. In this case either Ω or ω or both are undefined. In the case of the circular orbit ν is also undefined so it is necessary to measure the true anomaly from some arbitrary reference such as the ascending node or (if the orbit is also equatorial) from the unit vector **I.** Because of these and other difficulties the method of updating **r** and **v** via the classical orbital elements leaves much to be desired. Other more general methods of updating **r** and **v** that do not suffer from these defects will be presented in Chapter 4.

2.7 ORBIT DETERMINATION FROM A SINGULAR RADAR OBSERVATION

A radar installation located on the surface of Earth can measure the position and velocity of a satellite *relative to the radar site*. But the radar site is not located at the center of Earth so the position vector measured is not the **r** we need. Also, Earth is rotating, so the velocity of the satellite relative to the radar site is not the same as the velocity, **v**, relative to the center of the **IJK** frame, which we need to compute the orbital elements.

Before showing you how we can obtain **r** and **v** *relative to the center of Earth* from radar tracking data we must digress long enough to describe the coordinate system in which the radar site makes and expresses its measurements.

2.7.1 The Topocentric-Horizon Coordinate System

The origin of the topocentric-horizon system is the point on the surface of Earth (called the "topos") where the radar is located. The fundamental plane is the horizon and the X_h-axis points south. The Y_h-axis is east and the Z_h-axis is up. It seems hardly necessary to point out that the topocentric-horizon system is not an inertial reference frame since it rotates with Earth. The unit vectors **S**, **E** and **Z** shown in Figure 2-13 will aid us in expressing vectors in this system.

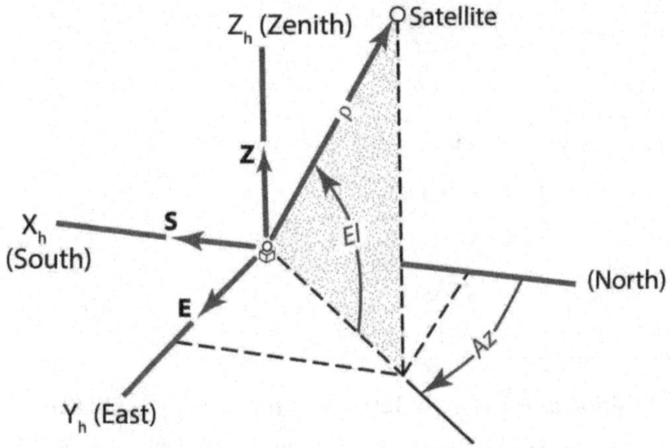

Figure 2-13 Topocentric-Horizon Coordinate System.

2.7.2 Expressing Position and Velocity Relative to the Topocentric-Horizon Reference Frame

The radar site measures the range and direction to the satellite. The range is simply the magnitude of the vector ρ (rho) shown in Figure 2-13. The direction to the satellite is determined by two angles that can be picked off the gimbal axes on which the radar antenna is mounted. The azimuth angle, Az, is measured clockwise from north; the elevation angle, El, is measured from the horizontal to the radar line of sight. If the radar is capable of detecting a shift in frequency in the returning echo (Doppler effect), the rate at which range is changing, $\dot{\rho}$, can also be measured. The sensors on the gimbal axes are capable of measuring the rate of change of the azimuth and elevation angles, $\dot{\text{Az}}$ and $\dot{\text{El}}$, as the radar antenna follows the satellite across the sky. Thus, we have the raw material for expressing the position and velocity of the satellite relative to the radar in the six measurements:

$$(\rho, \text{Az}, \text{El}, \dot{\rho}, \dot{\text{Az}}, \dot{\text{El}})$$

We will express the position vector as

$$\boldsymbol{\rho} = \rho_S\ \mathbf{S} + \rho_E\ \mathbf{E} + \rho_Z\ \mathbf{Z} \tag{2-45}$$

where, from the geometry of Figure 2-13,

$$\boxed{\begin{aligned} \rho_S &= -\rho \cos \text{El} \cos \text{Az} \\ \rho_E &= \rho \cos \text{El} \sin \text{Az} \\ \rho_Z &= \rho \sin \text{El} \end{aligned}} \tag{2-46}$$

The velocity relative to the radar site is just

$$\dot{\boldsymbol{\rho}} = \dot{\rho}_S\ \mathbf{S} + \dot{\rho}_E\ \mathbf{E} + \dot{\rho}_Z\ \mathbf{Z} \tag{2-47}$$

Differentiating Equations (2-46) we get the three components of $\dot{\boldsymbol{\rho}}$:

$$\boxed{\begin{aligned} \dot{\rho}_S &= -\dot{\rho} \cos \text{El} \cos \text{Az} + \rho \sin \text{El}\ (\dot{\text{El}}) \cos \text{Az} + \\ &\quad \rho \cos \text{El} \sin \text{Az}\ (\dot{\text{Az}}) \\ \dot{\rho}_E &= \dot{\rho} \cos \text{El} \sin \text{Az} - \rho \sin \text{El}\ (\dot{\text{El}}) \sin \text{Az} + \\ &\quad \rho \cos \text{El} \cos \text{Az}\ (\dot{\text{Az}}) \\ \dot{\rho}_Z &= \dot{\rho} \sin \text{El} + \rho \cos \text{El}\ (\dot{\text{El}}) \end{aligned}} \tag{2-48}$$

2.7.3 Position and Velocity Relative to the Geocentric Frame

There is an extremely simple relationship between the topocentric position vector $\boldsymbol{\rho}$ and the geocentric position vector $\mathbf{r}$. From Figure 2-14 we see that

$$\boxed{\mathbf{r} = \mathbf{R} + \boldsymbol{\rho}} \tag{2-49}$$

where $\mathbf{R}$ is the vector from the center of Earth to the origin of the topocentric frame. If Earth were perfectly spherical then the Z_h-axis, which defines the local vertical at the radar site, would pass through the center of Earth if extended downward and the vector $\mathbf{R}$ would be

$$\boxed{\mathbf{R} = r_{\oplus}\ \mathbf{Z}} \tag{2-50}$$

where $r_{\oplus}$ is the radius of Earth.

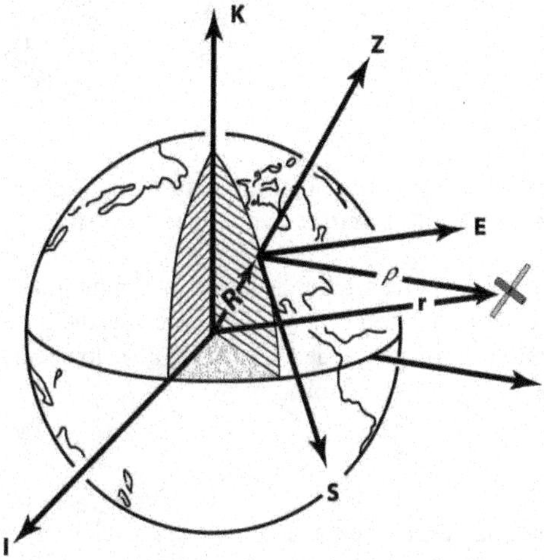

Figure 2-14 $\mathbf{r} = \mathbf{R} + \boldsymbol{\rho}$

Unfortunately, things are not that simple and to avoid errors of several miles in the position of the radar site relative to the geocenter it is necessary to use a more accurate model for the shape of Earth. A complete discussion of determining **R** on an oblate Earth is included later in this chapter. For the moment we will assume that Equation (2-50) is valid and proceed to the determination of **v**.

The general method of determining velocity relative to a "fixed" frame (hereafter referred to as the "true" velocity) when you are given the velocity relative to a moving frame may be stated in words as follows:

$$\begin{pmatrix}\text{Vel of object}\\ \text{rel to}\\ \text{fixed frame}\end{pmatrix} = \begin{pmatrix}\text{Vel of object}\\ \text{rel to}\\ \text{moving frame}\end{pmatrix} + \begin{pmatrix}\text{True vel of pt in}\\ \text{moving frame where}\\ \text{object is located}\end{pmatrix}$$

The sketches shown in Figure 2-15 illustrate this general principle for both a translating and a rotating frame.

If you visualize the three-dimensional volume of space (Figure 2-16) defined by the topocentric-horizon reference frame, you will see that every point in this frame moves with a different velocity relative to the center of Earth. If **r** is the position vector from the center of Earth to the satellite, the velocity of that point

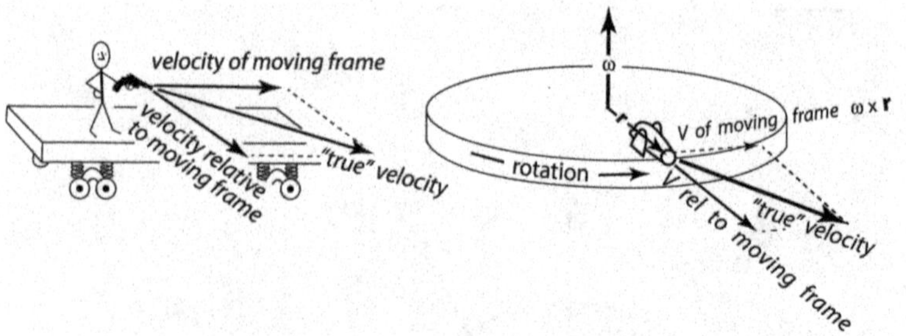

Figure 2-15 Relationship between Relative and True Velocity for Translating and Rotating Reference Frames.

in the topocentric frame where the satellite is located is simply ($\boldsymbol{\omega}_{\oplus} \times \mathbf{r}$) where $\boldsymbol{\omega}_{\oplus}$ is the angular velocity of Earth (hence, the angular velocity of the **SEZ** frame). It is the velocity that must be added vectorially to $\dot{\boldsymbol{\rho}}$ to obtain the "true" velocity **v**. Therefore,

$$\boxed{\mathbf{v} = \dot{\boldsymbol{\rho}} + \boldsymbol{\omega}_{\oplus} \times \mathbf{r}} \tag{2-51}$$

You may recognize this expression as a simple application of the Coriolis theorem, which will be derived in the next section as the general problem of derivatives in moving coordinate systems is discussed.

Example Problem. The position vector of a satellite relative to a radar site located at 169° W Long, 30° N Lat, is **2S** – **E** + 0.5**Z** (in units of **DU** to more easily show the calculations). The angle to Greenwich is 304°. Find the position vector of the satellite relative to fixed geocentric **IJK** coordinates. Assume the site is at sea level on a spherical Earth. We have

$$\mathbf{r} = \mathbf{R} + \boldsymbol{\rho} = 2\mathbf{S} - \mathbf{E} + 1.5\mathbf{Z}$$

Convert **r** to **IJK** coordinates using Equation (2-41):

$$L = 30^\circ \text{ (given)}$$

$$\theta = \theta_g + \lambda_E = 304^\circ + (-169^\circ) = 135^\circ$$

$$\tilde{\mathbf{D}}^{-1} = \begin{bmatrix} -0.3535 & -0.707 & -0.612 \\ 0.3535 & -0.707 & 0.612 \\ -0.866 & 0 & 0.5 \end{bmatrix}$$

$$\begin{bmatrix} r_I \\ r_J \\ r_K \end{bmatrix} = \tilde{\mathbf{D}}^{-1} \begin{bmatrix} r_S \\ r_E \\ r_Z \end{bmatrix} = \tilde{\mathbf{D}}^{-1} \begin{bmatrix} 2 \\ -1 \\ 1.5 \end{bmatrix} = \begin{bmatrix} -0.918 \\ 2.332 \\ -0.982 \end{bmatrix}$$

$$\mathbf{r} = \underline{\underline{-0.918\mathbf{I} + 2.332\mathbf{J} - 0.982\mathbf{K}}}$$

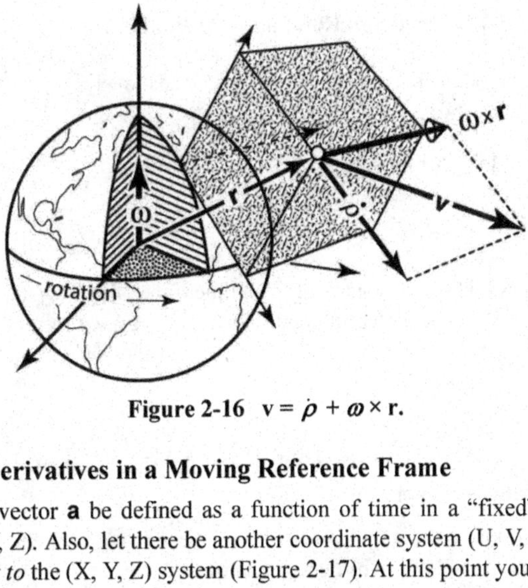

Figure 2-16 $\mathbf{v} = \dot{\rho} + \omega \times \mathbf{r}$.

2.7.4 Derivatives in a Moving Reference Frame

Let any vector **a** be defined as a function of time in a "fixed" coordinate frame (X, Y, Z). Also, let there be another coordinate system (U, V, W), rotating *with respect to* the (X, Y, Z) system (Figure 2-17). At this point you should note that this analysis applies to the *relative* motion between *any* two coordinate systems. It is not necessary that one system be fixed in inertial space!

The vector **a** may be *expressed* in either of these coordinate systems as

$$\underbrace{\mathbf{a} = a_I \mathbf{i} + a_J \mathbf{j} + a_K \mathbf{k}}_{\text{Fixed}} = \underbrace{a_U \mathbf{u} + a_V \mathbf{v} + a_W \mathbf{w}}_{\text{Rotating}} \tag{2-52}$$

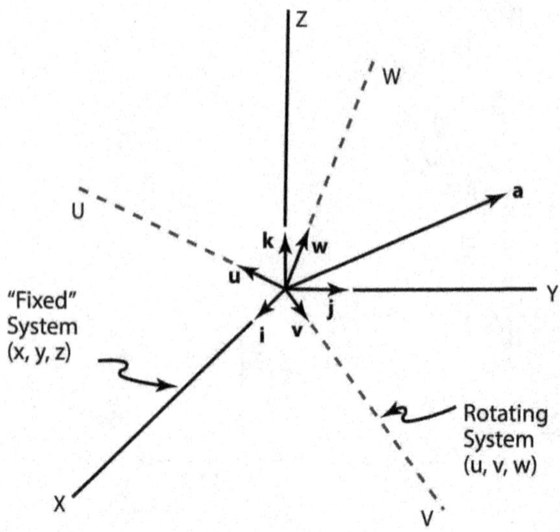

Figure 2-17 Fixed and Rotating Coordinate System.

Differentiating Equation (2-52) with respect to time leads to

$$\begin{aligned}\frac{d\mathbf{a}}{dt} &= \dot{a}_I\mathbf{i}+\dot{a}_J\mathbf{j}+\dot{a}_K\mathbf{k}+a_I\dot{\mathbf{i}}+a_J\dot{\mathbf{j}}+a_K\dot{\mathbf{k}} \\ &= \dot{a}_U\mathbf{u}+\dot{a}_V\mathbf{v}+\dot{a}_W\mathbf{w}+a_U\dot{\mathbf{u}}+a_V\dot{\mathbf{v}}+a_W\dot{\mathbf{w}}\end{aligned} \tag{2-53}$$

Now the unit vectors **i**, **j**, **k** may be moving with time if the "fixed" system is not inertial; however, if we specify the time derivative *with respect to the "fixed" coordinate system* then

$$\dot{\mathbf{i}} = \dot{\mathbf{j}} = \dot{\mathbf{k}} \equiv \mathbf{0} \tag{2-54}$$

The unit vectors **u**, **v** and **w** will move relative to the fixed system so that in general their derivatives are not zero.

Assume that we know the angular velocity of the moving reference frame relative to the fixed frame. We will denote this quantity by $\boldsymbol{\omega}$. Then using the results of the unit vector analysis we know that the time derivative of a unit vector is perpendicular to the unit vector and has magnitude equal to ω. You should prove to yourself that

$$\dot{\mathbf{u}} = \boldsymbol{\omega}\times\mathbf{u},\ \dot{\mathbf{v}} = \boldsymbol{\omega}\times\mathbf{v} \text{ and } \dot{\mathbf{w}} = \boldsymbol{\omega}\times\mathbf{w}$$

Hence, Equation (2-53) reduces to

$$\frac{d\mathbf{a}}{dt} = \dot{a}_I\mathbf{i} + \dot{a}_J\mathbf{j} + \dot{a}_K\mathbf{k} \tag{2-55}$$

$$= \dot{a}_U\mathbf{u} + \dot{a}_V\mathbf{v} + \dot{a}_W\mathbf{w} + a_U(\boldsymbol{\omega}\times\mathbf{u}) + a_V(\boldsymbol{\omega}\times\mathbf{v})a_W(\boldsymbol{\omega}\times\mathbf{w})$$

But

$$\dot{a}_U\mathbf{u} + \dot{a}_V\mathbf{v} + \dot{a}_W\mathbf{w} = \left.\frac{d\mathbf{a}}{dt}\right|_R$$

where $\left.\frac{d\mathbf{a}}{dt}\right|_R$ means the time derivative of **a** with respect to the rotating reference frame. Similarly,

$$\dot{a}_I\mathbf{i} + \dot{a}_J\mathbf{j} + \dot{a}_K\mathbf{k} = \left.\frac{d\mathbf{a}}{dt}\right|_F$$

where $\left.\frac{d\mathbf{a}}{dt}\right|_F$ means the time derivative of **a** with respect to the fixed reference frame. The remaining terms of Equation (2-55) may be rewritten as

$$\boldsymbol{\omega}\times(a_U\mathbf{u} + \boldsymbol{\omega}\times(a_V\mathbf{v}) + \boldsymbol{\omega}\times(a_W\mathbf{w}) = \boldsymbol{\omega}\times\mathbf{a}$$

Hence, Equation (2-55) reduces further to

$$\left.\frac{d\mathbf{a}}{dt}\right|_F = \left.\frac{d\mathbf{a}}{dt}\right|_R + \boldsymbol{\omega}\times\mathbf{a} \tag{2-56}$$

We now have a very general equation in which **a** is *any* vector. Equation (2-56) may be considered as a true operator:

$$\left.\frac{d(\ )}{dt}\right|_F = \left.\frac{d(\ )}{dt}\right|_R + \boldsymbol{\omega}\times(\ ) \tag{2-57}$$

where ω is the instantaneous angular velocity of the rotating frame with respect to the fixed reference frame. Equation (2-57) is referred to as the Coriolis theorem.

We will now apply the operator Equation (2-57) to the position vector **r**,

$$\left.\frac{d\mathbf{r}}{dt}\right|_F = \left.\frac{d\mathbf{r}}{dt}\right|_R + \boldsymbol{\omega}\times\mathbf{r} \tag{2-58}$$

or

$$\mathbf{v}_F = \mathbf{v}_R + \boldsymbol{\omega}\times\mathbf{r} \tag{2-59}$$

Again applying the operator equation to $\mathbf{v}_F$ we get

$$\left.\frac{d(\mathbf{v}_F)}{dt}\right|_F = \left.\frac{d(\mathbf{v}_F)}{dt}\right|_R + \boldsymbol{\omega} \times (\mathbf{v}_F) \tag{2-60}$$

Substituting Equation (2-59) into Equation (2-60) we get

$$\mathbf{a}_F = \mathbf{a}_R + \left[\left.\frac{d\boldsymbol{\omega}}{dt}\right|_R \times \mathbf{r}\right] + 2\boldsymbol{\omega} \times \left.\frac{d\mathbf{r}}{dt}\right|_R + \boldsymbol{\omega} \times (\boldsymbol{\omega} \times \mathbf{r}) \tag{2-61}$$

If we now solve Equation (2-61) for $\mathbf{a}_R$ we get

$$\mathbf{a}_R = \mathbf{a}_F - \left[\left.\frac{d\boldsymbol{\omega}}{dt}\right|_R \times \mathbf{r}\right] - 2\boldsymbol{\omega} \times \left.\frac{d\mathbf{r}}{dt}\right|_R - \boldsymbol{\omega} \times (\boldsymbol{\omega} \times \mathbf{r}) \tag{2-62}$$

Let us now apply this result to a problem of practical interest. Suppose we are standing on the surface of Earth. The angular velocity of our rotating platform is a constant $\boldsymbol{\omega}_\oplus$. Then Equation (2-62) reduces to

$$\mathbf{a}_R = \mathbf{a}_F - 2(\boldsymbol{\omega}_\oplus \times \mathbf{v}_R) - \boldsymbol{\omega}_\oplus \times (\boldsymbol{\omega}_\oplus \times \mathbf{r})$$

Inspection of this result indicates that the rotating observer sees the acceleration in the fixed system *plus* two others:

$$-2\boldsymbol{\omega}_\oplus \times \mathbf{v}_R = \text{coriolis acceleration}$$

and

$$-\boldsymbol{\omega}_\oplus \times (\boldsymbol{\omega}_\oplus \times \mathbf{r}) = \text{centrifugal acceleration}$$

The first is present only when there is relative motion between the object and the rotating frame. The second depends only on the position of the object from the axis of rotation.

2.8 SEZ TO IJK TRANSFORMATION USING AN ELLIPSOID EARTH MODEL

To fully complete the orbit determination problem of the previous section we need to convert the vectors we have expressed in **SEZ** components to **IJK** components. We could simply use a spherical Earth as a model and write the transformation matrix as discussed earlier. To be more accurate we will first discuss a nonspherical Earth model.

"Station coordinates" is the term used to denote the position of a tracking or launch site on the surface of Earth. If Earth were perfectly spherical, latitude and longitude could be considered as spherical coordinates with the radius being just Earth's radius plus the elevation above sea level. Earth is not a perfect geometric sphere. Therefore, to increase the accuracy of our calculations, a model for the geometric shape of Earth must be adopted. We will take as our approximate model an oblate spheroid. The first Vanguard satellite showed Earth to be slightly pear-shaped but this distortion is so small that the oblate spheroid is still an excellent representation.

We will discover that latitude can no longer be interpreted as a spherical coordinate and that Earth's radius is a function of latitude. We will find it most convenient to express the station coordinates of a point in terms of two rectangular coordinates and the longitude. (The interpretation of longitude is the same on an oblate Earth as it is on a spherical Earth.)

2.8.1 The Reference Ellipsoid

In the model that has been adopted, a cross section of Earth along a meridian is an ellipse whose semi-major axis, a_e, is just the equatorial radius and whose semi-minor axis, b_e, is just the polar radius of Earth. Sections parallel to the equator are, of course, circles. These radii and the consequent eccentricity of the elliptical cross section are as follows:

Equatorial radius (a_e) = 6,378.136 km

Polar radius (b_e) = 6,356.752 km

Eccentricity (e) = 0.08182

The reference ellipsoid is a good approximation to that hypothetical surface commonly referred to as "mean sea level." The actual mean sea level surface is called the "geoid" and it deviates from the reference ellipsoid slightly because of the uneven distribution of mass in Earth's interior. The geoid is a true equipotential surface and a plumb bob would hang perpendicular to the surface of the geoid at every point.

2.8.2 The Measurement of Latitude

An equatorial bulge is a difference between the equatorial and polar diameters of a planet, due to the centripetal force of its rotation. A rotating body tends to form an oblate spheroid rather than a sphere. An oblate spheroid is by definition a sphere with flattened poles. By convention the actual equatorial radius, including the equatorial bulge, of a planet is used as the definition of the radius and is accompanied by a flattening factor that defines the degree of oblateness in a spheroid. While an oblate Earth introduces no unique problems in the definition or measurements of terrestrial longitude, it does complicate the concept of latitude. Consider Figure 2-18. It illustrates the two most commonly used definitions of latitude.

The angle L' is called "geocentric latitude" and is defined as the angle between the equatorial plane and the radius from the geocenter.

The angle L is called "geodetic latitude" and is defined as the angle between the equatorial plane and the normal to the surface of the ellipsoid. The word "latitude" usually means geodetic latitude. This is the basis for most of the maps and charts we use. The normal to the surface is the direction that a plumb bob would hang were it not for local anomalies in Earth's gravitational field.

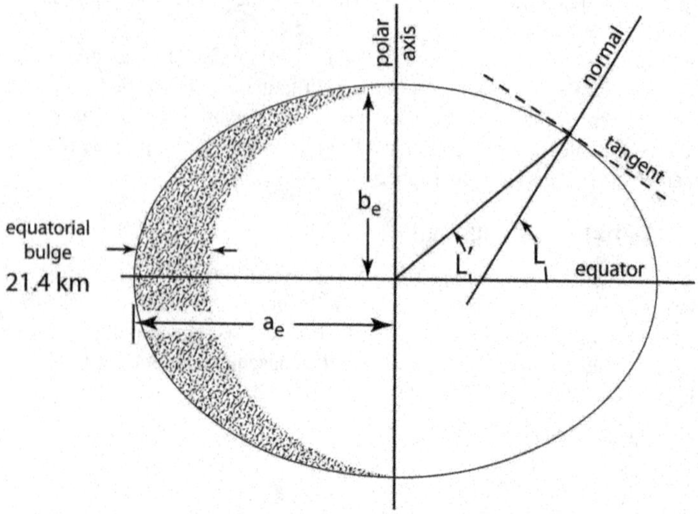

Figure 2-18 Geocentric and Geodetic Latitude.

The angle between the equatorial plane and the actual "plumb bob vertical" uncorrected for these gravitational anomalies is called L_a, the "astronomical latitude." Since the difference between the true geoid and our reference ellipsoid is slight, the difference between L and L_a is usually negligible.

When you are given the latitude of a place, it is safe to assume that it is the geodetic latitude L unless otherwise stated.

2.8.3 Station Coordinates

What we need now is a method of calculating the station coordinates of a point on the surface of our reference ellipsoid when we know the geodetic latitude and longitude of the point and its height above mean sea level (which we will take to be the height above the reference ellipsoid).

Consider an ellipse comprising a section of our adopted Earth model and a rectangular coordinate system as shown in Figure 2-19.

We will first determine the x and z coordinates of a point on the ellipse assuming that we know the geodetic latitude, L. It will then be a simple matter to adjust these coordinates for a point that is a know elevation above the surface of the ellipsoid in the direction of the normal.

It is convenient to introduce the angle β, the "reduced latitude," which is illustrated in Figure 2-19. The x and z coordinates can immediately be written in terms of β if we note that the ratio of the z coordinate of a point on the ellipse to the corresponding z coordinate of a point on the circumscribed circle is just b_e/a_e. Thus,

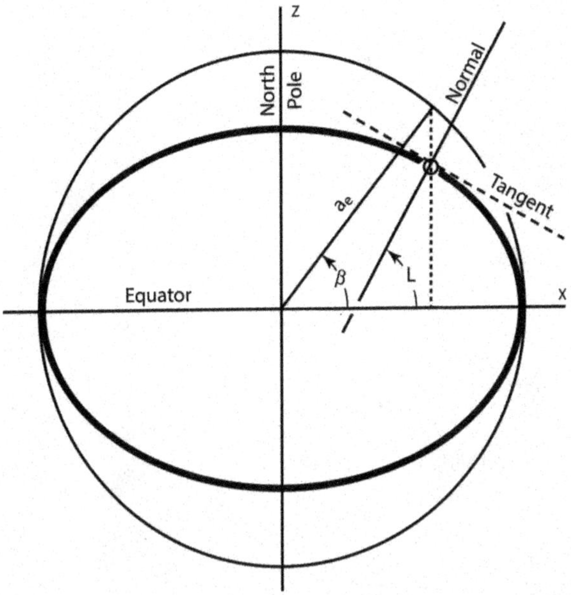

Figure 2-19 Station Coordinates.

$$x = a_e \cos \beta$$

$$z = \frac{b_e}{a_e} a_e \sin \beta \qquad (2\text{-}63)$$

But, for any ellipse, $a^2 = b^2 + c^2$ and $e = c/a$, so

$$b_e = a_e \sqrt{1-e^2}$$

and (2-64)

$$z = a_e \sqrt{1-e^2} \sin \beta$$

We must now express sin β in terms of the geodetic latitude L and the constants a_e and b_e. From elementary calculus we know that the slope of the tangent to the ellipse is just dz/dx and the slope of the normal is –dx/dz.

Since the slope of the normal is just tan L, we can write

$$\tan L = -\frac{dx}{dz}$$

The differentials dx and dz can be obtained by differentiating the expressions for x and z above. Thus,

$$dx = -a_e \sin\beta \, d\beta$$

$$dz = a_e\sqrt{1-e^2}\cos\beta \, d\beta$$

and

$$\tan L = \frac{\tan\beta}{\sqrt{1-e^2}}$$

or

$$\tan\beta = \sqrt{1-e^2}\tan L = \frac{\sqrt{1-e^2}\sin L}{\cos L}$$

Suppose we consider this last expression as the quotient

$$\tan\beta = \frac{A}{B}$$

where $A = \sqrt{1-e^2}\sin L$ and $B = \cos L$ with

$$\sin\beta = \frac{A}{\sqrt{A^2+B^2}} = \frac{\sqrt{1-e^2}\sin L}{\sqrt{1-e^2\sin^2 L}}$$

$$\cos\beta = \frac{B}{\sqrt{A^2+B^2}} = \frac{\cos L}{\sqrt{1-e^2\sin^2 L}} \tag{2-65}$$

We can now write the x and z coordinates for a point on the ellipse.

$$x = \frac{a_e \cos L}{\sqrt{1-e^2\sin^2 L}}$$

$$z = \frac{a_e\left(1-e^2\right)\sin L}{\sqrt{1-e^2\sin^2 L}} \tag{2-66}$$

For a point that is a height H above the ellipsoid (which we take to be mean sea level), it is easy to show that the x and z components of the elevation or height (H) normal to the adopted ellipsoid are

$$\Delta x = H\cos L$$

$$\Delta z = H\sin L \tag{2-67}$$

Adding these quantities to the relations for x and z, we get the following expressions for the two rectangular station coordinates of a point in terms of geodetic latitude, elevation above mean sea level and Earth equatorial radius and eccentricity:

$$x = \left[\frac{a_e}{\sqrt{1 - e^2 \sin^2 L}} + H \right] \cos L$$
$$z = \left[\frac{a_e(1 - e^2)}{\sqrt{1 - e^2 \sin^2 L}} + H \right] \sin L \qquad (2\text{-}68)$$

The third station coordinate is simply the east longitude of the point. If the Greenwich sidereal time, θ_g, is known, it can be combined with east longitude to find local sidereal time, θ. The x and z coordinates plus the angle θ completely locate the observer or launch site in the geocentric-equatorial frame as shown below.

From Figure 2-20 it is obvious that the vector **r** from the geocenter to the site on an oblate Earth is simply

$$\mathbf{r} = x \cos\theta \, \mathbf{i} + x \sin\theta \, \mathbf{j} + z \, \mathbf{k} \qquad (2\text{-}69)$$

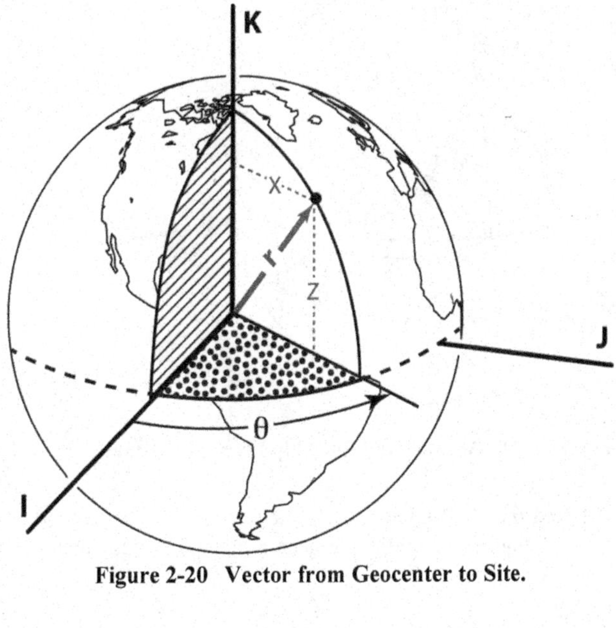

Figure 2-20 Vector from Geocenter to Site.

2.8.4 Transforming a Vector from SEZ to IJK Components

The only remaining problem is how to convert the vectors that we have expressed in **SEZ** components into the **IJK** components of the geocentric frame.

Recall that the geodetic latitude of the radar site, L, is the angle between the equatorial plane and the extension of the local vertical at the radar site (on a spherical or oblate Earth). Although Figure 2-21 shows a spherical Earth, the transformation matrix derived as follows is equally valid for an ellipsoid Earth model where L is the geodetic latitude for the site.

The angle between the unit vector **i** (vernal equinox direction) and the Greenwich meridian is called θ_g—the "Greenwich sidereal time." If we let λ_E be the geographic longitude of the radar site measured eastward from Greenwich, then

$$\boxed{\theta = \theta_g + \lambda_E} \tag{2-70}$$

where θ is called "local sidereal time."

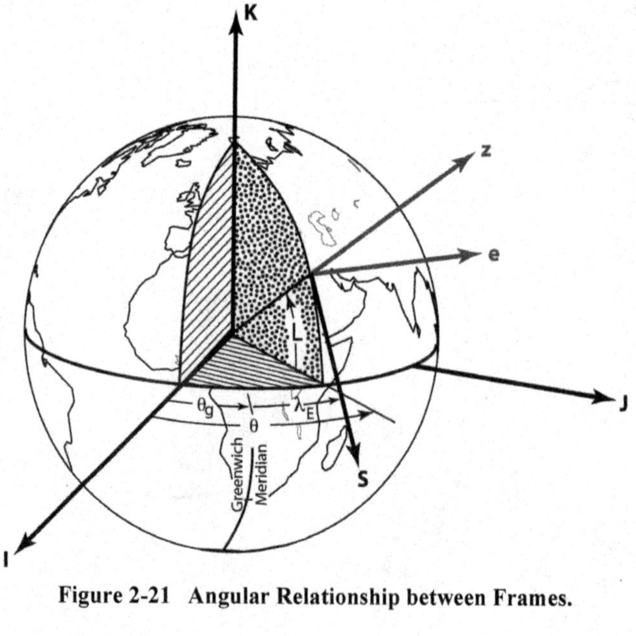

Figure 2-21 Angular Relationship between Frames.

The angles L and θ completely determine the relationship between the **IJK** frame and the **SEZ** frame. Obviously, we need a method of determining θ at some general time t.

If we knew θ_{go} at some particular time t_0 (say 0^h Universal Time on 1 Jan), we could determine θ at the time t from

$$\theta = \theta_{go} + \omega_{\oplus}(t - t_0) + \lambda_E \tag{2-71}$$

where $\omega_{\oplus}$ is the angular velocity of Earth.

The Astronomical Almanac[11] lists the value of θ_g at 0^h UT for every day of the year. For a more complete discussion of sidereal time see the next section.

We now have all we need to determine the rotation matrix ($\mathbf{D}^{-1}$) that transforms a vector from **SEZ** to **IJK** components. From Equation (2-41),

$$\tilde{\mathbf{D}}^{-1} = \begin{bmatrix} \sin L \cos\theta & -\sin\theta & \cos L \cos\theta \\ \sin L \sin\theta & \cos\theta & \cos L \sin\theta \\ -\cos L & 0 & \sin L \end{bmatrix} \tag{2-72}$$

If a_I, a_J, a_K and a_S, a_E, a_Z are the components of a vector **a** in each of the two systems then

$$\begin{bmatrix} a_I \\ a_J \\ a_K \end{bmatrix} = \tilde{\mathbf{D}}^{-1} \begin{bmatrix} a_S \\ a_E \\ a_Z \end{bmatrix} \tag{2-73}$$

2.9 THE MEASUREMENT OF TIME

Time is used as a fundamental dimension in almost every branch of science. When a scientist or layman uses the terms "hours, minutes or seconds" he or she is understood to mean units of *mean solar time*. This is the time kept by ordinary clocks. Since we will need to talk about another kind of time called "sidereal" time, it will help to understand exactly how each is defined. For more detailed discussions of several different dynamical measurement timescales beyond the scope of this book, refer to Vallado[12].

2.9.1 Solar and Sidereal Time

It is the Sun more than any other heavenly body that governs our daily activity cycle, so it is no wonder that ordinary time is reckoned by the Sun. The time between two successive upper transits of the Sun across our local meridian is called an *apparent solar day*. Earth has to turn through slightly more than one complete rotation on its axis relative to the "fixed" stars during this interval. The reason is that Earth travels about 1/365th of the way around its orbit in one day. This should be made clear by Figure 2-22.

A *sidereal day* consisting of 24 sidereal hours is defined as the time required for Earth to rotate once on its axis relative to the stars. This occurs in about 23^h 56^m 4^s of ordinary solar time and leads to the following relationship:

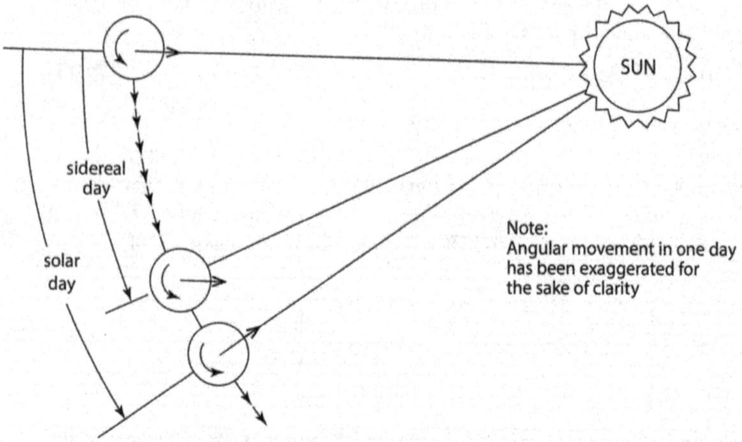

Figure 2-22 Solar and Sidereal Day.

$$
\begin{aligned}
1 \text{ day of mean solar time} &= 1.0027379093 \text{ days of mean sidereal time}\\
&= 24^h03^m56^s55536 \text{ of sidereal time}\\
&= 86{,}636.55536 \text{ mean sidereal seconds}
\end{aligned}
$$

$$
\begin{aligned}
1 \text{ day of mean sidereal time} &= 0.9972695664 \text{ days of solar time}\\
&= 23^h56^m04^s09054 \text{ of mean solar time}\\
&= 86{,}164.09054 \text{ mean solar seconds}
\end{aligned}
$$

So far, we have really defined only sidereal time and "apparent" solar time. Based on the definition of an apparent solar day illustrated in the figure, no two solar days would be exactly the same length because Earth's axis is not perpendicular to the plane of its orbit and because Earth's orbit is slightly elliptical. In early January when Earth is near perihelion, it moves farther around its orbit in 1 day then it does in early July when it is near aphelion. In order to avoid this irregularity in the length of a solar day, a mean solar day is defined based on the assumption that Earth is in a circular orbit whose period matches the actual period of Earth and that the axis of rotation is perpendicular to the orbital plane. An ordinary clock that ticks off 24 hours in one mean solar day would show the Sun arriving at our local meridian a little early at certain times of the year and a little late at other times of the year.

2.9.2 Julian Date

The concept of a Julian Date (JD) was introduced as a way to rationalize the keeping of calendar time across multiple centuries and multiple calendars. Astronomical days are numbered consecutively from an epoch that was chosen to be sufficiently far in the past to precede what is the current historical period. The Julian day number is defined to be 0 for the day starting at Greenwich mean noon on January 1,4713 B.C., on the Julian proleptic calendar. (The Julian proleptic calendar was produced by extending the Julian calendar backward to dates preceding 4 A.D., at which time the leap year was stabilized on a quadrennial basis.) The Julian Date at any instant is the Julian day number followed by the fraction of a day that has elapsed since the preceding noon. While a JD can be based on different timescales it is usually assumed to be based on UT1, which is described in the following section.

Some example Julian Dates for 2016 are:

J2000.0 = 2000 Jan 1.5 = JD 2,451,545.0

J2016.5 = 2016 Jul 2.125 = JD 2,457,571.625

A Modified Julian Date (MJD) is commonly used to simplify calculations and is defined as

$$\text{MJD} = \text{JD} - 2{,}400{,}000.5$$

2.9.3 Universal Time (UT) and International Atomic Time (TAI)

The modern standard for timekeeping is based on the concept of Universal Time (UT). UT conforms closely to the *mean* diurnal motion of the Sun while the *apparent* diurnal motion of the Sun is based on both the nonuniform diurnal rotation of Earth and the motion of Earth in its orbit around the Sun. Because of the nonuniform dynamics of the Earth–Sun system the mathematical relationship between UT and sidereal time is directly related by means of a mathematical formula. It is not precisely related to the hour angle of the Sun and does not refer to the motion of Earth because at any instant it can be derived from observations of the diurnal motion of the stars or radio sources.

UT0 is the uncorrected observed rotational timescale and that is dependent upon the location of the observation. The UT1 timescale is the measurement corrected for the polar motion of the longitude of the observing station and is now independent of observing location and influenced by the slightly variable rotation of Earth.

Since January 1, 1984, Greenwich mean sidereal time (GMST) has been related to UT1 by the following equation[13]:

$$\begin{aligned}\text{GMST of } 0^h\text{UT1} &= 24{,}110^s.54841 + 8{,}640{,}184^s.812866\ T_u \\ &+ 0^s.093104\ T_u^2 - 6.2 \times 10^{-6}\ T_u^3\end{aligned}$$

where d_u = the number of days of Universal Time elapsed since JD 2,451,545.5 UT1 (2000 January 1, 12^h UT1) and $T_u = d_u/36525$.

This is the definition of UT1 and it was based on the best available transit measurements of the FK5 star catalog. As new measurement techniques become operational for determining UT1, their arbitrary constants are adjusted to align with UT1 from the star catalog measurements.

Broadcast time services now provide time based on the redefined Coordinated Universal Time (UTC). UTC is maintained within 0.9 s of UT1 by introducing one-second leap seconds when necessary. The leap seconds are introduced at the end of June or December. DUT1 equals UT1 minus UTC and that number is transmitted over various broadcast bands.

International Atomic Time (TAI, from the French name Temps atomique international) is a high-precision atomic coordinate time standard based on the notional passage of proper time on Earth's geoid. It is the basis for UTC, which is used for civil timekeeping all over Earth's surface, and for Terrestrial Time, which is used for astronomical calculations. Since 31Dec 2016 when the last leap second was added, TAI has been exactly 37 s ahead of UTC. The 37 s results from the initial difference of 10 s at the start of 1972, plus 27 leap seconds in UTC since 1972. Time coordinates on the TAI scales are conventionally specified using traditional means of specifying days, carried over from non-uniform time standards based on the rotation of Earth. Specifically, both Julian Dates and the Gregorian calendar are used. TAI in this form was synchronized with Universal Time at the beginning of 1958, and the two have drifted apart ever since, due to the changing motion of Earth. TAI as a timescale is a weighted average of the time kept by over 200 atomic clocks in over 50 national laboratories worldwide. The clocks are compared using GPS signals and two-way satellite time and frequency transfer. Due to the averaging, it is far more stable than any clock would be alone (see signal averaging for a discussion). The majority of the clocks are cesium clocks; the definition of the SI second is written in terms of cesium.

If high precision is not required, the general term Universal Time (UT) may be used. The term Greenwich Mean Time (GMT) does not have a precise definition at the sub-second level, but it is often considered equivalent to UTC or UT1. Saying "GMT" often implies either UTC or UT1 when used within informal or casual contexts. In technical contexts, usage of "GMT" is avoided; the unambiguous terminology "UTC" or "UT1" is preferred.

2.9.4 Global Positioning System Time

The Global Positioning System (GPS) is a space-based satellite navigation system that provides location and time information in all weather conditions, anywhere on or near Earth where there is an unobstructed line of sight to four or more GPS satellites. The system provides critical capabilities to military, civil and commercial users around the world. It is maintained by the United States Air Force and is freely accessible to anyone with a GPS receiver.

Advances in technology and new demands on the existing system led to efforts to continuously modernize the GPS system and implement the next generation of GPS III satellites and Next Generation Operational Control System (OCX). In 2000, the U.S. Congress authorized the modernization effort, GPS Block III, and those satellites are past the demonstration phase and under continued production. Twenty-four GPS satellites in half-geosynchronous orbits at 55 degree inclination constitute a fully operational system and normally approximately 30 satellites are in orbit in an operational state at any time to provide on-orbit spares.

In addition to GPS, other systems are in use or under development. The Russian Global Navigation Satellite System (GLONASS) was developed contemporaneously with GPS, but it suffered from incomplete coverage of the globe until the mid-2000s. There are also the planned European Union Galileo positioning system, Chinese Compass navigation system, and Indian Regional Navigational Satellite System, which are in various stages of operational capability.

While most clocks derive their time from Coordinated Universal Time (UTC), the atomic clocks on the satellites are set to GPS time. The difference is that GPS time is not corrected to match the rotation of the Earth, so it does not contain leap seconds or other corrections that are periodically added to UTC. GPS time was set to match UTC in 1980, but has since diverged. The lack of corrections means that GPS time remains at a constant offset with International Atomic Time (TAI) (TAI – GPS = 19 s). Periodic corrections are performed to the on-board clocks to keep them synchronized with ground clocks.

The GPS navigation message includes the difference between GPS time and UTC. As of July 2015, GPS time was 17 s ahead of UTC because of the leap second added to UTC June 30, 2015. Receivers subtract this offset from GPS time to calculate UTC and specific time zone values. New GPS units may not show the correct UTC time until after receiving the UTC offset message. The GPS-UTC offset field can accommodate 255 leap seconds (eight bits). Users of GPS time must update the offset to properly calculate UTC.

The available GPS time signal for civilian use is theoretically accurate to about 14 ns. However, there are numerous sources of error (e.g., reflected signals, ionospheric refraction of signals and number of satellites in view) and time accuracy can drift to several times the nominal 14 ns value.

As opposed to the year, month and day format of the Gregorian calendar, the GPS date is expressed as a week number and a seconds-into-week number. The week number is transmitted as a 10-bit field in the civilian (C/A) and military [P(Y)] navigation messages, and so it becomes zero again every 1,024 weeks (19.6 years). GPS week zero started at 00:00:00 UTC (00:00:19 TAI) on January 6, 1980, and the week number became zero again for the first time at 23:59:47 UTC on August 21, 1999 (00:00:19 TAI on August 22, 1999). To determine the current Gregorian date, a GPS receiver must be provided with the approximate date (to within 3,584 days) to correctly translate the GPS date signal. To address this concern the modernized GPS navigation message uses a 13-bit field that only repeats every 8,192 weeks (157 years), thus lasting until the year 2137 (157 years after GPS week zero).

2.9.5 Local Mean Solar Time and Universal Time

Earth is divided into 24 time zones approximately 15° of longitude apart. The local mean solar time in each zone differs from the neighboring zones by 1 hour. (A few countries have adopted time zones that differ by only ½ hour from the adjacent zones.)

Table 2-1: Coordinated Universal Time (UTC) to Global Positioning System (GPS) Time Conversion.

Limits of Validity (at 0 h UTC)	TAI – UTC(s)	GPS – UTC(s)
1972-01-01 – 1972-07-01	10	--
1972-07-01 – 1973-01-01	11	--
1973-01-01 – 1974-01-01	12	--
1974-01-01 – 1975-01-01	13	--
1975-01-01 – 1976-01-01	14	--
1976-01-01 – 1977-01-01	15	--
1977-01-01 – 1978-01-01	16	--
1978-01-01 – 1979-01-01	17	--
1979-01-01 – 1980-01-01	18	--
1980-01-01 – 1981-07-01	19	0
1981-07-01 – 1982-07-01	20	1
1982-07-01 – 1983-07-01	21	2
1983-07-01 – 1985-07-01	22	3
1985-07-01 – 1988-01-01	23	4
1988-01-01 – 1990-01-01	24	5
1990-01-01 – 1991-01-01	25	6
1991-01-01 – 1992-07-01	26	7
1992-07-01 – 1993-07-01	27	8
1993-07-01 – 1994-07-01	28	9
1994-07-01 – 1996-01-01	29	10
1996-01-01 – 1997-07-01	30	11
1997-07-01 – 1999-01-01	31	12
1999-01-01 – 2006-01-01	32	13
2006-01-01 – 2009-01-01	33	14
2009-01-01 – 2012-07-01	34	15
2012-07-01 – 2015-07-01	35	16
2015-07-01 – 2017-01-01	36	17
2017-01-01 – TBD future correction	37	18

The local mean solar time on the Greenwich meridian is called Greenwich Mean Time (GMT), Universal Time (UT) or Zulu (Z) time.

A time given in terms of a particular time zone can be converted to Universal Time by simply adding or subtracting the correct number of hours. For example, if it is 1800 Eastern Standard Time (EST), the Universal Time is 2300. The conversion for time zones in the United States is as follows:

EST + 5 hr = UT
CST + 6 hr = UT
MST + 7 hr = UT
PST + 8 hr = UT

2.9.6 Finding the Greenwich Sidereal Time when Universal Time is Known

Often it is desired to relate observations made in the topocentric-horizon system to the **IJK** unit vectors of the geocentric equatorial system and vice versa. The geometrical relationship between these two systems depends on the latitude and longitude of the topos and Greenwich sidereal time, θ_g (expressed as an angle), at the date and time of the observation. This relationship is illustrated in Figure 2-21.

What we need is a convenient way to calculate the angle θ_g for any date and time of day. If we knew what θ_g was on a particular day and time we could calculate θ_g for any future time since we know that in one day Earth turns through 1.0027379093 complete rotations on its axis.

Suppose we take the value of θ_g at 0^h UT on the day before a particular month and call it θ_{g_0} (e.g., for February we use the value for 31 January). Then for the first day of the actual month (e.g., 01 February), we simply add 1 times the daily sidereal rotation of Earth to that initial value. For the 15th of a month, we add 15 times the daily sidereal rotation. For the local sidereal time at a particular location we add the local longitude measured from Greenwich (with East being positive), converted to hours or radians as needed.

If in addition, we express time in decimal fractions of a day, we can convert a particular date and time into a single number that indicates the number of days that have elapsed since our "time zero." If we call this number D, then

$$\theta_g = \theta_{g_0} + 1.0027379093 \times 360° \times D \text{ [degrees]}$$

or

$$\theta_g = \theta_{g_0} + 1.0027379093 \times 2\pi \times D \text{ [radians]}$$

2.9.7 Precession of the Equinoxes

To understand what the values of θ_{g_0} in Table2-2 represent, it is necessary to discuss the slow shifting of the vernal equinox direction known as precession (see Figure (2-23)).

Table 2-2: Mean Sidereal Time 2016. Greenwich mean sidereal time at 0^h UTC, θ_{g_0}.

Month	Day	Hour	h:m:s	Degrees	Radians
Jan	0	6.607	6:36:25	99.105	1.729709
Feb	0	8.644	8:38:38	129.66	2.262994
Mar	0	10.5496	10:32:59	158.244	2.761879
Apr	0	12.5866	12:35:12	188.799	3.295164
May	0	14.5579	14:33:28	218.3685	3.811249
Jun	0	16.5949	16:35:42	248.9235	4.344535
Jul	0	18.5662	18:33:58	278.493	4.86062
Aug	0	20.6032	20:36:12	309.048	5.393905
Sep	0	22.6402	22:38:25	339.603	5.92719
Oct	0	0.6115	0:36:41	9.1725	0.16009
Nov	0	2.6485	2:38:55	39.7275	0.693376
Dec	0	4.6198	4:37:11	69.297	1.209461

The direction of the equinox is determined by the line of intersection of the ecliptic plane (the plane of Earth's orbit) and the equatorial plane.

While the plane of the ecliptic is fixed relative to the stars, the equatorial plane is not. Due to the asphericity of Earth, the Sun produces a torque on Earth, which results in a wobbling or precessional motion similar to that of a simple top. Because Earth's equator is tilted ~23.5 deg to the plane of the ecliptic, the polar axis sweeps out a cone-shaped surface in space with a semi-vertex angle of ~23.5 deg. As Earth's axis precesses, the line of intersection of the equator and the ecliptic swings westward slowly. The period of the precession is about 26,000 years, so the equinox direction shifts westward about 50 arcseconds per year.

The Moon also produces a torque on Earth's equatorial bulge. However, the Moon's orbital plane precesses due to solar perturbation with a period of about 18.6 years, so the lunar-caused precession has this same period. The effect of the Moon is to superimpose a slight nodding motion call "nutation," with a period of 18.6 years on the slow westward precession caused by the Sun.

The *mean equinox* is the position of the equinox when solar precession alone is taken into account. The *apparent equinox* is the actual equinox direction when both precession and nutation are included.

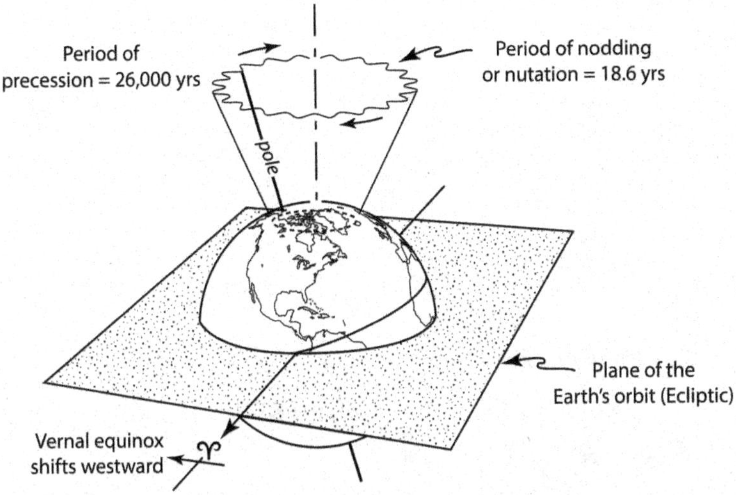

Figure 2-23 Precession of the Equinoxes.

The values of θ_{g_0} in the preceding table are referred to the mean equinox and equator of the dates shown.

Example Problem. What is the inertial position vector of a point 6.378 km above mean sea level on the equator, 57.296 degrees west longitude, at 0600 GMT, 2 January 1970?

The given information in consistent units is

UT = 0600 hr, Day = 1 (1 Jan 1970 = 0)
Long = $\lambda = -57.296° = -1$ radian (east longitude is positive)
H = 6.378 km = 0.001 $DU_\oplus$, Lat = L = 0°

From Section 2.8.2, θ_{g_0} for 1970 is 1.74933340 and

$$D = 1.25$$

$$\theta_g = \theta_{g_0} + 1.0027379093 \times 2\pi \times D = 9.6245 \text{ radians}$$

$$\theta = \theta_g + \lambda = 9.6245 - 1.0 = 8.6245 \text{ radians}$$

From Equation (2-68),

$$x = \left[\frac{a_e}{\sqrt{1 - e^2 \sin^2 L}} + H \right] \cos L = 1.001$$

$$z = \left[\frac{a_e \left(1 - e^2\right)}{\sqrt{1 - e^2 \sin^2 L}} + H \right] \sin L = 0$$

From Equation (2-69),

$$\mathbf{R} = x \cos\theta \, \mathbf{I} + x \sin\theta \, \mathbf{J} + Z\mathbf{K} \, DU_{\oplus}$$

$$= 1.001 \cos (8.6245)\mathbf{I} + 1.001 \sin (8.6245)\mathbf{J} + 0\mathbf{K} \, DU_{\oplus}$$

$$= -4{,}445.56\mathbf{I} + 4{,}579.50\mathbf{J} + 0\mathbf{K} \text{ km}$$

Example Problem. At 0600 GST a Ballistic Missile Early Warning System (BMEWS) tracking station (Lat 60° N, Long 150° W) detected a space object and obtained the following data:

Slant Range (ρ) = 0.4 $DU_{\oplus}$
Azimuth (Az) = 90°
Elevation (El) = 30°
Range Rate ($\dot{\rho}$) = 0
Azimuth Rate ($\dot{Az}$) = 10 rad/$TU_{\oplus}$
Elevation Rate ($\dot{El}$) = 5 rad/$TU_{\oplus}$

What were the velocity and position vectors of the space object at the time of observation?

From Equation (2-46),

$$\rho_S = -(0.4)(\cos 30°)(\cos 90°) = 0 \ DU_{\oplus}$$
$$\rho_E = (0.4)(\cos 30°)(\sin 90°) = 0.346 \ DU_{\oplus}$$
$$\rho_Z = (0.4) \sin 30° = 0.2 \ DU_{\oplus}$$

Hence

$$\boldsymbol{\rho} = 0.346E + 0.2Z \ (DU_{\oplus})$$

From Equation (2-48),

$$\dot{\rho}_S = (0)(\cos 30°)(\cos 90°) = (0.4)(\sin 30°)(5)\cos 90° + (0.4)(\cos 30°)(\sin 90°)(10) = 3.46 \text{ DU/TU}$$

$$\dot{\rho}_E = (0)(\cos 30°)(\sin 90°) - (0.4)(\sin 30°)(5)(\sin 90°) + (0.4)(\cos 30°)(\cos 90°)(10) = -1.0 \text{ DU/TU}$$

$$\dot{\rho}_Z = (0)(\sin 30°) + (0.4)(\cos 30°)(5) = 1.73 \text{ DU/TU}$$

Hence

$$\dot{\boldsymbol{\rho}} = 3.46\mathbf{S} - 1.0\mathbf{E} + 1.73\mathbf{Z} \text{ DU/TU}$$

From Equation (2-49), $\mathbf{r} = 0.346\mathbf{E} + 1.2\mathbf{Z}$ (DU)
From Equation (2-71),

$$\theta = (6)(15°/\text{hour}) - 150° = -60° \text{ (LST)}$$

The rotation matrix Equation (2-72) becomes

$$\tilde{\mathbf{D}}^{-1} = \begin{bmatrix} 0.433 & 0.866 & 0.25 \\ -0.75 & 0.5 & -0.433 \\ -0.5 & 0 & 0.866 \end{bmatrix}$$

and

$$\mathbf{r} = \tilde{\mathbf{D}}^{-1} \begin{bmatrix} 0 \\ 0.346 \\ 1.2 \end{bmatrix} = 0.6\mathrm{I} - 0.346\mathrm{J} + 1.04\mathrm{K}(\mathrm{DU}_\oplus)$$

Similarly,

$$\dot{\boldsymbol{\rho}} = \tilde{\mathbf{D}}^{-1} \begin{bmatrix} 3.46 \\ -1.0 \\ 1.73 \end{bmatrix} = 1.06\mathbf{I} - 3.84\mathbf{J} - 0.232\mathbf{K}(\text{DU/TU})$$

From Equation (2-51),

$$\mathbf{v} = \dot{\boldsymbol{\rho}} + (0.0588\mathbf{K}) \times \mathbf{r} \text{ DU/TU}$$

$$\mathbf{v} = 1.08\mathbf{I} - 3.8\mathbf{J} - 0.232\mathbf{K} \text{ DU/TU}$$

Thus the position and velocity vectors of the object at one epoch have been found, and the orbit is uniquely determined.

2.10 ORBIT DETERMINATION FROM THREE POSITION VECTORS[9]

In the preceding section we saw how to obtain **r** and **v** from a single radar measurement of $\rho, \dot{\rho}$, El, $\dot{\text{El}}$, Az, $\dot{\text{Az}}$. It may happen that a particular radar site is not equipped to measure Doppler phase shifts and so the rate information may be lacking. In this section we will examine a method for determining an orbit from three position vectors $\mathbf{r}_1$, $\mathbf{r}_2$ and $\mathbf{r}_3$ (assumed to be coplanar)[8]. See Figure (2-24). These three vectors may be obtained from successive measurements of ρ, El and Az at three times by the methods of the last section or by any other technique.

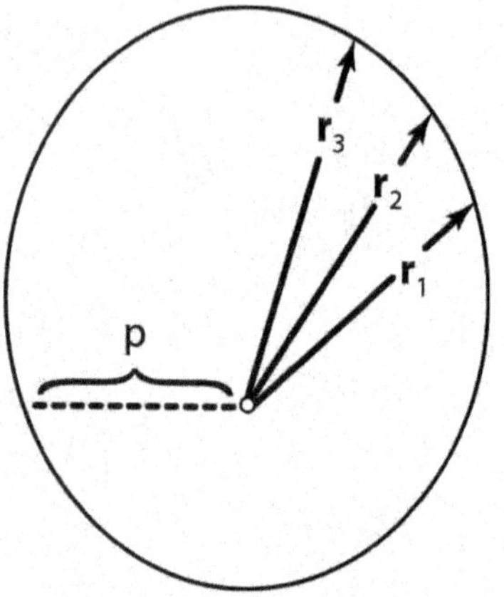

Figure 2-24 Orbit Through $\mathbf{r}_1$, $\mathbf{r}_2$ and $\mathbf{r}_3$.

The scheme to be presented is associated with the name of J.W. Gibbs and has come to be known as the Gibbsian method. As Baker[5] points out, it was developed using pure vector analysis and was historically the first contribution of an American scholar to celestial mechanics. Gibbs is, of course, well known for his contributions to thermodynamics, but the contributions to celestial mechanics of this fine scholar, who in 1863 received the USA's first PhD in engineering, is equally outstanding but less generally remembered.

The Gibbs problem can be stated as follows: Given three nonzero coplanar vectors $\mathbf{r}_1$, $\mathbf{r}_2$ and $\mathbf{r}_3$ that represent three sequential positions of an orbiting object on one pass, find the parameter p and the eccentricity e of the orbit and the perifocal base vectors **P**, **Q** and **W**.

Solution: Since the vectors $\mathbf{r}_1$, $\mathbf{r}_2$ and $\mathbf{r}_3$ are coplanar, there must exist scalars C_1, C_2 and C_3 such that

$$C_1\mathbf{r}_1 + C_2\mathbf{r}_2 + C_3\mathbf{r}_3 = \mathbf{0} \tag{2-74}$$

Using the polar equation of a conic section, Equation (1-43) and the definition of a dot product, we can show that

$$\mathbf{e} \cdot \mathbf{r} = p - r \tag{2-75}$$

Dotting Equation (2-74) by **e** and using the relation Equation (2-75) gives

$$C_1 (p - r_1) + C_2 (p - r_2) + C_3 (p - r_3) = 0. \tag{2-76}$$

Cross Equation (2-74) successively by $\mathbf{r_1}$, $\mathbf{r_2}$ and $\mathbf{r_3}$ to obtain

$$\begin{aligned} C_2 \mathbf{r}_1 \times \mathbf{r}_2 &= C_3 \mathbf{r}_3 \times \mathbf{r}_1 \\ C_1 \mathbf{r}_1 \times \mathbf{r}_2 &= C_3 \mathbf{r}_2 \times \mathbf{r}_3 \\ C_1 \mathbf{r}_3 \times \mathbf{r}_1 &= C_2 \mathbf{r}_2 \times \mathbf{r}_3 \end{aligned} \tag{2-77}$$

By multiplying Equation (2-76) by $\mathbf{r_3} \times \mathbf{r_1}$, then using Equation (2-77), we can eliminate $\mathbf{C_1}$ and $\mathbf{C_3}$ to obtain:

$$\begin{aligned} &C_2 \mathbf{r}_2 \times \mathbf{r}_3 \left(p - r_1\right) + C_2 \mathbf{r}_3 \times \mathbf{r}_1 \left(p - r_2\right) \\ &+ C_2 \mathbf{r}_1 \times \mathbf{r}_2 \left(p - r_3\right) = \mathbf{0} \end{aligned} \tag{2-78}$$

Notice that $\mathbf{C_2}$ may now be divided out. Multiplying the factors and collecting terms gives

$$\begin{aligned} &p\,(\mathbf{r}_1 \times \mathbf{r}_2 + \mathbf{r}_2 \times \mathbf{r}_3 + \mathbf{r}_3 \times \mathbf{r}_1) \\ &= r_3 \mathbf{r}_1 \times \mathbf{r}_2 + r_1 \mathbf{r}_2 \times \mathbf{r}_3 + r_2 \mathbf{r}_3 \times \mathbf{r}_1 \end{aligned} \tag{2-79}$$

Let the right side of Equation (2-79) be defined as a vector **N** and the coefficient of **p** be defined as a vector **D**, then

$$p\mathbf{D} = \mathbf{N} \tag{2-80}$$

Therefore, provided $\mathbf{N} \cdot \mathbf{D} = ND$, i.e., **N** and **D** have the same direction,

$$p = \frac{N}{D} \tag{2-81}$$

It can be shown that, for any set of vectors suitable for the Gibbsian method as described above, **N** and **D** do have the same direction and that this is the direction of the angular momentum vector **h**. This is also the direction of **W** in the perifocal coordinate system.

Since **P**, **Q** and **W** are orthogonal unit vectors,

$$\mathbf{Q} = \mathbf{W} \times \mathbf{P} \tag{2-82}$$

Since **W** is a unit vector in the direction of **N** and **P** is a unit vector in the direction of **e**, Equation (2-82) can be written as

$$\mathbf{Q} = \frac{1}{\mathrm{Ne}} (\mathbf{N} \times \mathbf{e}) \tag{2-83}$$

Now substituting for **N** from its definition in Equation (2-79) gives

$$\mathrm{Ne}\mathbf{Q} = r_3 (\mathbf{r}_1 \times \mathbf{r}_2) \times \mathbf{e} + r_1 (\mathbf{r}_2 \times \mathbf{r}_3) \times \mathbf{e} + r_2 (\mathbf{r}_3 \times \mathbf{r}_1) \times \mathbf{e} \tag{2-84}$$

Now use the general relationship for a vector triple product,

$$(\mathbf{a} \times \mathbf{b}) \times \mathbf{c} = (\mathbf{a} \cdot \mathbf{c})\, \mathbf{b} - (\mathbf{b} \cdot \mathbf{c})\, \mathbf{a} \tag{2-85}$$

to rewrite Equation (2-84) as

$$\begin{aligned} \mathrm{Ne}\mathbf{Q} &= r_3 (\mathbf{r}_1 \cdot \mathbf{e})\, \mathbf{r}_2 - r_3 (\mathbf{r}_2 \cdot \mathbf{e})\, \mathbf{r}_1 \\ &+ r_1 (\mathbf{r}_2 \cdot \mathbf{e})\, \mathbf{r}_3 - r_1 (\mathbf{r}_3 \cdot \mathbf{e})\, \mathbf{r}_2 \\ &+ r_2 (\mathbf{r}_3 \cdot \mathbf{e})\, \mathbf{r}_1 - r_2 (\mathbf{r}_1 \cdot \mathbf{e})\, \mathbf{r}_3 \end{aligned} \tag{2-86}$$

Again using the relationship Equation (2-75) and factoring p from the right side we have

$$\mathrm{Ne}\mathbf{Q} = \mathrm{p}\left[(r_2 - r_3)\mathbf{r}_1 + (r_3 - r_1)\mathbf{r}_2 + (r_1 - r_2)\mathbf{r}_3\right] \equiv \mathrm{p}\mathbf{S} \tag{2-87}$$

where **S** is defined by the bracketed quantity in Equation (2-87).

Therefore, since Ne**Q** = pS, **N** = pD and **Q** and **S** have the same direction,

$$\mathrm{e} = \frac{\mathrm{S}}{\mathrm{D}} \tag{2-88}$$

and

$$\mathbf{Q} = \frac{\mathbf{S}}{\mathrm{S}} \tag{2-89}$$

$$\mathbf{W} = \frac{\mathbf{N}}{\mathrm{N}} \tag{2-90}$$

Since **P, Q** and **W** are orthogonal,

$$\mathbf{P} = \mathbf{Q} \times \mathbf{W} \tag{2-91}$$

Thus, to solve this problem use the given **r** vectors to form the **N**, **D** and **S** vectors. Before solving the problem, check that $\mathbf{N} \neq 0$ and $\mathbf{D} \cdot \mathbf{N} > 0$. Then use Equation (2-81) to find p; Equation (2-88) to find **e**; Equation (2-89) to find **Q**; Equation (2-90) to find **W**; and Equation (2-91) to find **P**. The information thus obtained may be used in Equation (2-20) to obtain the velocity vector corresponding to any of the given position vectors. However, it is possible to develop an expression that gives the **v** vector directly in terms of the **D**, **N** and **S** vectors.

From Equation (1-38) we can write

$$\dot{\mathbf{r}} \times \mathbf{h} = \mu\left(\frac{\mathbf{r}}{r} + \mathbf{e}\right) \tag{2-92}$$

Cross **h** into Equation (2-92) to obtain

$$\mathbf{h} \times (\dot{\mathbf{r}} \times \mathbf{h}) = \mu\left(\frac{\mathbf{h} \times \mathbf{r}}{r} + \mathbf{h} \times \mathbf{e}\right) \tag{2-93}$$

Using the identity $\mathbf{a} \times (\mathbf{b} \times \mathbf{c}) = (\mathbf{a} \cdot \mathbf{c})\,\mathbf{b} - (\mathrm{a} \cdot \mathbf{b})\,\mathbf{c}$, the left side becomes $(\mathbf{h} \cdot \mathbf{h})\,\mathbf{v} - (\mathbf{h} \cdot \mathbf{v})\,\mathbf{h}$ and $\mathbf{h} \cdot \mathbf{v} = 0$.

Thus

$$h^2\mathbf{v} = \mu\left(\frac{\mathbf{h} \times \mathbf{r}}{r} + \mathbf{h} \times \mathbf{e}\right) \tag{2-94}$$

We can write $\mathbf{h} = h\mathbf{W}$ and $\mathbf{e} = e\mathbf{P}$, so

$$\mathbf{v} = \frac{\mu}{h}\left(\frac{\mathbf{W} \times \mathbf{r}}{r} + e\mathbf{W} \times \mathbf{P}\right)$$
$$\mathbf{v} = \frac{\mu}{h}\left(\frac{\mathbf{W} \times \mathbf{r}}{r} + e\mathbf{Q}\right) \tag{2-95}$$

Using the fact that $h = \sqrt{N\mu/D}$,

$$e = \frac{S}{D},\ \mathbf{Q} = \frac{\mathbf{S}}{S} \text{ and } \mathbf{W} = \frac{\mathbf{D}}{D} \tag{2-96}$$

we have $\mathbf{v} = \dfrac{1}{r}\sqrt{\dfrac{\mu}{ND}}\,\mathbf{D} \times \mathbf{r} + \sqrt{\dfrac{\mu}{ND}}\,\mathbf{S}$

To streamline the calculations let us define a vector and a scalar:

$$\mathbf{B} \triangleq \mathbf{D} \times \mathbf{r} \tag{2-97}$$

$$L \triangleq \sqrt{\frac{\mu}{DN}} \tag{2-98}$$

Then $\mathbf{v} = \dfrac{L}{r}\mathbf{B} + L\mathbf{S}$ (2-99)

Thus, to find any of the three velocities directly, proceed as follows: For example, to find $\mathbf{v}_2$:

1. Test $\mathbf{r}_1 \cdot \mathbf{r}_2 \times \mathbf{r}_3 = \mathbf{0}$ for coplanar vectors.
2. Form the **D**, **N** and **S** vectors.

3. Test $\mathbf{D} \neq 0$, $\mathbf{N} \neq 0$, $\mathbf{D} \cdot \mathbf{N} > 0$ to assure that the vectors describe a possible two-body orbit.
4. Form $\mathbf{B} = \mathbf{D} \times \mathbf{r_2}$
5. Form $L = \sqrt{\dfrac{\mu}{DN}}$
6. Finally, $\mathbf{v_2} = \dfrac{L}{r_2}\mathbf{B} + L\mathbf{S}$

There are a number of other derivations of the Gibbsian method of orbit determination, but all of them have some problems like quadrant resolution that makes computer implementation difficult. This method, using the stated tests, appears to be foolproof in that there are no known special cases.

There are several general features of the Gibbsian method worth noting that set it apart from other methods of orbit determination. Earlier in this chapter we noted that six independent quantities called "orbital elements" are needed to completely specify the size, shape and orientation of an orbit and the position of the satellite in that orbit. By specifying three position vectors we appear to have nine independent quantities—three components for each of the three vectors—from which to determine the six orbital elements. This is not exactly true. The fact that the three vectors must lie in the same plane means that they are not independent.

Another interesting feature of the Gibbsian method is that it is purely geometrical and vectorial and makes use of the theorem that "one and only one conic section can be drawn through three coplanar position vectors such that the focus lies at the origin of the three position vectors." Thus, the only feature of orbital motion that is exploited is the fact that the path is a conic section whose focus is the center of Earth. The time of flight between the three positions is not used in the calculations. If we make use of the dynamical equation of motion of the satellite it is possible to obtain the orbit from only two position vectors, $\mathbf{r}_1$ and $\mathbf{r}_2$, and the time of flight between these positions. This is such an important problem that we will devote Chapter 5 entirely to its solution.

Example Problem. Radar observations of an Earth satellite during a single pass yield in chronological order the following positions (canonical units):

$\mathbf{r}_1 = 1.000\mathbf{K}$

$\mathbf{r}_2 = -0.700\mathbf{J} - 0.8000\mathbf{K}$

$\mathbf{r}_3 = 0.9000\mathbf{J} + 0.5000\mathbf{K}$

Find **P**, **Q** and **W** (the perifocal basis vectors expressed in the **IJK** system), the semi-latus rectum, eccentricity, period and the velocity vector at position two.

Form the **D**, **N**, and **S** vectors:

$\mathbf{D} = 1.970\mathbf{I}$

$\mathbf{N} = 2.047\mathbf{I}$ DU

$$\mathbf{S} = -0.0774\mathbf{J} - 0.0217\mathbf{K}$$

Test: $\mathbf{r_1} \cdot \mathbf{r_2} \times \mathbf{r_3} = 0$ to verify that the observed vectors are coplanar.
Since $\mathbf{D} \neq 0$, $\mathbf{N} \neq 0$ and $\mathbf{D} \cdot \mathbf{N} > 0$ we know that the given data present a solvable problem:

$$p = \frac{N}{D} = \frac{2.047}{1.97} = 1.039 \text{ DU } (6627 \text{ km})$$

$$e = \frac{S}{D} = \frac{0.0804}{1.97} = 0.04081$$

$$a = 1.041 \text{ DU } (6639 \text{ km})$$

$$\mathbb{TP} = 2\pi\sqrt{\frac{a^3}{\mu}} = 6.67 \text{ TU } (89.7 \text{ minutes})$$

$$r_p = p / (1 + e) = 0.9983 \text{ DU } (6367 \text{ km}) = \text{suborbital!}$$

$$\mathbf{Q} = \frac{\mathbf{S}}{S} = -0.963\mathbf{J} - 0.270\mathbf{K}$$

$$\mathbf{W} = \frac{\mathbf{N}}{N} = 1.000\mathbf{I}$$

$$\mathbf{P} = \mathbf{Q} \times \mathbf{W} = -0.270\mathbf{J} + 0.963\mathbf{K}$$

Now form the **B** vector:

$$\mathbf{B} = \mathbf{D} \times r_2 = 1.576\mathbf{J} - 1.379\mathbf{K}$$

and the scalar

$$L = 1/\sqrt{DN} = 0.4979$$

then $v_2 = \frac{L}{r_2}\mathbf{B} + L\mathbf{S} = 0.700\mathbf{J} - 0.657\mathbf{K}$ DU/TU

$$v_2 = 0.960 \text{ DU/TU } (7.5932 \text{ km/s})$$

The same equations used above are very efficient for use in a computer solution to problems of this type. Another approach is to immediately solve for $\mathbf{v_2}$ and then use $\mathbf{r_2}$, $\mathbf{v_2}$ and the method of Section 2.4 to solve for the elements.

2.11 ORBIT DETERMINATION FROM OPTICAL SIGHTINGS

The modern orbit determination problem is made much simpler by the availability of radar range and range-rate information. However, the angular pointing accuracy and resolution of radar sensors is far below that of optical sensors.

Six independent quantities suffice to completely specify a satellite's orbit. These may be the six classical orbital elements or they may be the six components of the vectors **r** and **v** at some epoch. In either case, an optical observation yields only two independent quantities such as El and Az or right ascension and declination, so a minimum of three observations is required at three different times to determine the orbit.

Since astronomers had to determine the orbits of comets and minor planets (asteroids) using angular data only, the method presented below has been in long use and was first suggested by Laplace in 1780.[6]

2.11.1 Determining the Line-of-Sight Unit Vectors

Let us assume that we have the topocentric right ascension and declination of a satellite at three separate times: α_1, δ_1, α_2, δ_2, α_3, δ_3 (see Figure 2-25). These could easily be obtained from a photograph of the satellite against the star background. If we let $\mathbf{L_1}$, $\mathbf{L_2}$ and $\mathbf{L_3}$ be unit vectors along the line of sight to the satellite at the three observation times then

$$\mathbf{L}_i = \begin{bmatrix} L_I \\ L_J \\ L_K \end{bmatrix}_i = \begin{bmatrix} \cos\delta_i \cos\alpha_i \\ \cos\delta_i \sin\alpha_i \\ \sin\delta_i \end{bmatrix}_i, \quad i = 1, 2, 3, \tag{2-100}$$

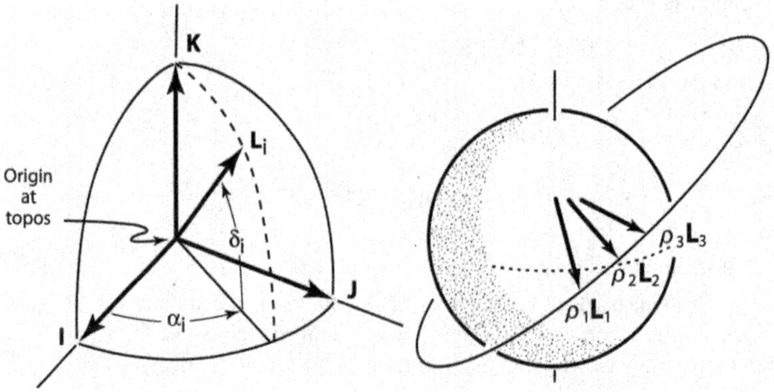

Figure 2-25 Line-of-Sight Vectors.

Now since $\mathbf{L_i}$ are unit vectors directed along the slant range vector $\boldsymbol{\rho}$ from the observation site to the satellite, we may write

$$\mathbf{r} = \rho\mathbf{L} + \mathbf{R} \tag{2-101}$$

where subscripts have been omitted for simplicity and where ρ is the slant range to the satellite, $\mathbf{r}$ is the vector from the center of Earth to the satellite and $\mathbf{R}$ is the vector from the center of Earth to observation site (see Figure 2-14).

We may differentiate Equation (2-101) twice to obtain

$$\dot{\mathbf{r}} = \dot{\rho}\mathbf{L} + \rho\dot{\mathbf{L}} + \dot{\mathbf{R}} \tag{2-102}$$

$$\ddot{\mathbf{r}} = 2\dot{\rho}\dot{\mathbf{L}} + \ddot{\rho}\mathbf{L} + \rho\ddot{\mathbf{L}} + \ddot{\mathbf{R}} \tag{2-103}$$

From the equation of motion we have the dynamical relationship

$$\ddot{\mathbf{r}} = -\mu\frac{\mathbf{r}}{r^3}$$

Substituting this into Equation (2-103) and simplifying yields

$$\boxed{\mathbf{L}\,\ddot{\rho} + 2\dot{\mathbf{L}}\,\dot{\rho} + \left(\ddot{\mathbf{L}} + \frac{\mu}{r^3}\mathbf{L}\right)\rho = -\left(\ddot{\mathbf{R}} + \mu\frac{\mathbf{R}}{r^3}\right)} \tag{2-104}$$

At a specified time, say the middle observation, the above vector equation represents three component equations in 10 unknowns. The vectors **L**, **R** and $\ddot{\mathbf{R}}$ are known at time t_2; $\dot{\mathbf{L}}$, $\ddot{\mathbf{L}}$, ρ, $\dot{\rho}$, $\ddot{\rho}$ and r, however, are not known.

2.11.2 Derivatives of the Line-of-Sight Vector

Since we have the value of **L** at three times, t_1, t_2 and t_3, we can numerically differentiate to obtain $\dot{\mathbf{L}}$ and $\ddot{\mathbf{L}}$ at the central time, t_2, provided the three observations are not too far apart in time. We may use the Lagrange interpolation formula to write a general analytical expression for **L** as a function of time:

$$\mathbf{L}(t) = \frac{(t-t_2)(t-t_3)}{(t_1-t_2)(t_1-t_3)}\mathbf{L}_1 + \frac{(t-t_1)(t-t_3)}{(t_2-t_1)(t_2-t_3)}\mathbf{L}_2 + \frac{(t-t_1)(t-t_2)}{(t_3-t_1)(t_3-t_2)}\mathbf{L}_3$$

Note that this second-order polynomial in t reduces to $\mathbf{L}_1$ when $t = t_1$, $\mathbf{L}_2$ when $t = t_2$ and $\mathbf{L}_3$ when $t = t_3$. Differentiating this equation twice yields $\dot{\mathbf{L}}$ and $\ddot{\mathbf{L}}$, thus

$$\boxed{\dot{\mathbf{L}}(t) = \frac{2t-t_2-t_3}{(t_1-t_2)(t_1-t_3)}\mathbf{L}_1 + \frac{2t-t_1-t_3}{(t_2-t_1)(t_2-t_3)}\mathbf{L}_2 + \frac{2t-t_1-t_2}{(t_3-t_1)(t_3-t_2)}\mathbf{L}_3} \tag{2-105}$$

$$\boxed{\begin{aligned}\ddot{\mathbf{L}}(t) = {} & \frac{2}{(t_1 - t_2)(t_1 - t_3)}\mathbf{L}_1 + \frac{2}{(t_2 - t_1)(t_2 - t_3)}\mathbf{L}_2 \\ & + \frac{2}{(t_3 - t_1)(t_3 - t_2)}\mathbf{L}_3\end{aligned}} \tag{2-106}$$

By setting $t = t_2$ in Equations (2-105) and (2-106) we can obtain numerical values for $\dot{\mathbf{L}}$ and $\ddot{\mathbf{L}}$ at the central time. It should be noted, however, that if more than three observations are available, more accurate values of $\dot{\mathbf{L}}$ and $\ddot{\mathbf{L}}$ at the central date may be obtained by fitting higher order polynomials with the Lagrange interpolation formula to the observations or, better yet, making a least-squares polynomial fit to the observations. This, in fact, must be done if $\dddot{\mathbf{L}}$ and higher order derivatives are not negligible.[4]

Equation (2-104) written for the central time now represents three component equations in four unknowns, ρ, $\dot{\rho}$, $\ddot{\rho}$ and r.

2.11.3 Solving for the Vector r

For the time being, let us assume that we know **r** and solve Equation (2-104) for ρ using Cramer's rule.

The determinant of the coefficients is

$$D = \begin{vmatrix} L_I & 2\dot{L}_I & \ddot{L}_I + \mu L_I/r^3 \\ L_J & 2\dot{L}_J & \ddot{L}_J + \mu L_J/r^3 \\ L_K & 2\dot{L}_K & \ddot{L}_K + \mu L_K/r^3 \end{vmatrix}$$

Since the value of the determinant is not changed if we subtract μ/r^3 times the first column from the third column, **D** reduces to

$$D = 2\begin{vmatrix} L_I & \dot{L}_I & \ddot{L}_I \\ L_J & \dot{L}_J & \ddot{L}_J \\ L_K & \dot{L}_K & \ddot{L}_K \end{vmatrix} \tag{2-107}$$

By applying Cramer's rule to Equation (2-104) it is evident that

$$D\rho = -\begin{vmatrix} L_I & 2\dot{L}_I & \ddot{R}_I + \mu R_I/r^3 \\ L_J & 2\dot{L}_J & \ddot{R}_J + \mu R_J/r^3 \\ L_K & 2\dot{L}_K & \ddot{R}_K + \mu R_K/r^3 \end{vmatrix}$$

This determinant can be conveniently split to produce

$$D\rho = -2\begin{vmatrix} L_I & \dot{L}_I & \ddot{R}_I \\ L_J & \dot{L}_J & \ddot{R}_J \\ L_K & \dot{L}_K & \ddot{R}_K \end{vmatrix} - 2\frac{\mu}{r^3}\begin{vmatrix} L_I & \dot{L}_I & R_I \\ L_J & \dot{L}_J & R_J \\ L_K & \dot{L}_K & R_K \end{vmatrix} \tag{2-108}$$

For convenience, let us call the first determinant D_1 and the second D_2. Then

$$\rho = \frac{-2\,D_1}{D} - \frac{2\mu\,D_2}{r^3 D},\ D \neq 0 \tag{2-109}$$

Provided that the determinant of the coefficients, D, is not zero, we have succeeded in solving for ρ as a function of the still unknown **r**. The conditions that result in D being zero will be discussed in a later section.

From geometry we know that ρ and **r** are related by

$$\boxed{\mathbf{r} = \rho\mathbf{L} + \mathbf{R}} \tag{2-101}$$

Dotting this equation into itself yields

$$r^2 = \rho^2 + 2\rho\,\mathbf{L}\cdot\mathbf{R} + R^2 \tag{2-110}$$

Equations (2-109) and (2-110) represent two equations in two unknowns, ρ and r. Substituting Equation (2-109) into (2-110) leads to an eighth-order equation in r that may be solved by iteration.

Once the value of r at the central time is known, Equation (2-109) may be solved for ρ and the vector **r** obtained from Equation (2-101).

2.11.4 Solving for Velocity

Applying Cramer's rule again to Equation (2-104), we may solve for $\dot{\rho}$ in a manner exactly analogous to that of the preceding section. If we do this we find that

$$D\dot{\rho} = -\begin{vmatrix} L_I & \ddot{R}_I & \ddot{L}_I \\ L_J & \ddot{R}_J & \ddot{L}_J \\ L_K & \ddot{R}_K & \ddot{L}_K \end{vmatrix} - \frac{\mu}{r^3}\begin{vmatrix} L_I & R_I & \ddot{L}_I \\ L_J & R_J & \ddot{L}_J \\ L_K & R_K & \ddot{L}_K \end{vmatrix} \tag{2-111}$$

For convenience, let us call the first determinant D_3 and the second D_4. Then

$$\dot{\rho} = -\frac{D_3}{D} - \frac{\mu}{r^3}\frac{D_4}{D},\ D \neq 0 \tag{2-112}$$

Since we already know r we can solve Equation (2-112) for $\dot{\rho}$. To obtain the velocity vector, **v**, at the central time we only need to differentiate **r** in Equation (2-104), so

$$\boxed{\mathbf{v} = \dot{\mathbf{r}} = \dot{\rho}\mathbf{L} + \rho\dot{\mathbf{L}} + \dot{\mathbf{R}}} \tag{2-113}$$

2.11.5 Vanishing of the Determinant, D

In the preceding analysis we have assumed that the determinant D is not zero so that Cramer's rule may be used to solve for ρ and $\dot{\rho}$. Moulton[7] has shown that D will be zero only if the three observations lie along the arc of a great circle as viewed from the observation site at time t_2. This is another way of saying that, if the observer lies in the plane of the satellite's orbit at the central time, Laplace's method fails.

A somewhat better method of orbit determination from optical sightings will be presented in Chapter 5.

2.12 IMPROVING A PRELIMINARY ORBIT BY DIFFERENTIAL CORRECTION

The preceding sections dealt with the problem of determining a preliminary orbit from a minimum number of observations. This preliminary orbit may be used to predict the position and velocity of the satellite at some future date. As a matter of fact, one of the first things that is done when a new satellite is detected is to compute an ephemeris for the satellite so that a downrange tracking station can acquire the satellite and thus improve the accuracy of the preliminary orbit by making further observations.

There are two ways in which further observations of a satellite may be used to improve the orbital elements. If the downrange station can get another "six-dimensional fix" on the satellite, such as ρ, $\dot{\rho}$, El, $\dot{\text{El}}$, Az, $\dot{\text{Az}}$ or three optical sightings of α_1, δ_1, α_2, δ_2, α_3, δ_3, then a complete redetermination of the orbital elements can be made and the new or "improved" orbital elements taken as the average of all preceding determinations of the elements.

It may happen, however, that the downrange station cannot obtain the type of six-dimensional fix required to redetermine the orbital elements. For example, a future observation may consist of six closely spaced observations of range-rate only. The question then arises, "can this information be used to improve the accuracy of the preliminary orbit?" The answer is "yes." By using a technique known as "differential correction" any six or more subsequent observations may be used to improve our knowledge of the orbital elements.

2.12.1 Computing Residuals

Differential correction is based upon the concept of residuals. A residual is the difference between an actual observation and what the observation would have been if the satellite traveled exactly along the nominal orbit. Because such a nominal orbit will not be exactly correct owing to sensor errors or uncertainty in the original observation station's geographical coordinates, the observational data collected by a downrange station (e.g., Doppler range-rate, $\dot{\rho}$) will differ from the computed data.

Suppose that the six components of $\mathbf{r}_0$ and $\mathbf{v}_0$ are taken as the preliminary orbital elements of a satellite at some epoch t_0. Now assume that by some analytical method based on two-body orbital mechanics or some numerical method based on perturbation theory we can predict what the range-rates relative to some downrange observing site will be at six times, t_1, t_2, t_3, t_4, t_5 and t_6. These predictions are, of course, based on the assumption that the six nominal elements $[r_I, r_J, r_K, v_I, v_J, v_K]$ at $t = t_0$ are correct.

The downrange station now makes its observations of $\dot{\rho}$ at the six prescribed times and forms a set of six residuals based on the difference between the predicted values of $\dot{\rho}$ and the actual observations. The six residuals are $\Delta\dot{\rho}_1$, $\Delta\dot{\rho}_2$, $\Delta\dot{\rho}_3$, $\Delta\dot{\rho}_4$, $\Delta\dot{\rho}_5$ and $\Delta\dot{\rho}_6$.

2.12.2 The Differential Correction Equations

Assuming that the residuals are small, we can write the following six first-order equations:

$$\begin{aligned}
\Delta\dot{\rho}_1 &= \frac{\partial\dot{\rho}_1}{\partial r_I}\Delta r_I + \frac{\partial\dot{\rho}_1}{\partial r_J}\Delta r_J + \frac{\partial\dot{\rho}_1}{\partial r_K}\Delta r_K + \frac{\partial\dot{\rho}_1}{\partial v_I}\Delta v_I + \frac{\partial\dot{\rho}_1}{\partial v_J}\Delta v_J + \frac{\partial\dot{\rho}_1}{\partial v_K}\Delta v_K \\
&\quad\vdots \qquad\qquad \vdots \qquad\qquad \vdots \qquad\qquad \vdots \qquad\qquad \vdots \qquad\qquad \vdots \\
\Delta\dot{\rho}_6 &= \frac{\partial\dot{\rho}_6}{\partial r_I}\Delta r_I + \frac{\partial\dot{\rho}_6}{\partial r_J}\Delta r_J + \frac{\partial\dot{\rho}_6}{\partial r_K}\Delta r_K + \frac{\partial\dot{\rho}_6}{\partial v_I}\Delta v_I + \frac{\partial\dot{\rho}_6}{\partial v_J}\Delta v_J + \frac{\partial\dot{\rho}_6}{\partial v_K}\Delta v_K
\end{aligned} \tag{2-114}$$

By assuming that the partial derivatives can be numerically evaluated, Equations (2-114) constitute a set of six simultaneous linear equations in six unknowns, Δr_I, Δr_J, Δr_K, Δv_I, Δv_J and Δv_K. By using matrix methods these equations can be inverted and solved for the correction terms, which will then be added to preliminary orbital elements, yielding a "corrected" or "improved" set of orbital elements

$$[r_I + \Delta r_I,\ r_J + \Delta r_J \ldots\ldots,\ v_K + \Delta v_K]$$

These corrected elements are then used to recompute the predicted range-rates at the six observation times. New residuals are formed and the whole process is repeated until the residuals cease to become smaller with further iterations. In essence, the differential correction process is just a six-dimensional Newton iteration where we are trying by trial and error, to find the value of the orbital elements at time t_0 that will correctly predict the observations, i.e., reduce the residuals to zero.

2.12.3 Evaluation of Partial Derivatives

The inversion of Equations (2-114) requires knowledge of all 36 partial derivatives such as $\partial\dot{\rho}_1/\partial r_I$, etc. Usually it is impossible to obtain such derivatives analytically. With the aid of a computer, however, it is a simple matter to obtain them numerically. All that is required is to introduce a small variation, such as Δr_I, to each of the original orbital elements in turn and compute the resulting variation in each of the predicted $\dot{\rho}$'s. (A variation of 1% or 2% in the original elements is usually sufficient.) Then, for example,

$$\frac{\partial \dot{\rho}_1}{\partial r_I} \approx \frac{\dot{\rho}_1(r_I + \Delta r_I, r_J, r_K, \ldots, v_K) - \dot{\rho}_1(r_I, r_J, \ldots, v_K)}{\Delta r_I}$$

Although the preceding analysis was based on using range-rate data to differentially correct an orbit, the general method is valid no matter what type of data the residuals are based upon. The only requirement is that at least six independent observations are necessary.

The example problems that follow will demonstrate the differential correction method and will use the following notation:

n = number of observations.

p = number of elements (parameters in the equation, usually six for orbit problems).

$\tilde{\mathbf{W}}$ is an n × n diagonal matrix whose elements consist of the square of the confidence placed in the corresponding measurements (e.g., for 90% confidence, use 0.81). In this way we can reduce the effect of questionable data without completely disregarding it.

α and β are the elements (two here, since p = 2 below).

$\tilde{\mathbf{A}}$ is an n × p matrix of partial derivatives of each quantity with respect to each of the elements measured.

$\tilde{\mathbf{b}}$ is the n × 1 matrix of residuals based on the previous estimate of the elements.

$\Delta\tilde{\mathbf{z}}$ is the p × 1 matrix computed corrections to the estimates of the elements.

x_i is an independent variable measurement.

y_i is a dependent variable measurement corresponding to x_i.

$\bar{\mathsf{y}}_i$ is a computed (predicted) value of the dependent variable using previous values for the elements (α and β).

If $\mathsf{p} > \mathsf{n}$ we do not have enough data to solve the problem.

If $\mathsf{p} = \mathsf{n}$ the result is a set of simultaneous equations that can be solved explicitly for the exact solution.

If $\mathsf{p} < \mathsf{n}$ we have more equations than we do unknowns so that there is no unique solution. In this case we seek the best solution in "least-squares" sense. This means we find the curve that causes the sum of the squares of the residuals to be a minimum or equivalently we minimize the square root of the arithmetic mean of the squared residuals called the root mean square (RMS). This can occur even though the curve may not pass *through* any points.

The solution to weighted least-squares iterative differential correction is given by the following equation:

$$\Delta\tilde{\mathbf{z}} = (\tilde{\mathbf{A}}^{\mathrm{T}}\tilde{\mathbf{W}}\tilde{\mathbf{A}})^{-1}\tilde{\mathbf{A}}^{\mathrm{T}}\tilde{\mathbf{W}}\tilde{\mathbf{b}} \tag{2-115}$$

Follow this procedure:

1. Solve Equation (2-115) for the changes to the elements.
2. Correct the elements ($\alpha_{new} = \alpha_{old} + \Delta\alpha$).
3. Compute new residuals using the same data with new elements.
4. Repeat steps 1 through 3 until the residuals are:
   a. zero, for the exactly determined case ($p = n$) or
   b. a minimum as described above

Example Problem.[10] To make this example simple let us first use two observations and two elements. In this case the problem is exactly determined ($n = p = 2$).

Let the elements be α and β ($\Delta\tilde{\alpha} = \Delta\tilde{z}_1$ and $\Delta\tilde{\beta} = \Delta\tilde{z}_2$) in the relationship $\bar{y} = \alpha + \beta x$.

Assume equal confidence in all data. Therefore, **W** is the identity matrix and will not be carried through the calculations. Choose $\alpha = 2$ and $\beta = 3$ for the first estimates.

Given:

Observation	x_i	y_i	$\bar{y}_i$	Residual
1	2	1	8	–7
2	3	2	11	–9

where we have predicted the values of the variable y for the two values of x,

$$x_1 = 2,\ \bar{y}_1 = 2 + 3x_1 = 8$$

$$x_2 = 3,\ \bar{y}_2 = 2 + 3x_2 = 11$$

and computed the residuals for the (2 × 1) matrix

$$\tilde{\mathbf{b}} = \begin{bmatrix} y_1 - \bar{y}_1 \\ y_2 - \bar{y}_2 \end{bmatrix} = \begin{bmatrix} -7 \\ -9 \end{bmatrix},\ \tilde{\mathbf{W}} = \begin{bmatrix} 1 & 0 \\ 0 & 1 \end{bmatrix}$$

From the fitting equation $y = \alpha + \beta x$ the partial derivatives are:

$$\frac{\partial y_i}{\partial \alpha} = 1 \qquad \frac{\partial y_i}{\partial \beta} = x_i$$

The matrix $\tilde{\mathbf{A}}_{(2 \times 2)}$ is

$$\tilde{\mathbf{A}} = \begin{bmatrix} \frac{\partial y_1}{\partial \alpha} & \frac{\partial y_1}{\partial \beta} \\ \frac{\partial y_2}{\partial \alpha} & \frac{\partial y_2}{\partial \beta} \end{bmatrix} = \begin{bmatrix} 1 & 2 \\ 1 & 3 \end{bmatrix}$$

We will now solve Equation (2-115) for the elements of the (2 × 1) matrix

$$\Delta\tilde{\mathbf{z}} = \begin{bmatrix} \Delta\alpha \\ \Delta\beta \end{bmatrix}$$

$$\tilde{\mathbf{A}}^T\tilde{\mathbf{A}} = \begin{bmatrix} 2 & 5 \\ 5 & 13 \end{bmatrix}, (\tilde{\mathbf{A}}^T\tilde{\mathbf{A}})^{-1} = \begin{bmatrix} 13 & -5 \\ -5 & 2 \end{bmatrix}$$

$$\tilde{\mathbf{A}}^T\tilde{\mathbf{b}} = \begin{bmatrix} -16 \\ -41 \end{bmatrix}$$

$$\Delta\tilde{\mathbf{z}} = \begin{bmatrix} \Delta\alpha \\ \Delta\beta \end{bmatrix} = \begin{bmatrix} 13 & -5 \\ -5 & 2 \end{bmatrix} \begin{bmatrix} -16 \\ -41 \end{bmatrix} = \begin{bmatrix} -3 \\ -2 \end{bmatrix}$$

Thus,

$$\alpha_{new} = \alpha_{old} + \Delta\alpha = -1, \ \beta_{new} = \beta_{old} + \Delta\beta = 1$$

The fitting equation is now

$$y = -1 + x$$

which yields the following:

Observation	x_i	y_i	$\bar{y}_i$	Residual
1	2	1	1	0
2	3	2	2	0

Notice that we achieved the exact result in one iteration. This is not surprising since our equation is that of a straight line and we used only two points.

The relationship between the orbit elements and components of position and velocity are very nonlinear. In practice, large numbers of measurements are used to determine the orbit. This means that in a realistic problem 100 × 6 or larger matrices (100 observations and 6 orbit elements) would be used. This obviously implies the use of a computer for the solution. To illustrate the *method* involved, we will use a simple, yet still nonlinear relationship for our "elements" and a small, yet still overspecified number of measurements.

Example Problem. Let the elements be α and β in the relationship:

$$y = \alpha \text{ x raised to power } \beta \qquad (p = 2)$$

Given the following measurements of equal confidence, let us assume that our curve passes *through* points 3 and 4. This gives us a first guess of

$$\alpha = 0.474 \text{ and } \beta = 3.360$$

Thus using

$$\bar{y}_i = 0.474\, x_i^{3.36}$$

we predict values of y_i corresponding to the given x_i and compute residuals.

x_i	y_i	$\bar{y}_i$	$y_i - \bar{y}_i$	$(y_i - \bar{y}_i)^2$
1	2.500	0.474	2.026	4.105
2	8.000	4.865	3.135	9.828
3	19.000	19.007	−0.007	0
4	50.000	49.969	0.031	0.001
				13.934

$$\text{RMS residual} = \sqrt{\frac{\sum_{i=1}^{n} (y_i - \bar{y}_i)^2}{n}} = \sqrt{\frac{13.934}{4}} = 1.866$$

The partial derivatives for matrix $\tilde{\mathbf{A}}$ are given by

$$\frac{\partial y_i}{\partial \alpha} = x_i^{\beta} = x_i^{3.36}$$

$$\frac{\partial y_i}{\partial \beta} = \alpha x_i^{\beta} \ln\left(x_i\right) x_i = 0.474 x_i^{3.36} \log_e x_i$$

The residuals for matrix (**b**) are given by

$$\tilde{b}_i = y_i - 0.474 x_i^{3.36}$$

Since the data are of equal confidence

$$\tilde{W} = \begin{bmatrix} 1 & 0 & 0 & 0 \\ 0 & 1 & 0 & 0 \\ 0 & 0 & 1 & 0 \\ 0 & 0 & 0 & 1 \end{bmatrix}$$

After computing the elements of the matrices we have

$$\tilde{A} = \begin{bmatrix} 1.00 & 0.00 \\ 10.27 & 3.37 \\ 40.10 & 20.88 \\ 105.41 & 69.27 \end{bmatrix}, \quad \tilde{b} = \begin{bmatrix} 2.026 \\ 3.133 \\ -0.007 \\ 0.031 \end{bmatrix}$$

$$\tilde{A}^T \tilde{W} \tilde{A} = \begin{bmatrix} 12{,}830 & 8{,}175 \\ 8{,}175 & 5{,}246 \end{bmatrix}$$

$$(\tilde{A}^T \tilde{W} \tilde{A})^{-1} = \begin{bmatrix} 0.011 & -0.017 \\ -0.017 & 0.027 \end{bmatrix}, \quad \tilde{A}^T \tilde{W} \tilde{b} = \begin{bmatrix} 37.21 \\ 12.58 \end{bmatrix}$$

Now using Equation (2-115) we have

$$\tilde{\Delta z} = (\tilde{A}^T \tilde{W} \tilde{A})^{-1}\, \tilde{A}^T \tilde{W} \tilde{b} = \begin{bmatrix} 0.196 \\ -0.304 \end{bmatrix}$$

Thus

$$\alpha_{new} = 0.474 + 0.196, \quad \beta_{new} = 3.36 - 0.304$$

$$\alpha_{new} = 0.670, \quad \beta_{new} = 3.056$$

Computation using the new residuals (based on the new elements and the observed data) yields an RMS residual of 2.360, which is larger than what we started with, so we iterate.

The second iteration yields

$$\Delta\alpha = 0.062, \quad \Delta\beta = -0.018$$

Thus

$$\alpha_{new} = 0.733, \quad \beta_{new} = 3.039$$

Recalculation of residuals using these newer parameters yields an RMS residual of 1.582.

We will consider our RMS residual to be a minimum if its value on two successive iterations differs by no more than 0.001. This is not yet the case so we iterate again. The next values of the parameter are:

$\alpha = 0.735, \qquad \beta = 3.038$

This time the RMS residual is 1.581, which means that to three significant figures we have converged on the best values of α and β.

This best relationship between the given data and our fitting equation is given by

$$y = 0.735\, x^{3.038}$$

2.12.4 Unequally Weighted Data

If the data are not weighted equally we can observe several things:

First: If a piece of data that is actually exact should be weighted less than other good data, the algorithm above will still converge to the same best value but it will take more iterations.

Second: Weighting incorrect data less than the other data will cause the final values of the elements to be much closer to the values obtained using correct data than if the bad data had been weighted equally. However, the lower the weight given the bad data, the larger the RMS residual will be. The number of iterations needed for convergence increases the lower the weight.

2.13 SPACE SURVEILLANCE

In the preceding sections we have seen how, in theory, we can determine the orbital elements of a satellite from only a few observations. In practice, however, a handful of observation on new orbiting objects cannot secure the degree of precision needed for orbital surveillance and prediction. Typical requirements are for 100–200 observations per object per day during the first few days of orbit, 20–50 observations per object per day to update already established orbits and finally, during orbital decay, 200–300 observations to confirm and locate reentry.[8]

In 1975 there were nearly 3,500 detected objects in orbit around Earth. By 2015 the total number of tracked objects in the near-Earth (NE) catalog of space objects had increased to approximately 22,000 items and that number will increase by an order of magnitude as new optical and radar tracking sites become operational and the data sets are continually integrated.

2.13.1 The Spacetrack System

The task of keeping track of this growing space population belongs to the U.S. Strategic Command (USSTRATCOM). Space surveillance is a critical part of USSTRATCOM's mission and involves detecting, tracking, cataloging and identifying human-made objects orbiting Earth, i.e., active and inactive satellites, spent rocket bodies or fragmentation debris. Space surveillance comprises several critical functions:

a. predict when and where a decaying space object will reenter Earth's atmosphere;

b. prevent a returning space object, which to radar looks like a missile, from triggering a false alarm in missile-attack warning sensors of the U.S.A. and other countries;

c. chart the present position of space objects and plot their anticipated orbital paths;

d. detect new human-made objects in space;

e. produce a running catalog of human-made space objects;

f. determine which country owns a reentering space object;

g. inform NASA whether or not objects may interfere with the International Space Station and other operational space assets.

The command accomplishes these tasks through its Space Surveillance Network (SSN) of the U.S. Army, Navy and Air Force operated, ground-based radars and optical sensors at 25 sites worldwide.

2.13.2 Space Surveillance Network

The SSN has been tracking space objects since 1957 when the Soviets opened the space age with the launch of Sputnik-1. While many objects and pieces of space debris have reentered Earth's atmosphere and disintegrated, or survived reentry and impacted Earth, there approximately 22,000 objects larger than a baseball that are tracked frequently enough to be maintained in the space objects catalog. The space objects now orbiting Earth range from satellites weighing several tons to pieces of spent rocket bodies weighing a few kilograms down to minuscule paint flakes. About 7% of the space objects are operational satellites; the rest are debris. USSTRATCOM is primarily interested in the active satellites, but it also tracks space debris.

2.13.3 Space Surveillance Network Sensors

The SSN uses a "predictive" technique to monitor space objects; i.e., it spot checks them rather than tracking them continually. This technique is used because of the limits of the SSN (number of sensors, geographic distribution, capability and availability). Below is a brief description of each type of sensor.

Phased-array radars can maintain tracks on multiple satellites simultaneously and scan large areas of space in a fraction of a second. These radars have no moving mechanical parts to limit the speed of the radar scan—the radar beam is steered electronically.

Conventional radars use a fixed detection antenna and a steerable tracking antenna. The detection antenna transmits radar energy into space in the shape of a large fan. When a satellite intersects the fan the energy is reflected back to the antenna, triggering the tracking antenna. The tracking antenna, then, locks its narrow beam on the target and follows it in order to establish orbital data.

The **Ground-Based Electro-Optical Deep Space Surveillance System (GEODSS)** use a worldwide system of cameras, each consisting of three one-meter telescopes with low-light sensors that capture images of space objects. As an optical system, it relies on nighttime and good weather, but, unlike the Baker–Nunn cameras that they replaced, it uses electronic sensors instead of film, directly producing computer-compatible data for analysis. There are three operational GEODSS sites that report to the 21st Operations Group:

- Socorro, New Mexico, at 33.8172°N 106.6599°W,
- AMOS, Maui, Hawaii, at 20.7088°N 156.2578°W, and
- Diego Garcia, British Indian Ocean Territory, at 7.41173°S 72.45222°E.

A site at Choe Jong San, South Korea, was closed in 1993 because of nearby smog from the town, weather and cost concerns. Originally, the fifth GEODSS was planned to be operated from a site in Portugal, but this was never built.

There are multiple space-based measurements that are fed into the SSN to provide additional space object tracking data. The Space-Based Space Surveillance (SBSS) system is a planned constellation of satellites and supporting ground infrastructure that will improve the ability of USSTRATCOM to detect and track space objects in orbit around Earth. The first "pathfinder" satellite of the SBSS system was successfully placed into orbit on board a Minotaur IV rocket on September 25, 2010. It is designed to examine every spacecraft in geosynchronous orbit at least once a day. Sapphire, Canada's first military satellite, is a small spacecraft designed to monitor space debris and satellites that orbit between 6,000 and 40,000 km above Earth. The satellite has been providing data to the SSN since January 2014.

Combined, these types of sensors make up to 80,000 satellite observations each day. The data are transmitted directly to USSTRATCOM's Space Control Center (SCC) via satellite, ground wire, microwave and phone. Every available means of communications is used to ensure a backup is readily available if necessary.

2.13.4 Orbital Space Debris

About 22,000 items are in the space objects catalog but somewhere between 200,000 and 500,000 are detectable by various sources. The catalog may grow to around 200,000 in the near future if various sensor platform and computing upgrades take place. Most debris is out between 700 and 900 km—roughly twice the normal altitude of the International Space Station (ISS), which orbits at about 300 km. Only a small amount of debris exists where the ISS orbits.

The likelihood of a significant collision between a piece of debris (of 5 cm and larger) and the ISS is extremely remote. The statistical estimate is one chance in 10,000 years, in the worst case. The probability is higher for objects smaller than baseball size, which cannot be tracked well with currently available sensors. Although 22,000 larger than baseball-sized space objects seems like a large number, in the 800 km band there are normally only a few items in an area roughly equivalent to the commercial airspace over the continental U.S.A. Therefore, the likelihood of collision between objects is very small.

However, on February 10, 2009, 16:56 UTC an Iridium 33 and Kosmos-2251 collided at a speed of 11.70 km/s and an altitude of 789 km above the Taymyr Peninsula in Siberia. Kosmos-2251 was a 950 kg Russian Strela military communications satellite that had been deactivated prior to the collision, and it remained in orbit as space debris. Iridium 33 was a 560 kg commercial U.S.-built satellite and was part of the commercial satellite phone Iridium constellation of 66 communications satellites. The collision destroyed both Iridium 33 and Kosmos 2251. The Iridium satellite was operational at the time of the collision—and had the ability to perform minimal thrust maneuvers for orbit maintenance and collision avoidance.

NASA estimated that the satellite collision created approximately 1,000 pieces of debris larger than ten centimeters (four inches), in addition to many smaller ones. By July 2011, the SSN had cataloged over 2,000 large debris fragments. NASA determined the risk to the International Space Station to be low, as was any threat to the shuttle launch (STS-119) then planned for late February 2009. However, Chinese scientists have said that the debris does pose a threat to Chinese satellites in Sun-synchronous orbits, and the ISS did have to perform an avoidance maneuver because of collision debris in March 2011.

By December 2011, many pieces of debris were in a steady orbital decay toward Earth and were expected to burn up in the atmosphere within one or two years. By January 2014, 24% of the known debris had decayed and in 2016 the collision was characterized as the fourth biggest fragmentation event in history, with Iridium 33 producing 628 pieces of cataloged debris, of which 364 pieces of tracked debris remain in orbit as of January 2016.[14]

A small piece of Kosmos 2251 satellite debris safely passed by the International Space Station at 2:38 a.m. EDT, Saturday, March 24, 2012. As a precaution, the six crew members on board the orbiting complex took refuge inside the two docked Soyuz rendezvous spacecraft until the debris had passed.[15]

Events where two satellites approach within several kilometers of each other occur numerous times each day. Sorting through the large number of potential collisions to identify those that are high risk presents a challenge. Precise, up-to-date information regarding current satellite positions is difficult to obtain. Calculations made by CelesTrak had expected these two satellites to miss by 584 meters.[16]

Planning an avoidance maneuver with due consideration of the risk, the fuel consumption required for the maneuver, and its effects on the satellite's normal functioning can also be challenging. Reference (14) discusses the difficulty of handling all the notifications the satellite operators were receiving regarding close approaches, which numbered 400 per week (for approaches within 5 km) for the entire Iridium constellation. The estimated risk of collision per conjunction was one in 50 million.

This collision and numerous near misses have renewed calls for mandatory disposal of defunct satellites (typically by deorbiting them or at a minimum sending them in to a graveyard orbit), but no such international law exists yet. The U.S. Air Force updated its space safety policy in 2014[17] to require that all new space objects in low Earth orbit have a maximum on-orbit lifetime of 25 years. The U.S. Federal Communications Commission (FCC) requires all geostationary satellites launched after March 18, 2002, to commit to moving to a graveyard orbit at the end of their operational life.[32]

It is clear that the challenges of tracking space objects and space debris, maintaining an ever-growing space object catalog, calculating innumerable potential collision scenarios (conjunction analyses) and providing warning to satellite operators while the amount of space debris increases are substantial and will continue to grow.

2.14 GROUND TRACK OF A SATELLITE

While knowing the orbital elements of a satellite enables you to visualize the orbit and its orientation in the **IJK** inertial reference frame, it is often important to know what the ground track of a satellite is. One of the most valuable characteristics of an artificial Earth satellite is its ability to pass over large portions of Earth's surface in a relatively short time. As a result, it has tremendous potential as an instrument for scientific, commercial or military surveillance.

2.14.1 Ground Track on a Non-rotating Earth

The orbit of an Earth satellite always lies in a plane passing through the center of Earth. The track of this plane on the surface of a non-rotating spherical Earth is a great circle. On this non-rotating Earth, the satellite would retrace the same ground track over and over. The maximum latitude north or south of the equator that the satellite passes over is just equal to the inclination, i, of the orbit. For a retrograde orbit the most northerly or southerly latitude on the ground track is $180° - i$.

2.14.2 Effect of Launch Site Latitude and Launch Azimuth on Orbit Inclination

We can determine the effect of launch site latitude and launch azimuth on orbit inclination by studying Figure 2-26. Suppose a satellite is launched from point C on Earth whose latitude and longitude are L_0 and λ_0, respectively, with a launch azimuth of β_0. The ground track of the resulting orbit crosses the equator at a point A at an angle equal to the orbital inclination. The arc CB forms the third side of a spherical triangle and is formed by the meridian passing through the launch site and subtends the angle L_0 at the center of Earth. Since we know two angles and the included side of this triangle we can solve for the third angle, i:

$$\cos i = -\cos 90° \cos \beta_0 + \sin 90° \sin \beta_0 \cos L_0$$

$$\boxed{\cos i = \sin \beta_0 \cos L_0} \tag{2-116}$$

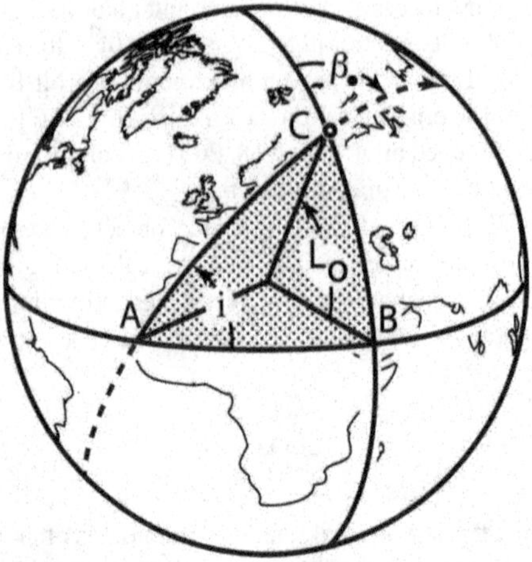

Figure 2-26 Effect of Launch Azimuth and Latitude on Inclinations.

There is a tremendous amount of interesting information concealed in this innocent-looking equation. For a direct orbit ($0 < i < 90°$) cos i must be positive. Since L_0 can range between 0° and 90° for launch sites in the northern hemisphere, and between 0° and –90° for launch sites in the southern hemisphere, cos L_0 must always be positive. A direct orbit requires, therefore, that the launch azimuth, β_0, be easterly, i.e., between 0° and 180°.

Suppose we now ask "what is the minimum orbital inclination we can achieve from a launch site at latitude L_0?" If i is to be minimized, cos i must be maximized, which implies that launch azimuth, β_0, should be 90°. For a due east launch, Equation (2-116) tells us that the orbital inclination will be the minimum possible from a launch site at latitude L_0 and i will be precisely equal to L_0!

Among other things, this tells us that a satellite cannot be put directly into an equatorial orbit ($i = 0°$) from a launch site that is not on the equator. Russia is at a particular disadvantage in this regard because none of its launch sites are closer than 45° to the equator, so it cannot launch a satellite whose inclination is less than 45°. If the Russians wish to establish an equatorial orbit it requires a plane change of at least 45° after the satellite is established in its initial orbit. This is an expensive maneuver, as we shall see in the next chapter.

2.14.3 Effect of Earth Rotation on the Ground Track

The orbital plane of a satellite remains fixed in space while Earth turns under the orbit. The net effect of Earth rotation is to displace the ground track westward on each successive revolution of the satellite by the number of degrees Earth turns during one orbital period. The result is illustrated in Figure 2-27.

Instead of retracing the same ground track over and over, a satellite eventually covers a swath around Earth between latitudes north and south of the equator equal to the inclination. Figure 2-27 shows the ground track for a satellite in a circular orbit with a 500 km altitude and a 45° inclination. A global surveillance satellite would have to be in a polar orbit to overfly Earth's entire surface.

If the time required for one complete rotation of Earth on its axis (23 hr 56 min) is an exact multiple of the satellite's period then eventually the satellite will retrace exactly the same path over Earth as it did on its initial revolution. This is a desirable property for a reconnaissance satellite where you wish to have it overfly a specific target once each day. It is also desirable in human spaceflight to overfly the primary astronaut recovery areas at least once each day. Figure 2-28 shows the ground track from 15 orbits (approximately 24 hours) from a satellite at 500 km altitude and 45° inclination.

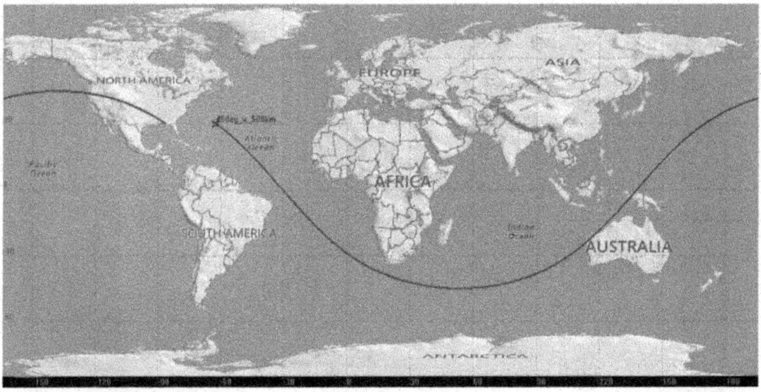

Figure 2-27 Westward Displacement of Ground Track due to Earth Rotation.

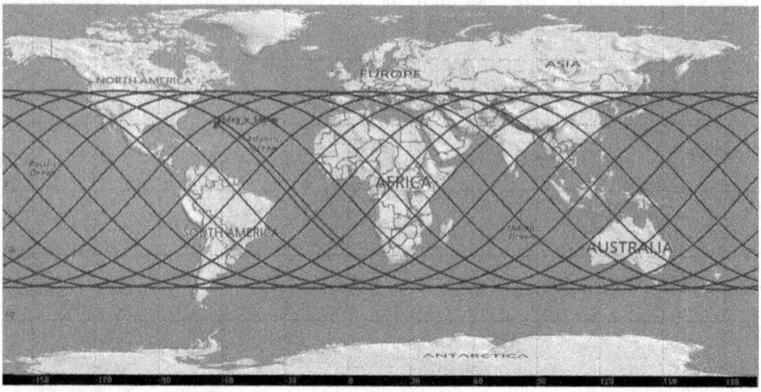

Figure 2-28 Ground Track Displacement for 15 Orbits.

Exercises

2.1 Determine the orbital elements for an Earth orbit that has the following positions and velocity vectors:

$\mathbf{r} = 1\mathbf{K}$ DU

$\mathbf{v} = 1\mathbf{I}$ DU/TU

(Partial answers: a = 1 DU, i = 90°, ω = undefined)

2.2 Given $\mathbf{r} = -4500\mathbf{I} + 4500\mathbf{J}$ km and

$\mathbf{v} = 4.0\mathbf{J}$ km/s

Determine the orbital elements and sketch the orbit.

(Partial answers: p = 812.844 km, e = 0.881576, $\nu_0 = 171.67°$)

2.3 Answer the following:

a. Which takes longer, a solar day or sidereal day?

b. What causes an apparent solar day to be different from a mean solar day?

c. What was the local sidereal time (radians) of Greenwich, England, at 0448 hours (local) on 4 July 2016?

d. What was the local sidereal time (radians) of the U.S. Air Force Academy, Colorado (104.89° W Long) at that same time?

e. Does it make a difference whether Colorado is on standard or daylight saving time?

2.4 What was the Greenwich sidereal time in radians on 3 June 2015 at 17 h 00 m 00 s UT? What is the remainder over an integer number of revolutions?

(Answer: 2.5627 radians = GST)

2.5 Determine by inspection if possible the orbital elements for the following objects:

a. Object A is crossing the negative **J** axis in a direct equatorial circular orbit at an altitude of 6,378.136 km.

b. Object B departs from a point **r** = –6,378.136**K** km with local escape speed in the –**I** direction.

c. Object C departs from point **r** = 6,378.136**K** km with **v** = 11.1799**I** + 11.1799**J** km/s.

Object	p	e	i	Ω	ω	ν_0
A	____	____	____	____	____	____
B	____	____	____	____	____	____
C	____	____	____	____	____	____

2.6 Radar readings determine that an object is located at 7,653.76**K** km with a velocity of 3.162**I** – 2.3716**K** km/s. Determine p, e, u_0, Ω, ω, ν_0, (ℓ_0) and the latitude of impact.

(Answer: [partial] e = 0.820748, ℓ_0 = 270°)

2.7 A radar site reduces a set of observed quantities such that:

$\mathbf{r}_0$ = –6,378.14**I** –6,378.14**J** –6,378.14**K** (km)

$\mathbf{v}_0$ = 2.63234**I** –2.632334**J** +2.63234**K** (km/s)

In the geocentric equatorial coordinate system. Determine the orbital elements.

(Partial answers: p = 5657.48 km, ω = 104.51°)

2.8 For the following orbital elements:

p = 1,466.97 km	Ω = 180°
e = 0.82	ω = 260°
i = 90°	ν_0 = 190°
u_0 = 90°	ℓ_0 = 270°

a. Express the **r** and **v** vectors for the satellite in the perifocal system along the unit vectors **P**, **Q**, **W**.

b. By a suitable coordinate transformation technique express the **r** and **v** vectors in the geocentric equatorial system in the **IJK** system.

(Answer: **r** = 7,622**K**)

2.9 A radar station at Sunnyvale, California, makes an observation on an object at 2048 hours, PST, 10 January 2016. Site longitude is 121.5° West. What is the local sidereal time?

(Answer: LST = 0.848 radians)

2.10 A basis IJK requires three rotations before it can be lined up with another basis UVW; the 1st rotation is 30° about the first axis; the 2nd rotation is 60° about the second axis; the final rotation is 90° about the third axis.

a. Find the matrix required to transform a vector from the IJK basis to the UVW basis.

b. Transform **r** = 2**I** – **J** + 4**K** to the UVW basis.

2.11 The values for θ_{g0} for 1968 through 1971 given in the text cluster near a value of approximately 100°. Explain why you would expect this to be so.

2.12 Determine the orbital elements by inspection for an object crossing the positive Y axis in a retrograde, equatorial, circular orbit at the altitude or 1 DU.

2.13 Determine the orbital elements of the following objects using the Gibbs method. Use of a computer is suggested, but not necessary. Be sure to make all the tests since all the orbits may not be possible. Units are Earth canonical units.

		I	J	K
a.	$\mathbf{r}_1$	1.41422511	0	1.414202
	$\mathbf{r}_2$	1.81065659	1.06066883	0.3106515
	$\mathbf{r}_3$	1.35353995	1.41422511	–0.6464495
	(Partial answer: e = 0.171, p = 1.76)			
b.	$\mathbf{r}_1$	0.70711255	0	0.70710101
	$\mathbf{r}_2$	–0.89497879	0.56568081	–0.09496418
	$\mathbf{r}_3$	–0.09497879	–0.56568681	–0.89497724
c.	$\mathbf{r}_1$	1.0	0	0
	$\mathbf{r}_2$	–0.8	0.6	0
	$\mathbf{r}_3$	0.8	–0.6	0

d.	$\mathbf{r}_1$	0.20709623	3.53552813	1.2071255
	$\mathbf{r}_2$	0.91420062	4.9497417	1.91423467
	$\mathbf{r}_3$	1.62130501	6.36395526	2.62134384
	(Partial answer: Straight line orbit—hyperbola with infinite eccentricity)			
e.	$\mathbf{r}_1$	1.0	0	0
	$\mathbf{r}_2$	0	1.0	0
	$\mathbf{r}_3$	−1.0	0	0
f.	$\mathbf{r}_1$	7.0	2.0	0
	$\mathbf{r}_2$	1.0	1.0	0
	$\mathbf{r}_3$	2.0	7.0	0
g.	$\mathbf{r}_1$	0	2.7	0
	$\mathbf{r}_2$	2.97	0	0
	$\mathbf{r}_3$	−2.97	0	0

2.14 * An SSN site determines that an object is located at 1.2**K** DU with a velocity of 0.4**I** −0.3**K** DU/TU. Determine the orbital elements of the orbit and the latitude of impact of the object.

2.15 * Given the orbital elements for objects A, B, C and D fill in the blanks to correctly complete the following statements:

Object	i	Ω	Π	ℓ_o
A	0°	undefined	210°	30°
B	4°	180°	260°	90°
C	110°	90°	110°	140°
D	23°	60°	260°	160°

a. Object _______ is in retrograde motion.

b. Object _______ has a true anomaly at epoch of 180°.

c. Object _______ has its perigee south of the equatorial plane.

d. Object _______ has a line of nodes that coincides with the vernal equinox direction.

e. Object _______ has an argument of perigee of 200°.

2.16 * A radar site located in Greenland observes an object that has components of the position and velocity vectors only in the **K** direction of the geocentric-equatorial coordinate system. Draw a sketch of the orbit and discuss the orbit type.

2.17 * A radar tracking site located at 30° N, 97.5° W obtains the following data at 0930 GST for a satellite passing directly overhead:

$\rho = 637.814$ km $\qquad \dot{\rho} = 0$

$\text{Az} = 30^\circ \qquad \dot{\text{Az}} = 0$

$\text{El} = 90^\circ \qquad \dot{\text{El}} = 0.7094 \text{ deg/ s}$

a. Determine the rectangular coordinates of the object in the topocentric-horizon system.

b. What is the velocity of the satellite relative to the radar site in terms of south, east and zenith (up) components?

c. Express the vector **r** in terms of topocentric-horizon coordinates.

d. Transform the vector **r** into geocentric-equatorial coordinates.

e. Determine the velocity **v** in terms of geocentric-equatorial coordinates.

(Answer: $\mathbf{v} = 4.8969\mathbf{I} - 0.06159\mathbf{J} - 5.92276\mathbf{K}$ km/s)

List of References

1. Gauss, Carl Friedrich. *Theory of the Motion of the Heavenly Bodies Moving About the Sun in Conic Sections.* A translation of *Theoria Motus* (1857) by Charles H. Davis. New York, Dover Publications, Inc., 1963.

2. Armitage, Angus. *Edmond Halley.* London, Thomas Nelson and Sons, Ltd., 1966.

3. Newman, James R. Commentary on "An Ingenious Army Captain and on a Generous and Many-sided Man," in *The World of Mathematics.* Vol. 3, New York, NY, Simon and Schuster, 1956.

4. Escobal, Pedro Ramon. *Methods of Orbit Determination.* New York, NY, John Wiley & Sons, Inc., 1965.

5. Baker, Robert M.L. Jr. *Astrodynamics: Applications and Advanced Topics.* New York and London, Academic Press, 1967.

6. Laplace, Pierre Simon. *Memoire de l' Academie Royale des Sciences de Paris,* 1780.

7. Moulton, Forest Ray. *An Introduction to Celestial Mechanics.* Second Revised Edition, New York, NY, Macmillan, 1914.

8. Thomas, Paul G., "Space Traffic Surveillance," *Space/Aeronautics*, Vol. 48, No. 6, November, 1967.

9. Bate, Roger R. and Eller, Thomas J. *An Improved Approach to the Gibbsian Method.* Department of Astronautics and Computer Science Report A-70-2, USAF Academy, Colorado, 1970.

10. Carson, Gerald C. *Computerized Satellite Orbit Determination.* Satellite Control Facility, Air Force Systems Command, Sunnyvale, California, 1966.

11. *The Astronomical Almanac.* http://asa.usno.navy.mil/.

12. Vallado, David A. *Fundamentals of Astrodynamics and Applications.* 2nd edition. El Segundo, California, Microcosm Press, 2004.

13. *Explanatory Supplement to the Astronomical Almanac.* U.S. Naval Observatory, University Science Books, Mill Valley, California, 1992.

14. "10 breakups account for 1/3 of catalogued debris," Space News April 25 2016, page 21.

15. Orbital Debris Safely Passes International Space Station (Web Broadcast). National Aeronautics and Space Association. 2012-03-23. Event occurs at 23 minutes 30 seconds. Retrieved 25 July 2016.

16. "Iridium 33/Cosmos 2251 Collision." CelesTrak. Archived from the original on 17 March 2009. Retrieved 18 July 2016.

17. Space Safety and Mishap Prevention Program, Air Force Instruction 91-217, 11 April 2014.

Chapter 3

Basic Orbital Maneuvers

> But if we now imagine bodies to be projected in the directions of lines parallel to the horizon from greater heights, as of 5, 10, 100, 1,000 or more miles, or rather as many semidiameters of Earth, those bodies, according to their different velocity, and the different force of gravity in different heights will describe arcs either concentric with Earth, or variously eccentric, and go on revolving through the heavens in those orbits just as the planets do in their orbits.
>
> -Isaac Newton[1]

3.1 Historical Background

The concept of artificial satellites circling Earth was introduced to scientific literature by Sir Isaac Newton in 1686. After that, for the next 250 years, the idea seems to have been forgotten. The great pioneers of rocketry—Tsiolkovsky, Goddard and Oberth—were the first to predict that high-performance rockets together with the principle of "staging" would make such artificial satellites possible. It is a curious fact that these pioneers foresaw and predicted manned satellites but none of them could see a use for unmanned Earth satellites. Ley[2] suggests that the absence of reliable telemetering techniques in the early 1930s would explain this oversight.

Who was the first to think of an unmanned artificial satellite after the concept of a manned space station had been introduced by Hermann Oberth in 1923 still remains to be established. It very likely may have been Wernher von Braun or one of the others at Peenemünde—the first place on Earth where a group of people with space-travel inclinations was paid to devote all of their time, energy and imagination to rocket research.[8] Walter Dornberger in his book *V-2* mentions that the discussions about future developments at Peenemünde included the rather macabre suggestion that the space-travel pioneers be honored by placing their embalmed bodies into glass spheres that would be put into orbit around Earth.[3]

The earliest plans to actually fire a satellite into Earth orbit were proposed by Von Braun at a meeting of the Space-Flight Committee of the American Rocket

Society in the Spring of 1954. His plan was to use a Redstone rocket with successive clusters of a small solid propellant rocket called "Loki" on top. The scheme was endorsed by the Office of Naval Research and dubbed "Project Orbiter." The launch date was tentatively set for midsummer of 1957.

However, on 29 July 1955, the White House announced that the United States would orbit an artificial Earth satellite called "Vanguard" as part of its International Geophysical Year Program.

Project Orbiter, since it contemplated the use of military hardware, was not considered for launching a "strictly scientific" satellite and the plan was shelved.

On 4 October 1957, the Soviet Union successfully orbited Sputnik I. After several Vanguard failures, the von Braun team was at last given its chance and put up Explorer I on 1 February 1958. Vanguard I was finally launched successfully a month and a half later.

In this chapter we will describe the methods used to establish Earth satellites in both low- and high-altitude orbits and the techniques for maneuvering them from one orbit to another.

3.2 LOW-ALTITUDE EARTH ORBITS

Manned flight, still in its infancy, has largely been confined to regions of space very near the surface of Earth—with the exception of the U.S. missions to the lunar surface. The reason for this is neither timidity nor lack of large booster rockets. Rather, the environment of near-Earth space conspires to limit the altitude of an artificial satellite, particularly if it is manned, to a very narrow region just above Earth's sensible atmosphere.

Altitudes below 120 km are not possible because of atmospheric drag, and the Van Allen radiation belts limit manned flights to altitudes below about 560 km.

3.2.1 Effect of Orbital Altitude on Satellite Lifetimes

The exact relationship between orbital altitude and satellite lifetime depends on several factors. For circular orbits of a manned satellite about the size of the Gemini or Apollo spacecraft, Figure 3-1 shows the limits imposed by drag, radiation and meteorite damage considerations.

For elliptical orbits, the limits are slightly different but you should keep in mind that the perigee altitude cannot be much less than 200 km nor the apogee altitude much greater than 560 km for the reasons just stated.

The eccentricity of an elliptical orbit whose perigee altitude is 200 km and whose apogee altitude is 560 km is less than 0.03! Because it is difficult to imagine just how close such an orbit is to the surface of Earth, we have drawn one to scale in Figure 3-2. The potential of a low-altitude satellite for reconnaissance is obvious from this sketch.

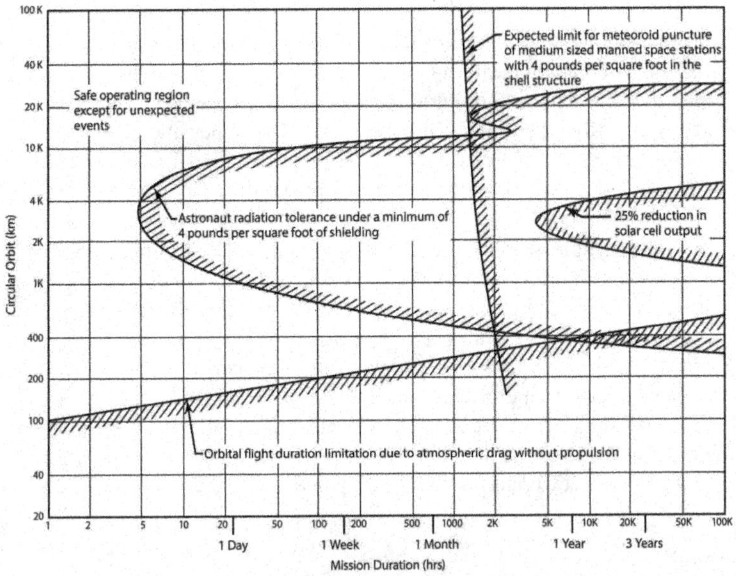

Figure 3-1 Satellite Lifetime vs. Orbital Altitude.[4]

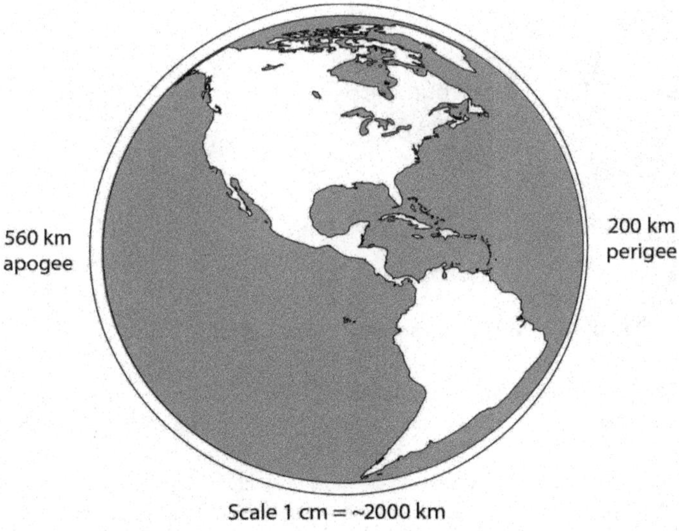

Figure 3-2 Typical Low-altitude Earth Orbit.

3.2.2 Direct Ascent to Orbit

It is possible to inject a satellite directly into a low-altitude orbit by having its booster rockets burn continuously from liftoff to a burnout point somewhere on the desired orbit. The injection or burnout point is usually planned to occur at the perigee with a flight-path angle at burnout of 0°. Any deviation from the correct burnout speed or flight-path angle could be catastrophic for, as you can see from Figure 3-2, there is very little clearance between the orbit and the surface of Earth.

It normally takes at least a two-stage booster to inject a two-or three-human vehicle into low Earth orbit. The vehicle is not allowed to coast between first stage booster separation and second stage ignition. The powered flight trajectory looks something like what is shown in Figure 3-3. The vehicle rises vertically from the launch pad, immediately beginning a roll to the correct azimuth. The pitch program—a slow tilting of the vehicle to the desired flight-path angle—normally begins about 15 s after liftoff and continues until the vehicle is traveling horizontally at the desired burnout altitude. The final burnout point is usually about 560 km downrange of the launch point.

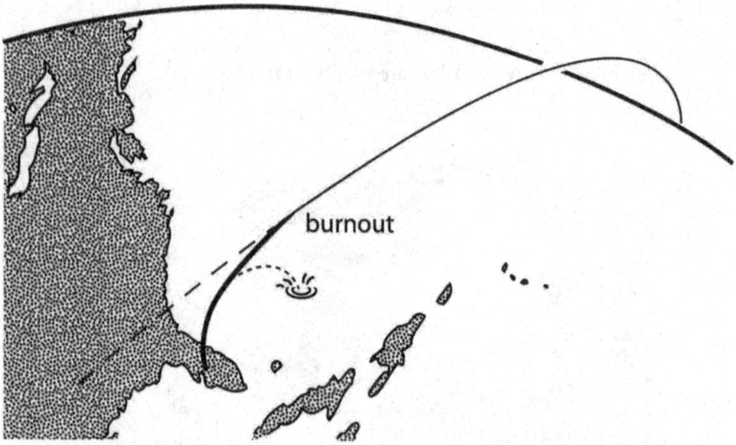

Figure 3-3 Direct Ascent into a Low-altitude Orbit.

The first stage booster falls to Earth several hundred kilometers downrange but the final stage booster, since it has essentially the same speed and direction at burnout as the satellite itself, may orbit Earth for several revolutions before atmospheric drag causes its orbit to decay and it reenters. The lower ballistic coefficient (mass-to-area ratio) of the booster causes it to be affected more by drag than the vehicle itself would be.

3.2.3 Perturbations of Low-altitude Orbits Owing to the Oblate Shape of Earth

Earth is not spherical as we assumed it to be and, therefore, its center of gravity is not coincident with its center of mass. If you are very far from the center of Earth the difference is not significant, but for low-altitude Earth orbits the effects are not negligible. To accurately determine the perturbations during a single orbit the osculating elements should be used rather than the mean elements. For mission analysis applications it is more useful to determine an average change in the node on a daily basis. In addition, the mass distribution of Earth is not constant.

The two principal effects are regression of the line of nodes and rotation of the line-of-apsides (major axis). Nodal regression is a rotation of the plane of the orbit about Earth's axis of rotation at a rate that depends on both orbital inclination and altitude. As a result, successive ground traces of direct orbits are displaced westward farther than would be the case owing to Earth rotation alone.

The gravitational effect of an oblate Earth can more easily be visualized by picturing a spherical Earth surrounded by a belt of excess matter representing the equatorial bulge. When a satellite is in the positions shown in Figure 3-4, the net effect of the bulges is to produce a slight torque on the satellite about the center of Earth. This torque will cause the plane of the orbit to precess just as a gyroscope would under similar torque. The result is that the nodes move westward for direct orbits and eastward for retrograde orbits.

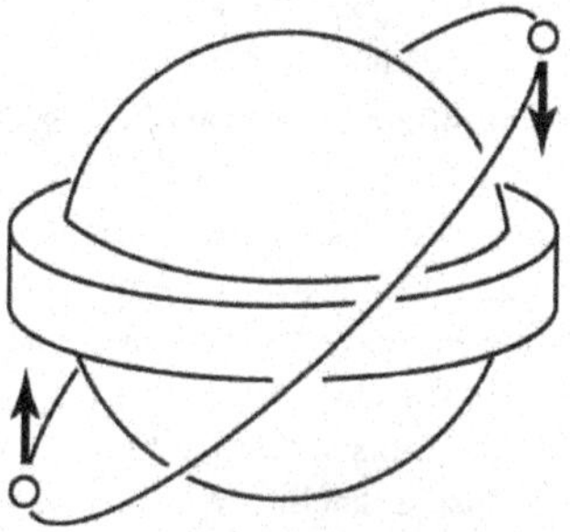

Figure 3-4 Perturbative Torque Caused by Earth's Equatorial Bulge.

Earth can be more accurately modeled by using a more complete gravitational model that incorporates the varying mass distribution around the oblate spheroid shape variation with latitude and the "orange slices" variation with longitude. Detailed derivation of the perturbative effects of gravitational irregularities from a perfect sphere can be found in Vallado[6] and Prussing.[7] The oblate spheroid term is the J_2 parameter in the gravity field model and that is the cause of node precession and perigee wandering. The nodal regression rate in degrees per day is

$$\dot{\Omega}_{J_2} = -2.06474 \times 10^{14} a^{-3.5} (\cos i)(1 - e^2)^{-2} \tag{3-1}$$

and is shown in Figure 3-5. Note that for low-altitude orbits of low inclination the rate approaches 9° per day. You should be able to look at Figure 3-5 and see why the nodal regression is greatest for orbits whose inclination is near 0° or 180° and goes to zero for polar orbits.

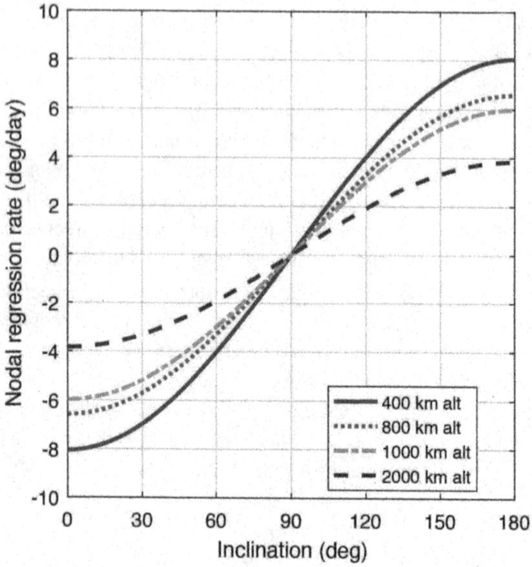

Figure 3-5 Nodal Regression Rate in Deg/Day.

There are satellite orbits specifically designed to match Earth's rotation rate on its axis as it moves around the Sun with the nodal regression rate so that the satellite passes over the equator every day at exactly the same local time through the life of the satellite. These orbits are known as "Sun synchronous" orbits because the orientation of the orbital plane relative to the Sun remains constant during the year. The result is that a satellite with a remote sensing payload (typically optical in the visible part of the spectrum) passes over Earth's surface with consistent shadowing effects.

The rotation of the line of apsides is only applicable to eccentric orbits. With this perturbation, which also is due to oblateness, the major axis of an elliptical trajectory will rotate in the direction of motion of the satellite if the orbital inclination is less than 63.4° or greater than 116.6° and opposite to the direction of motion for inclinations between 63.4° and 116.6°. The rate at which the major axis rotates is a function of both orbit altitude and inclination angle, and mean orbital elements result in the following daily average change:

$$\dot{\omega}_{J_2} = (1.03237 \times 10^{14} a^{-3.5})(4 - 5\sin^2 i)(1 - e^2)^{-2} \qquad (3\text{-}2)$$

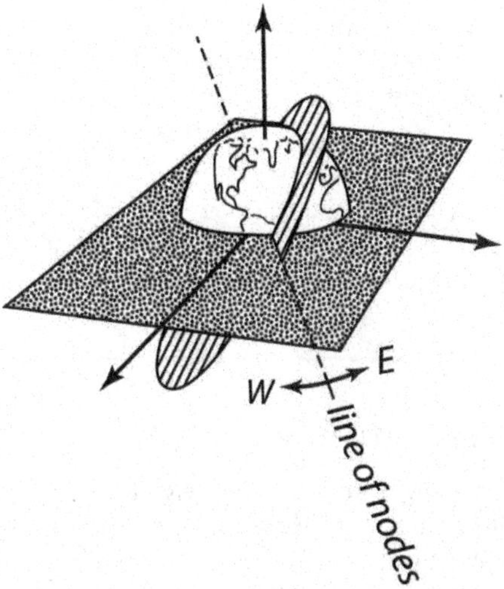

Figure 3-6 Nodal Regression.

Figure 3-7 shows the apsidal rotation rate versus inclination angle for a perigee altitude of 400 km and various apogee altitudes.

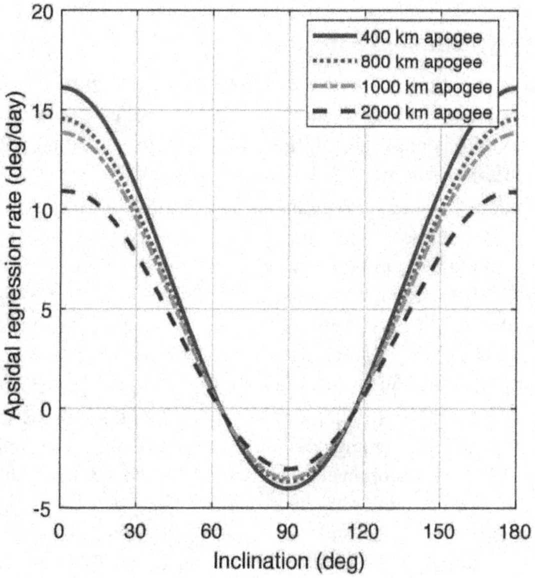

Figure 3-7 Apsidal Rotation Rate.

Example Problem. A satellite with an optical imaging payload is designed to operate at an altitude of 800 km in a circular orbit and needs to keep the sunlight illumination angle constant.

- What is the inclination of the orbit?

Earth completes a rotation about the Sun every 365.2421897 days so that a nodal regression rate of +0.98564736 deg/day will maintain a constant orientation between the orbit plane of the satellite and the Sun. Solving Equation 3-1 with that nodal rate gives i = 98.6032° (slightly retrograde).

3.3 HIGH-ALTITUDE EARTH ORBITS

As satellite usage has evolved from the beginning of the space age there are several classes of satellite orbits that have common meanings. Low-Earth orbit (LEO) satellites have an altitude less than 2,000 km. Mid-Earth orbit (MEO) satellites are generally in the range of 2,000 km to 30,000 km and are exposed to significantly higher levels of radiation than LEO satellites. Geosynchronous orbits at approximately 35,786 km in altitude (a = 42,164 km) match their orbital period to the rotation of Earth. If they also have an inclination of 0°, then they appear to be stationary to an observer on Earth's surface. Global Positioning System (GPS) satellites are at a semi-synchronous altitude with a 12 hr period.

3.3.1 The Geosynchronous Satellite

If the period of a *circular direct equatorial* orbit is exactly 23 hr 56 min it will appear to hover motionless over a point on the equator. This, of course, is an illusion since the satellite is not at rest in the inertial IJK frame but only in the noninertial topocentric-horizon frame.

The correct altitude for a synchronous circular orbit is 35,786 km or about 5.6 Earth radii above the surface. The great utility of such a satellite for communications is obvious. It can be useful to detect missile launches by their infrared "signatures." High-resolution photography of Earth's surface would be difficult from that altitude, but it is very useful as a location for geostationary weather satellites that can image a large section of Earth constantly.

There is often misunderstanding even among otherwise well-informed persons concerning the ground trace of a synchronous satellite. Many think it possible to "hang" a synchronous satellite over any point on Earth—the People's Republic of China, for example. This is, unfortunately, not the case. The satellite can only appear to be motionless over a point on the equator. If the synchronous circular orbit is inclined to the equator its ground trace will be a "figure-eight" curve that carries it north and south approximately along a meridian. Figure 3-8 shows why.

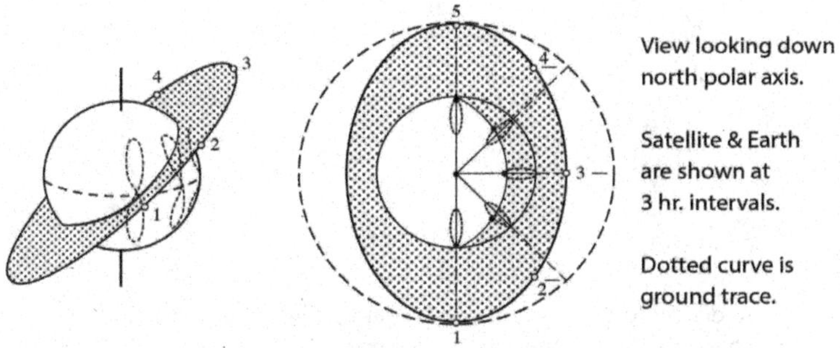
View looking down north polar axis.
Satellite & Earth are shown at 3 hr. intervals.
Dotted curve is ground trace.

Figure 3-8 Ground Trace of an Inclined Synchronous Satellite.

3.3.2 Launching a High-Altitude Satellite—The Ascent Ellipse

Launching a high-altitude satellite is a two-step operation requiring two burnouts separated by a coasting phase. The initial stage booster climbs almost vertically, reaching burnout with a flight-path angle of 45° or more. This first phase places the final stage in an elliptical orbit (called an "ascent ellipse") that has its apogee at the altitude of the desired orbit. At apogee of the ascent ellipse, the final stage booster engines are fired to increase the speed of the satellite and establish it in its final orbit. This burn-coast-burn technique is normally used any time the altitude of the orbit injection point exceeds 800 km. The ascent ellipse is a section of the geo-transfer orbit (GTO) and the perigee is typically between 150 and 300 km so that the upper stage of the booster rocket will lose energy via atmospheric drag during each subsequent orbit and eventually burn up in the atmosphere.

3.4 IN-PLANE ORBIT CHANGES

Because of small errors in burnout altitude, speed, and flight-path angle, the exact orbit desired may not be achieved. Sometimes this is not serious, but, if a rendezvous is contemplated or if, for some other reason, a very precise orbit is required, it may be necessary to make small corrections in the orbit. This may be done by applying small speed changes or Δv's, as they are called, at appropriate points in the orbit. In the next two sections we shall consider both small in-plane corrections to an orbit and large changes from one circular orbit to a new one of different size.

3.4.1 Adjustment of Perigee and Apogee Height

In Chapter 1 we derived the following energy relationship that is valid for all orbits:

$$E = \frac{v^2}{2} - \frac{\mu}{r} = -\frac{\mu}{2a} \tag{3-3}$$

If we solve for v^2 we get

$$v^2 = \mu\left(\frac{2}{r} - \frac{1}{a}\right) \tag{3-4}$$

Suppose we decide to change the speed, v, at a point in an orbit, leaving r unchanged. What effect would this Δv have on the semi-major axis, a? We can find out by taking the differential of both sides of Equation (3-4), considering r as fixed and Δv in the velocity direction. We then have

$$2v\,dv = \frac{\mu}{a^2}\,da \tag{3-5}$$

or

$$da = \frac{2a^2}{\mu}\,v\,dv \tag{3-6}$$

For an infinitesimally small change in speed dv, we get a change in semi-major axis, da, of the orbit given by Equation (3-6). Since the major axis is 2a, the size of the orbit changes by twice this amount or 2da.

But suppose we make the speed change in perigee? The resulting change in major axis will actually be a change in the height of apogee. Similarly, a Δv applied at apogee will result in a change in perigee height. We can specialize the general relationship shown by Equation (3-6) for small but finite Δv's applied at perigee and apogee as

$$\boxed{\begin{aligned} \Delta h_a &\approx \frac{4a^2}{\mu} v_p\,\Delta v_p \\ \Delta h_p &\approx \frac{4a^2}{\mu} v_a\,\Delta v_a \end{aligned}} \tag{3-7}$$

This method of evaluating a small change in one of the orbital elements as a result of a small change in some other variable is illustrative of one of the techniques used in perturbation theory. In technical terms what we have just done is called "variation of parameters."

3.4.2 The Hohmann Transfer

Transfer between two circular coplanar orbits is one of the most useful maneuvers we have. It represents an alternate method of establishing a satellite in a high-altitude orbit. For example, we could first make a direct ascent to a low-altitude "parking orbit" and then transfer to a higher circular orbit by means of an elliptical transfer orbit, which is just tangent to both of the circular orbits.

The least velocity change (Δv) required for a transfer between two circular orbits is achieved by using such a doubly-tangent transfer ellipse. The first recognition of this principle was by Hohmann in 1925 and such orbits are, therefore, called *Hohmann transfer orbits*.[5] Consider the two circular orbits shown in Figure 3-9. Suppose we want to travel from the small orbit, whose radius is r_1, to the large orbit, whose radius is r_2, along the transfer ellipse.

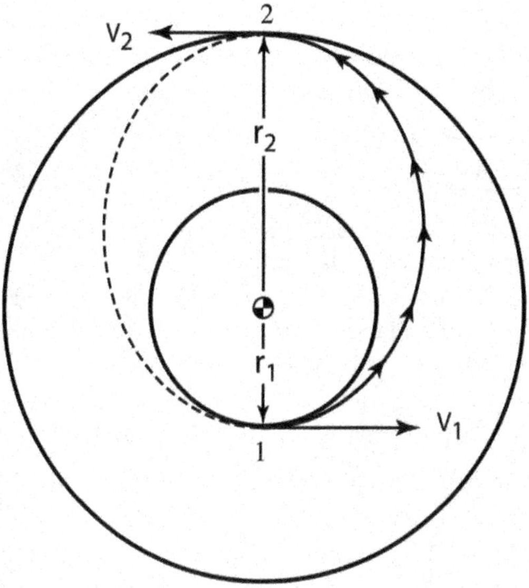

Figure 3-9 Hohmann Transfer.

We call the speed at point 1 on the transfer ellipse v_1. Since we know r_1, we could compute v_1 if we knew the energy of the transfer orbit E_t. From the geometry of Figure 3-9,

$$2a_t = r_1 + r_2 \tag{3-8}$$

and, since $E = -\mu/2a$,

$$E_t = -\mu/(r_1 + r_2) \tag{3-9}$$

We can now write the energy equation for point 1 of the elliptical orbit and solve it for $\mathbf{v}_1$:

$$v_1 = \sqrt{2\left[\frac{\mu}{r_1} + E_t\right]} \tag{3-10}$$

Since our satellite already has circular speed at point 1 of the small orbit, its speed is

$$v_{cs_1} = \sqrt{\frac{\mu}{r_1}} \tag{3-11}$$

To make our satellite go from the small circular orbit to the transfer ellipse we need to increase its speed from $\mathbf{v}_{cs}$ to $\mathbf{v}_1$. So,

$$\Delta v_1 = v_1 - v_{cs_1} \tag{3-12}$$

The speed change required to transfer from the ellipse to the large circle at point 2 can be computed in a similar fashion.

Although, in our example, we went from a smaller orbit to a larger one, the same principles may be applied to a transfer in the opposite direction. The only difference would be that two speed decreases would be required instead of two speed increases.

The time of flight for a Hohmann transfer is obviously just half the period of the transfer orbit. Since $\mathbb{P}_t = 2\pi\sqrt{a_t^3/\mu}$ and we know $\mathbf{a}_t$,

$$TOF = \pi\sqrt{\frac{a_t^3}{\mu}} \tag{3-13}$$

While the Hohmann transfer is the *most economical* from the standpoint of Δv required, it also *takes longer* than any other possible transfer orbit between the same two circular orbits. The other possible transfer orbits between coplanar circular orbits are discussed in the next section.

Example Problem. A remote sensing satellite is in a circular orbit of altitude 400 km. Find the minimum Δv required to double the altitude of the satellite.

Minimum Δv implies a Hohmann transfer. For the transfer trajectory,

$$r_1 = 400 \text{ km} + 6{,}378.137 \text{ km} = 6{,}778.137 \text{ km}$$

$$r_2 = 800 \text{ km} + 6{,}378.137 \text{ km} = 7{,}178.137 \text{ km}$$

$$a_t = (r_1 + r_2)/2 = 6{,}978.137 \text{ km}$$

$$E_t = -\mu / 2a_t = -28.5607 \text{ km}^2/\text{s}^2$$

$$v_1 = \sqrt{2\left[\frac{\mu}{r_1} + E_t\right]} = 7.7777 \text{ km/s}$$

$$v_{cs_1} = \sqrt{\frac{\mu}{r_1}} = 7.6686 \text{ km/s}$$

$$\Delta v_1 = v_1 - v_{cs_1} = 0.1091 \text{ km/s}$$

$$v_2 = \sqrt{2\left[\frac{\mu}{r_2} + E_t\right]} = 7.3443 \text{ km/s}$$

$$v_{cs_2} = \sqrt{\frac{\mu}{r_2}} = 7.4518 \text{ km/s}$$

$$\Delta v_2 = v_{cs_2} - v_2 = 0.1076 \text{ km/s}$$

$$\Delta v_{TOT} = \Delta v_1 + \Delta v_2 = \underline{\underline{0.2167 \text{ km/s}}}$$

3.4.3 General Coplanar Transfer Between Circular Orbits

Transfer between circular coplanar orbits merely requires that the transfer orbit intersect or at least be tangent to both of the circular orbits. Figure 3-10 shows transfer orbits that are either possible or impossible.

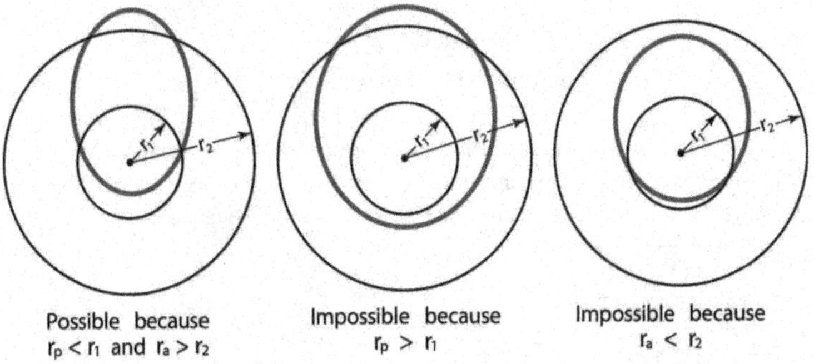

Figure 3-10 Transfer Orbit Must Intersect Both Circular Orbits.

It is obvious from Figure 3-10 that the periapsis radius of the transfer orbit must be equal to or less than the radius of the inner orbit *and* the apoapsis radius must be equal to or exceed the radius of the outer orbit if the transfer orbit is to touch both circular orbits. We can express this condition mathematically as

$$r_p = \frac{p}{1+e} \leq r_1 \tag{3-14}$$

$$r_a = \frac{p}{1-e} \geq r_2 \tag{3-15}$$

where p and e are the parameter and eccentricity of the transfer orbit and where r_1 and r_2 are the radii of the inner and outer circular orbits, respectively.

Orbits that satisfy both of these equations will intersect or at least be tangent to both circular orbits. We can plot these two equations (see Figure 3-11) and interpret them graphically. To satisfy both conditions, p and e of the transfer orbit must specify a point that lies in the shaded area. Values of (p, e) that fall on the limit lines correspond to orbits that are just tangent to one or the other of the circular orbits.

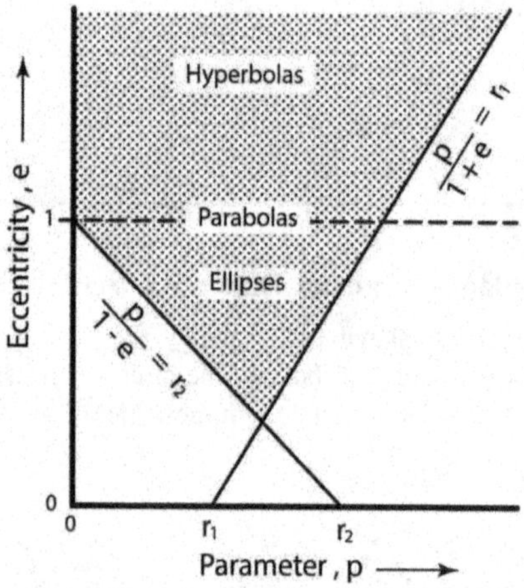

Figure 3-11 p vs. e.

Suppose we have picked values of p and e for our transfer orbit that satisfy the conditions above. Knowing p and e, we can compute the energy, E_t, and the angular momentum, h_t, in the transfer orbit. Since $p = a(1-e^2)$ and $E = -\mu/2a$,

$$\boxed{E_t = -\mu(1-e^2)/2p} \tag{3-16}$$

Since $p = h^2/\mu$

$$\boxed{h_t = \sqrt{\mu p}} \tag{3-17}$$

We now can proceed just as we did in the case of Hohmann transfer. Solving the energy equation for the speed at point 1 in the transfer orbit, we get

$$\boxed{v_1 = \sqrt{2\left(\frac{\mu}{r_1} + E_t\right)}} \tag{3-18}$$

Since our satellite already has circular speed at point 1 of the small orbit, its speed is

$$v_{cs_1} = \sqrt{\frac{\mu}{r_1}} \tag{3-19}$$

The angle between $\mathbf{v_1}$ and $\mathbf{v_{cs_1}}$ is just the flight-path angle, ϕ_1. Since $h = rv \cos\phi$,

$$\cos \phi_1 = h_t/r_1\, v_1 \tag{3-20}$$

Now, since we know two sides and the included angle of the vector triangle shown in Figure 3-12, we can use the law of cosines to solve for the third side, $\Delta\mathbf{v}$:

$$\boxed{\Delta v_1^2 = v_1^2 + v_{cs_1}^2 - 2v_1\, v_{cs_1} \cos \phi_1} \tag{3-21}$$

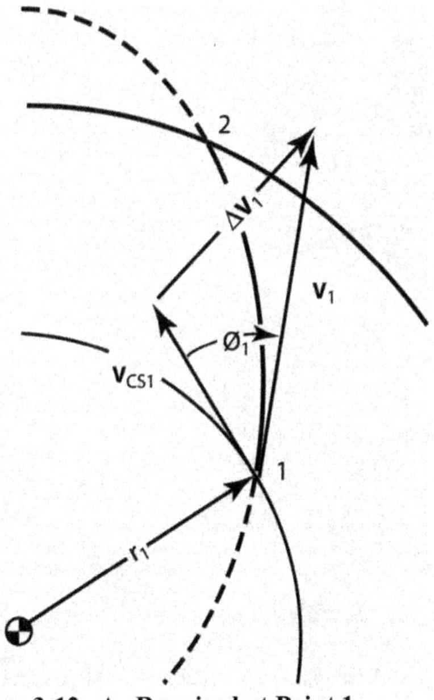

Figure 3-12 Δv Required at Point 1.

The speed change required at point 2 can be computed in a similar fashion.

Since the Hohmann transfer is just a special case of the problem illustrated above, it is not surprising that Equation (3-21) reduces to Equation (3-12) when the flight-path angle, ϕ_1, is zero.

3.5 OUT-OF-PLANE ORBIT CHANGES

A velocity change, $\Delta\mathbf{v}$, that lies in the plane of the orbit can change the orbit's size or shape, or it can rotate the line of apsides. To change the orientation of the orbital plane in space requires a $\Delta\mathbf{v}$ component perpendicular to the plane of the orbit.

3.5.1 Simple Plane Change

If, after applying a finite $\Delta\mathbf{v}$ the speed and flight-path angle of the satellite are unchanged, then only the plane of the orbit has been altered. This is called a *simple plane change*.

An example of a simple plane change would be changing an inclined orbit to an equatorial orbit as shown in Figure 3-13. The plane of the orbit has been changed through an angle θ. The initial velocity and final velocity are identical in magnitude and, together with the $\Delta\mathbf{v}$ required, form an isosceles vector triangle. We can solve for the magnitude of $\Delta\mathbf{v}$ using the law of cosines, assuming that we know v and θ. Or, more simply, we can divide the isosceles triangle into two right triangles, as shown at the right of Figure 3-13, and obtain directly

$$\boxed{\Delta v = 2v \sin \frac{\theta}{2}} \qquad (3\text{-}22)$$

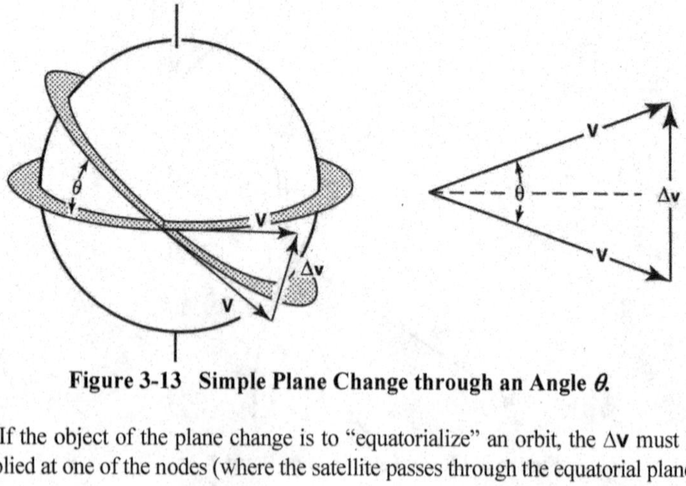

Figure 3-13 Simple Plane Change through an Angle θ.

If the object of the plane change is to "equatorialize" an orbit, the $\Delta\mathbf{v}$ must be applied at one of the nodes (where the satellite passes through the equatorial plane).

Large plane changes are prohibitively expensive in terms of the velocity change required. A plane change of 60°, for example, makes $\Delta \mathbf{v}$ equal to $\mathbf{v}$! Russia must pay this high price if it wants equatorial satellites. Since it has no launch sites south of latitude 45° N, it cannot launch a satellite whose inclination is less than 45°. Therefore, a turn of at least 45° must be made at the equator if the satellite is to be equatorialized.

Example Problem. As part of an automated rendezvous and resupply mission, a satellite in a parking orbit of 650 km must travel to a communication satellite in a geosynchronous coplanar circular orbit. The transfer will be accomplished via an elliptical orbit tangent to the lower orbit and crossing the high orbit at the end of the minor axis of the transfer orbit.

a. Determine the total $\Delta \mathbf{v}$ required to accomplish the mission.
b. If the communication satellite orbit is inclined at an angle of 10° to the low parking orbit, calculate the additional $\Delta \mathbf{v}$ required for the simple plane change necessary to accomplish the transfer. Assume the plane change is performed *after* you have established that the refueling satellite is in a geosynchronous circular orbit.
c. The given information is

$h_1 = 650$ km
$h_2 = 35{,}780$ km

Therefore, $r_1 = r_\oplus + h_1 = 7{,}028.137$ km
$r_2 = r_\oplus + h_2 = 42{,}158.137$ km

Since the transfer ellipse is tangent to the low orbit, $\Delta v_1 = v_1 - v_{cs_1}$.

We need to know $\mathcal{E}_t$ for the transfer orbit. It is given that the transfer orbit intersects the high orbit at the end of its (transfer orbit) minor axis. Therefore, $\mathbf{a}_t = \mathbf{r}_2$. Hence,

$$\mathcal{E}_t = -\frac{\mu}{2a_t} = -\frac{\mu}{2r_1} = -4.7275 \text{ km}^2/\text{s}$$

$$v_1 = \sqrt{2\left(\mathcal{E}_t + \frac{\mu}{r_1}\right)} = 10.1968 \text{ km/s}$$

$$v_{cs_1} = \sqrt{\frac{\mu}{r_1}} = 7.5309 \text{ km/s}$$

$$\Delta v_1 = v_1 - v_{cs_1} = 2.6659 \text{ km/s}$$

Since $\mathbf{v}_2$ and $\mathbf{v}_{cs_2}$ are not tangent,

$$|\Delta \mathbf{v}_2| = |\mathbf{v}_{cs_2} - \mathbf{v}_2|$$

and we must use Equation (3-22) to determine Δv_2:

$$h_t = r_1 v_1 = 71{,}665 \text{ km}^2/\text{s}$$

$$v_2 = \sqrt{2\left(E_t + \frac{\mu}{r_2}\right)} = 3.0749 \text{ km/s}$$

Therefore,

$$\cos\phi_2 = \frac{h_t}{r_2 v_2} = 0.5528$$

$$v_{cs_2} = \sqrt{\frac{\mu}{r_2}} = 3.0749 \text{ km/s}$$

(Why does $v_2 = v_{cs_2}$?)

From Equation (3-21), we have

$$\Delta v_2^2 = v_2^2 + v_{cs_2}^2 - 2 v_2 v_{cs_2} \cos\phi_2$$

$$\text{so } \Delta v_2 = 2.9079 \text{ km/s}$$

Therefore,

$$\Delta v_{Total} = \Delta v_1 + \Delta v_2 = \underline{\underline{5.5738 \text{ km/s}}}$$

d. For a simple plane change, use of Equation (3-22) gives

$$\Delta v = 2v \sin\frac{\theta}{2} = \underline{\underline{0.5360 \text{ km/s}}}$$

Exercises

3.1 What is the inclination of a circular orbit of period 100 minutes designed such that the trace of the orbit moves eastward at the rate of 3° per day?

3.2 Calculate the total Δv required to transfer between two coplanar circular orbits of radii $r_1 = 12{,}750$ km and $r_2 = 31{,}890$ km, respectively, using a transfer ellipse having parameters $p = 13{,}457$ km and $e = 0.76$.

(Answer: 7.086 km/s)

3.3 Two satellites are orbiting Earth in circular orbits—not at the same altitude or inclination. What sequence of orbit changes and plane changes would most effectively place the lower satellite in the same orbit as that of the higher one? Assume only one maneuver can be performed at a time, i.e., a plane change *or* an orbit transfer.

3.4 Refer to Figure 3-8. What would be the effect on the ground trace of a synchronous satellite if:

a. $e > 0$ but the period remained 23 hours 56 minutes?

b. i changed—all other parameters remained the same?

c. $\mathbb{P} < 23$ hours 56 minutes?

3.5 You are in a circular Earth orbit with a velocity of 7 km/s. Your Service Module is in another circular orbit with a velocity of 3.5 km/s. What is the minimum Δv needed to transfer to the Service Module's orbit?

(Answer: 3.140 km/s)

3.6 Determine which of the following orbits could be used to transfer between two circular-coplanar orbits with radii 1.2 DU and 4 DU, respectively.

a. $r_p = 1$ DU

$e = 0.5$

b. $a = 2.5$ DU

$e = 0.56$

c. $\mathcal{E} = -0.1$ DU^2/TU^2

$h = 1.34$ DU^2/TU

d. $p = 1.95$ DU

$e = 0.5$

3.7 Design a satellite orbit that could provide a continuous communications link between Moscow and points in Siberia for at least 1/3 to 1/2 a day.

3.8 Compute the minimum Δv required to transfer between two coplanar elliptical orbits that have their major axes aligned. The parameters for the ellipses are given by

$r_{p_1} = 7{,}000$ km $\qquad r_{p_2} = 32{,}000$ km

$e_1 = 0.290 \qquad e_2 = 0.412$

Assume both perigees lie on the same side of Earth.

(Answer: 1.562 km/s)

3.9 Calculate the total Δv required to transfer from a circular orbit of radius 1 DU to a circular orbit of infinite radius and then back to a circular orbit of 15 DU, using Hohmann transfers. Compare this with Δv required to make a Hohmann transfer from the 1 DU circular orbit directly to the 15 DU circular orbit. At least 5 digits of accuracy are needed for this calculation.

3.10 *Note that in Problem 3.9 it is more economical to use the three-impulse transfer mode. This is often referred to as a bi-elliptical transfer. Find the ratio between circular orbit radii (outer to inner) beyond which it is more economical to use the bi-elliptical transfer mode.

List of References

1. Newton, Sir Isaac. *Principia.* Motte's translation revised by Cajori. Vol. 2, Berkeley and Los Angeles, University of California Press, 1962.

2. Ley, Willy. *Rockets, Missiles and Space Travel.* 2nd revised ed. New York, NY, Viking Press, 1961.

3. Dornberger, Walter. *V-2*. New York, NY, Viking Press, 1955.

4. *Space Planners Guide*. Washington, DC, Headquarters, Air Force Systems Command, 1 July 1965.

5. Baker, Robert M.L., and Maud W. Makemson. *An Introduction to Astrodynamics*. New York and London, Academic Press, 1960.

6. Vallado, David A. *Fundamentals of Astrodynamics and Applications.* 2nd edition. El Segundo, California, Microcosm Press, 2004.

7. Prussing, John E. and Conway, Bruce A. *Orbital Mechanics*. New York, Oxford University Press, Inc., 1993.

8. Koestler, Arthur. *The Sleepwalkers*. New York, NY, Macmillan, 1959.

Chapter 4

Position and Velocity as a Function of Time

> ...the determination of the true movement of the planets, including Earth...This was Kepler's first great problem. The second problem lay in the question: What are the mathematical laws controlling these movements? Clearly, the solution of the second problem, if it were possible for the human spirit to accomplish it, presupposed the solution of the first. For one must know an event before one can test a theory related to this event.
>
> -Albert Einstein[1]

4.1 Historical Background

The year Tycho Brahe died, 1601, Johannes Kepler, who had worked with Tycho in the 18 months preceding his death, was appointed as Imperial Mathematician to the Court of Emperor Rudolph II. Recognizing the gold mine of information locked up in Tycho's painstaking observations, Kepler packed them up and moved them to Prague with him. In a letter to one of his English admirers he calmly reported:

> "I confess that when Tycho died, I quickly took advantage of the absence, or lack of circumspection, of the heirs, by taking the observations under my care, or perhaps usurping them..." [2]

Kepler stayed in Prague from 1601 to 1612. It was the most fruitful period of his life and saw the publication of *Astronomia Nova* in 1609 in which he announced his first two laws of planetary motion. The manner in which Kepler arrived at these two laws is fascinating and can be told only because in *New Astronomy* Kepler leads his reader into every blind alley, detour, trap or pitfall that he himself encountered.

> "What matters to me," Kepler points out in his Preface, "is not merely to impart to the reader what I have to say, but above all to convey to him the reasons, subterfuges, and lucky hazards which

led me to my discoveries. When Christopher Columbus, Magellan and the Portuguese relate how they went astray on their journeys, we not only forgive them, but would regret to miss their narration because without it the whole grand entertainment would be lost."[2]

Kepler selected Tycho's observations of Mars and tried to reconcile them with some simple geometrical theory of motion. He began by making three revolutionary assumptions: (a) that the orbit was a circle with the Sun slightly off-center (Figure 4-1), (b) that the orbital motion took place in a plane that was fixed in space and (c) that Mars did not necessarily move with uniform velocity along this circle. Thus, Kepler immediately cleared away a vast amount of rubbish that had obstructed progress since Ptolemy.

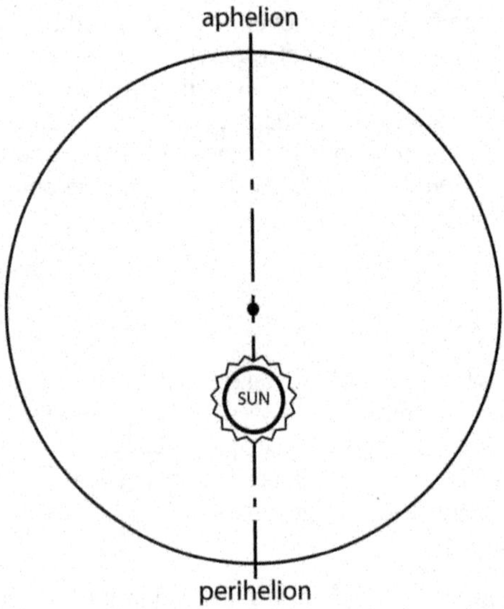

Fig. 4-1 Kepler First Assumed a Circular Orbit with the Sun Off-Center.

Kepler's first task was to determine the radius of the circle and the direction of the axis connecting perihelion and aphelion. At the very beginning of a whole chapter of excruciating trial-and-error calculations, Kepler absentmindedly put down three erroneous figures for three vital longitudes of Mars, never noticing his error. His results, however, were nearly correct because of several mistakes of simple arithmetic committed later in the chapter that happened very nearly to cancel out his earlier errors.

At the end, he seemed to have achieved his goal of representing within 2 arc-minutes the position of Mars at all 10 oppositions recorded by Tycho. But then, without a word of transition, in the next two chapters Kepler explains almost with masochistic delight, how two other observations from Tycho's collection did not fit; there was a discrepancy of 8 minutes of arc. Others might have shrugged off this minor disparity between fact and hypothesis. It is to Kepler's everlasting credit that he made it the basis for a complete reformation of astronomy. He decided that the sacred concept of circular motion had to go.

Before Kepler could determine the true shape of Mars' orbit, without benefit of any preconceived notions, he had to determine precisely Earth's motion around the Sun. For this purpose he designed a highly original method and when he had finished his computations he was ecstatic. His results showed that Earth did not move with uniform speed but moved faster or slower according to its distance from the Sun. Moreover, at the extrema of the orbit (perihelion and aphelion) Earth's velocity proved to be inversely proportional to distance.

At this point, Kepler could contain himself no longer and becomes airborne, as it were, with the warning: "Ye physicists, prick your ears, for now we are going to invade your territory." He was convinced that here was "a force in the Sun" that moved the planets. What could be more beautifully simple than that the force should vary inversely with distance? He had proved the inverse ratio of speed to distance for only *two points* in the orbit, perihelion and aphelion, yet he made the patently incorrect generalization that this "law" held true for the *entire* orbit. This was the first of the critical mistakes that would cancel itself out "as if by a miracle" and lead him by faulty reasoning to the correct result.

Forgetting his earlier resolve to abandon circular motion, he reasoned, again incorrectly that, since speed was inversely proportional to distance, the line joining the Sun (which was off-center in the circle) and the planet swept out equal areas in the orbit in equal times (Figure 4-2).

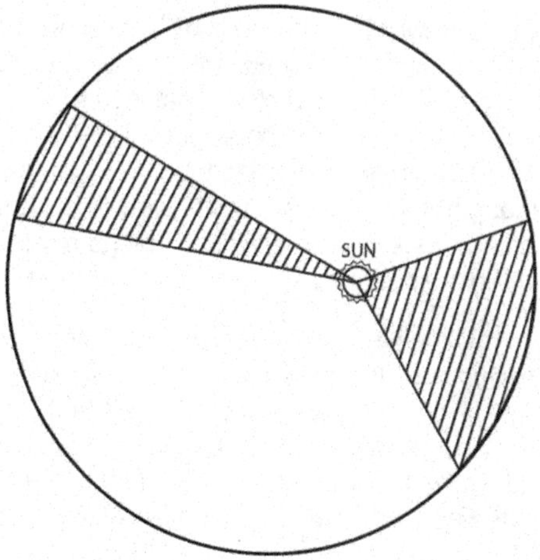

Fig. 4-2 Kepler's Law of Equal Areas.

This was his famous Second Law—discovered before the First—a law of amazing simplicity, arrived at by a series of faulty steps that he himself later recognized with observation: "But these two errors—it is like a miracle—cancel out in the most precise manner, as I shall prove further down."[2]

The correct result is even more miraculous than Kepler realized since his explanation of why the errors canceled was also erroneous!

Kepler now turned again to the problem of replacing the circle with another geometric shape; he settled on the oval. But a very special oval: it had the shape of an egg with the narrow end at the perihelion and the broad end at aphelion. As Koestler[2] observed: "No philosopher had laid such a monstrous egg before."

Finally, a kind of snowblindness seemed to descend upon him: he held the solution in his hands but was unable to see it. On 4 July 1603 he wrote a friend that he was unable to solve the problem of computing the area of his oval; but "if only the shape were a perfect ellipse all the answers could be found in Archimedes' and Apollonius' work."[2] Finally, after struggling with his egg for more than a year he stumbled onto the secret of the Martian orbit. He was able to express the distance from the Sun by a simple mathematical formula: but he did not recognize that this formula specifically defined the orbit as an ellipse. Any student of analytical geometry today would have recognized it immediately; but analytical geometry came after Kepler. He had reached his goal, but he did not realize that he had reached it!

He tried to construct the curve represented by his equation, but he did not know how, made a mistake in geometry and ended up with a "chubby-faced" orbit. The climax to his comedy of errors came when, in a moment of despair, Kepler threw out his equation (which denoted an ellipse) because he wanted to try an entirely new hypothesis: to wit, an elliptic orbit! When the orbit fit and he realized what had happened, he frankly confessed:

> "Why should I mince my words? The truth of Nature, which I had rejected and chased away, returned by stealth through the backdoor, disguising itself to be accepted. That is to say, I laid [the original equation] aside and fell back on ellipses, believing that this was a quite different hypothesis, whereas the two . . . are one and the same . . . I thought and searched, until I went nearly mad, for reason why the planet preferred an elliptical orbit [to mine] . . . Ah, what a foolish bird I have been!"[2]

In the end, Kepler was able to write an empirical expression for the time of flight of a planet from one point in its orbit to another—although he still did not know the true reason why it should move in an orbit at all. In this chapter we will, with the benefit of hindsight and concepts introduced by Newton, derive the Kepler time-of-flight equation in much the same way that Kepler did. We will then turn our attention to the solution of what has come to be known as "the Kepler problem"—predicting the future position and velocity of an orbiting object as a function of some known initial position and velocity and the time of flight. In doing this we will introduce a very important concept in the field of orbital mechanics—a universal formulation of the time-of-flight relationships valid for all conic orbits.

4.2 TIME OF FLIGHT AS A FUNCTION OF ECCENTRIC ANOMALY

Many of the concepts introduced by Kepler, along with the names he used to describe them, have persisted to this day. You are already familiar with the term "true anomaly" used to describe the angle from periapsis to the orbiting object measured in the direction of motion. In this section we will encounter a new term called "eccentric anomaly," which was introduced by Kepler in connection with elliptical orbits. Although he was not aware that parabolic and hyperbolic orbits existed, the concept can be extended to these orbits also as we shall see.

It is possible to derive time-of-flight equations analytically using only the dynamical equation of motion and integral calculus. We will pursue a lengthier, but more motivated, derivation in which the eccentric anomaly arises quite naturally in the course of geometrical arguments. This derivation is presented more for its historical value than for actual use. The universal variable approach is strongly recommended as the best method for general use.

4.2.1 Time of Flight on the Elliptical Orbit

We have already seen that in one orbital period the radius vector sweeps out an area equal to the total area of an ellipse; i.e., in going part way around an orbit, say from periapsis to some general point, P, where the true anomaly is ν, the radius vector sweeps out the shaded area, A_1, in Figure 4-3. Because area is swept out at a constant rate in an orbit (Kepler's Second Law) we can say that

$$\frac{t-T}{A_1} = \frac{\mathbb{P}}{\pi ab} \tag{4-1}$$

where T is the time of periapsis passage and $\mathbb{P}$ is the period.

The only unknown in Equation (4-1) is the area A_1. The geometrical construction illustrated in Figure 4-4 will enable us to write an expression for A_1.

A circle of radius a has been circumscribed about the ellipse. A dotted line, perpendicular to the major axis, has been extended through P to where it intersects the "auxiliary circle" at Q. The angle E is called the *eccentric anomaly.*

Before proceeding further we must derive a simple relationship between the ellipse and its auxiliary circle. From analytical geometry, the equations of the curves in Cartesian coordinates are

$$\text{ellipse: } \frac{x^2}{a^2} + \frac{y^2}{b^2} = 1, \qquad \text{circle: } \frac{x^2}{a^2} + \frac{y^2}{a^2} = 1$$

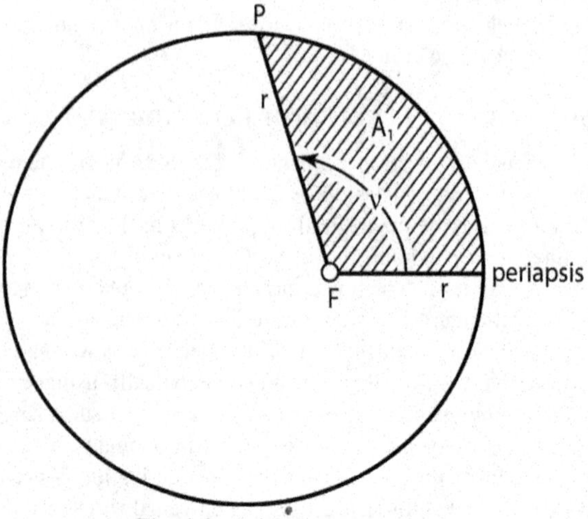

Fig. 4-3 Area Swept out by r.

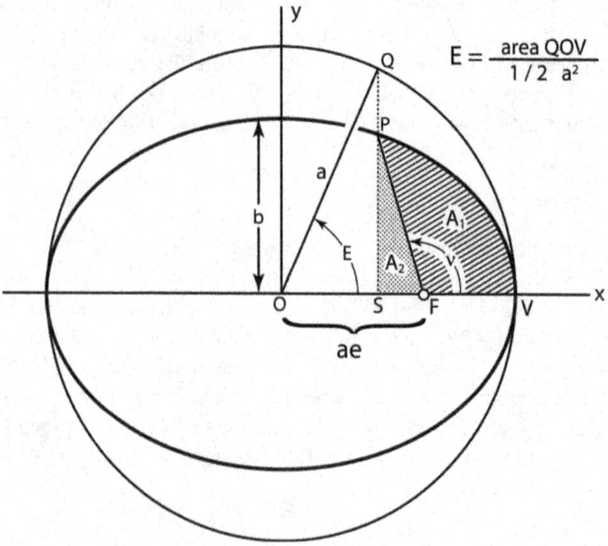

Fig. 4-4 Eccentric Anomaly, E.

From these relationships, we obtain

$$y_{ellipse} = \sqrt{\frac{a^2b^2 - b^2x^2}{a^2}} = \frac{b}{a}\sqrt{a^2 - x^2}$$

$$y_{circle} = \sqrt{a^2 - x^2}$$

and hence

$$\frac{y_{ellipse}}{y_{circle}} = \frac{b}{a} \tag{4-2}$$

This simple relationship between the y-ordinates of the two curves will play a key role in subsequent area and length comparisons.

From Figure 4-4 we note that the area swept out by the radius vector is Area PSV minus the dotted area, A_2:

$$A_1 = \text{Area PSV} - A_2$$

Since A_2 is the area of a triangle whose base is $ae - a\cos E$ and whose altitude is $b/a\,(a \sin E)$, we can write

$$A_2 = \frac{ab}{2}(e \sin E - \cos E \sin E) \tag{4-3}$$

Area PSV is the area under the ellipse; it is bounded by the dotted line and the major axis. Area QSV is the corresponding area under the auxiliary circle. It follows directly from Equation (4-2) that

$$\text{Area PSV} = \frac{b}{a}(\text{Area QSV})$$

Area QSV is just the area of the sector QOV, which is $(1/2)\,a^2 E$ (where E is in radians), minus the triangle, whose base is $(a \cos E)$ and whose altitude is $(a \sin E)$. Hence

$$\text{Area PSV} = \frac{ab}{2}(E - \cos E \sin E)$$

Substituting into the expression for area A_1 yields

$$A_1 = \frac{ab}{2}[E - e \sin E]$$

Finally, substituting into Equation (4-1) and expressing the period as $2\pi\sqrt{a^3/\mu}$ we get

$$t - T = \sqrt{\frac{a^3}{\mu}}(E - e \sin E) \tag{4-4}$$

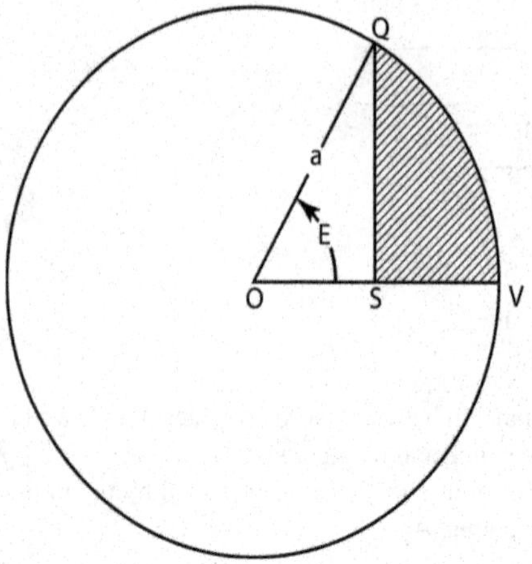

Fig. 4-5 Area QSV.

Kepler introduced the definition

$$M = E - e \sin E \tag{4-5}$$

where M is called the "mean anomaly." If we also use the definition

$$n \equiv \sqrt{\mu / a^3}$$

where n is called the "mean motion," then the mean anomaly may be written as

$$M = n(t - T) = E - e \sin E \tag{4-6}$$

which is often referred to as *Kepler's equation.*

Obviously, to use Equation (4-4), we must be able to relate the eccentric anomaly, E, to its corresponding true anomaly, ν.
From Figure 4-4,

$$\cos E = \frac{ae + r \cos \nu}{a} \tag{4-7}$$

Since $r = \dfrac{a(1-e^2)}{1 + e\cos \nu}$ Equation (4-7) reduces to

$$\cos E = \frac{e + \cos \nu}{1 + e \cos \nu} \tag{4-8}$$

The eccentric anomaly may be determined from Equation (4-8). The correct quadrant for E is obtained by noting that ν and E are always in the same half-plane; when ν is between 0 and π, so is E.

Suppose we want to find the time of flight between a point defined by ν_0 and some general point defined by ν when the initial point is not periapsis. Provided the object does not pass through periapsis en route from ν_0 to ν (Figure 4-6 left), we can say that

$$t - t_0 = (t - T) - (t_0 - T)$$

If the object does pass through periapsis (which is the case whenever ν_0 is greater than ν) then, from Figure 4-6 (right panel),

$$t - t_0 = \mathbb{P} + (t - T) - (t_0 - T_0)$$

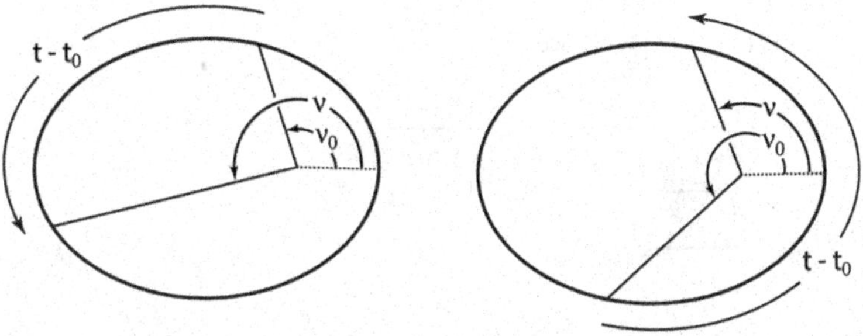

Fig. 4-6 Time of flight Between Arbitrary Points.

In general we can say that

$$\boxed{t - t_0 = \sqrt{\frac{a^3}{\mu}}\left[2k\pi + \left(E - e\sin E\right) - \left(E_0 - e\sin E_0\right)\right]} \tag{4-9}$$

where k is the number of times the object passes through periapsis en route from ν_0 to ν.

At this point it is instructive to note that this same result can be derived analytically. In this case the eccentric anomaly appears as a convenient variable transformation to enable integration. Only the skeleton of the derivation will be shown here.

From equations in Section 1.7.2 we can write

$$\int_T^t h\,dt = \int_0^\nu r^2\,d\nu \tag{4-10}$$

or

$$h\,(t - T) = \int_0^\nu \frac{p^2 d\nu}{(1 + e\cos\nu)^2} \tag{4-11}$$

Now let us introduce the eccentric anomaly as a variable change to make Equation (4-11) easily integrable. From Equation (4-8) and geometry the following relationships can be derived:

$$\cos \nu = \frac{e - \cos E}{e \cos E - 1} \tag{4-12}$$

$$\sin \nu = \frac{a\sqrt{1-e^2}}{r} \sin E \tag{4-13}$$

$$r = a(1 - e \cos E) \tag{4-14}$$

Differentiating Equation (4-12), we obtain

$$d\nu = \frac{\sin E\,(1 + e\cos \nu)}{\sin \nu (1 - e\cos E)} dE = \frac{\sin E\,(p/r)}{\sin \nu\,(r/a)}$$

$$= \frac{a\sqrt{1-e^2}}{r} dE \tag{4-15}$$

Then

$$h(t-T) = \frac{p}{\sqrt{1-e^2}} \int_0^E r\,dE$$

$$= \frac{pa}{\sqrt{1-e^2}} \int_0^E (1 - e\cos E)\,dE$$

$$= \frac{pa}{\sqrt{1-e^2}} (E - e\sin E) \tag{4-16}$$

since

$$h = \sqrt{\mu p}$$

$$(t-T) = \sqrt{\frac{a^3}{\mu}}(E - e\sin E) \tag{4-17}$$

which is identical to the geometrical result.

4.2.2 Time of Flight on Parabolic and Hyperbolic Orbits

In a similar manner, the analytical derivation of the parabolic time of flight can be shown to be

$$\boxed{t - T = \frac{1}{2\sqrt{\mu}}\left[pD + \frac{1}{3}D^3\right]} \tag{4-18}$$

or

$$\boxed{t - t_0 = \frac{1}{2\sqrt{\mu}}\left[\left(pD + \frac{1}{3}D^3\right) - \left(pD_0 + \frac{1}{3}D_0^3\right)\right]} \tag{4-19}$$

where

$$D = \sqrt{p}\tan\frac{\nu}{2}$$

is the"parabolic eccentric anomaly."

From either a geometrical or analytical approach the hyperbolic time of flight, using the "hyperbolic eccentric anomaly," F, can be derived as

$$\boxed{t - T = \sqrt{\frac{(-a)^3}{\mu}}(e\sinh F - F)} \tag{4-20}$$

or

$$t - t_0\sqrt{\frac{(-a)^3}{\mu}}[(e\sinh F - F) - (e\sinh F_0 - F_0)] \tag{4-21}$$

where

$$\boxed{\cosh F = \frac{e + \cos\nu}{1 + e\cos\nu}} \tag{4-22}$$

or

$$F = \ln\left[y + \sqrt{y^2 - 1}\right]$$

for $y = \cosh F$. Whenever ν is between 0 and π, F should be taken as positive; whenever ν is between π and 2π, F should be taken as negative. Figure 4-7 illustrates the hyperbolic variables.

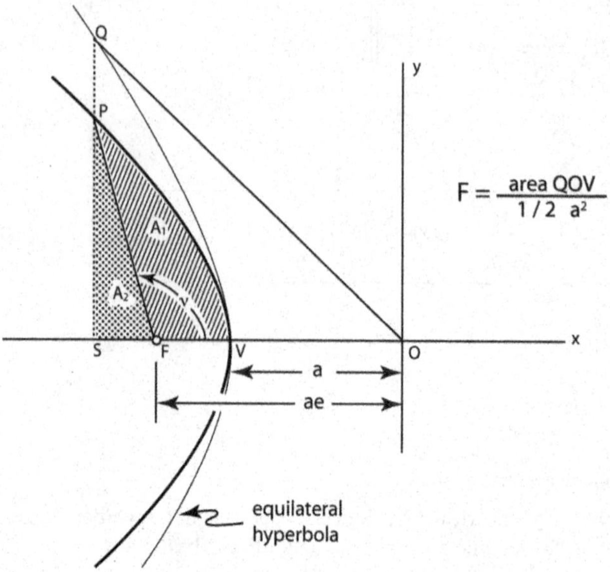

Fig. 4-7 Hyperbolic Eccentric Anomaly, F.

Example Problem. A space probe is in an elliptical orbit around the Sun. The perihelion distance is 0.5 AU and aphelion is 2.5 AU. How many days during each orbit is the probe closer than 1 AU to the Sun?

The given information is:

$$r_p = 0.5 \text{ AU}, \; r_a = 2.5 \text{ AU}$$

$$r_1 = 1 \text{ AU}, \; r_2 = 1 \text{ AU}$$

Since the portion of the orbit in question is symmetrical, we can compute the time of flight from periapsis to point 2 and then double it.

From Equation (1-64), $e = \dfrac{r_a - r_p}{r_a + r_p} = \dfrac{2}{3}$.

From Equation (1-62), $a = \dfrac{r_p + r_a}{2} = \dfrac{3}{2}$.

From $r_2 = \dfrac{a(1-e^2)}{1 + e\cos\nu_2}$, $\cos\nu_2 = \dfrac{1(1-e^2) - r_2}{e r_2} = -\dfrac{1}{4}$,

$$\cos E_2 = \frac{e + \cos\nu_2}{1 + e\cos\nu_2} = \frac{1}{2}$$

$$E_2 = 60° = 1.048 \text{ radians}, \quad \sin E_2 = 0.866$$

From Equation (4-4),

$$t_2 - T = \sqrt{\frac{(1.5)^3}{1}}\,[1.048 - \frac{2}{3}\,(0.866)] = 0.862 \text{ TU}_\odot$$

$$\text{TOF} = 1.724 \text{ TU}_\odot \left(58.133 \frac{\text{days}}{\text{TU}_\odot}\right) = \underline{\underline{100 \text{ days}}}$$

4.2.3 Loss of Numerical Accuracy for Near-Parabolic Orbits

The Kepler time-of-flight equations suffer from a severe loss in computational accuracy near $e = 1$. The nature of the difficulty can best be illustrated by a numerical example:

Suppose we want to compute the time of flight from periapsis to a point where $\nu = 60°$ on an elliptical orbit with $a = 100$ DU and $e = 0.999$. The first step is to compute the eccentric anomaly. From Equation (4-8),

$$\cos E = \frac{0.999 + 0.5}{1 + 0.999(0.5)} = 0.99967$$

Therefore, $E = 0.02559$, $\sin E = 0.02560$ and $(e \sin E) = 0.02557$. Substituting these values into Equation (4-4) gives

$$t - T = \sqrt{\frac{100^3}{1}}(0.02559 - 0.02557)$$

$$= 1{,}000\,(0.00002) = 0.02 \text{ TU}$$

There is a loss of significant digits in computing E from cos E when E is near zero. There is a further loss in subtracting two nearly equal numbers in the last step. As a result, the answer is totally unreliable.

This loss of computational accuracy near $e = 1$ and the inconvenience of having a different equation for each type of conic orbit will be our principal motivation for developing a universal formulation for time of flight in Section 4.3.

4.3 A UNIVERSAL FORMULATION FOR TIME OF FLIGHT

The classical formulations for time of flight involving the eccentric anomalies, E or F, don't work very well for near-parabolic orbits. We have already seen the loss of numerical accuracy that can occur near $e = 1$ in the

Kepler equation. Also, solving for E or F when a, e, ν_o and $t - t_o$ are given is difficult when e is nearly 1 because the trial-and-error solutions converge too slowly or not at all. Both of these defects are overcome in a reformulation of the time-of-flight equations made possible by the introduction of a new auxiliary variable different from the eccentric anomaly. Furthermore, the introduction of this new auxiliary variable allows us to develop a single time-of-flight equation valid for all conic orbits.

The change of variable is known as the "Sundman transformation" and was first proposed in 1912.[6] In the middle of the 20th century, this was used to develop a unified time-of-flight equation. Goodyear,[7] Lemmon,[8] Herrick,[9] Stumpff,[10] Sperling[11] and Battin[3] have all presented formulas for computation of time of flight via "generalized" or "universal" variables. The original derivation presented below was suggested by Bate[12] and partially makes use of notation introduced by Battin.

4.3.1 Definition of the Universal Variable, x

Angular momentum and energy are related to the geometrical parameters p and a by the familiar equations

$$h = r^2\dot{\nu} = \sqrt{\mu p}$$

$$E = \frac{1}{2}v^2 - \frac{\mu}{r} = \frac{-\mu}{2a}$$

As shown in Figure 4-8 we can resolve **v** into its radial component, $\dot{r}$, and its transverse component, $r\dot{\nu}$; the energy equation can then be written as

$$\frac{1}{2}\dot{r}^2 + \frac{1}{2}(r\dot{\nu})^2 - \frac{\mu}{r} = \frac{-\mu}{2a}$$

Solving for $\dot{r}^2$ and setting $(r\dot{\nu})^2 = \frac{\mu p}{r^2}$, we get

$$\dot{r}^2 = -\frac{\mu p}{r^2} + \frac{2\mu}{r} - \frac{\mu}{a} \tag{4-23}$$

Since the solution of this equation is not obvious, we introduce a new independent variable, x, whose derivative is defined as

$$\boxed{\dot{x} = \frac{\sqrt{\mu}}{r}} \tag{4-24}$$

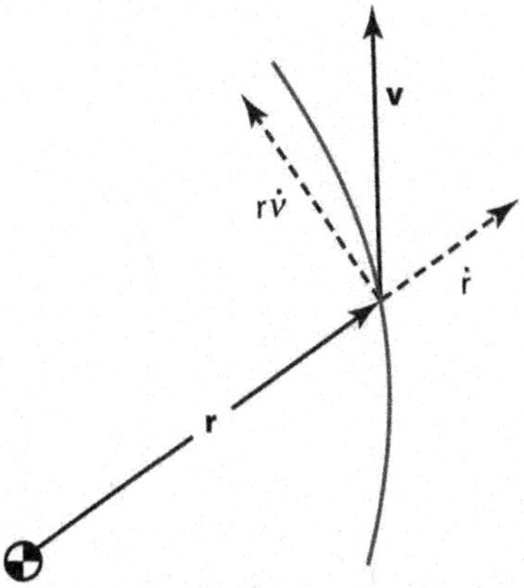

Fig. 4-8 Radial and Transverse Components of the Velocity Vector, v.

First we will develop a general expression for r in terms of x. If we divide Equation (4-23) by equation (4-24) squared, we obtain

$$\left(\frac{dr}{dx}\right)^2 = -p + 2r - \frac{r^2}{a}$$

Separating the variables yields

$$dx = \frac{dr}{\sqrt{-p + 2r - r^2/a}} \tag{4-25}$$

For, $e \neq 1$, the indefinite integral is

$$x + c_0 = \sqrt{a}\ \sin^{-1}\frac{(r/a - 1)}{\sqrt{1 - p/a}}$$

where c_0 is the integration constant. But $p = a(1 - e^2)$, so $e = \sqrt{1 - p/a}$ and so we may write

$$x + c_0 = \sqrt{a}\ \sin^{-1}\frac{(r/a - 1)}{e}$$

Finally, solving this equation for r gives

$$\boxed{r = a\left(1 + e \sin\frac{x + c_0}{\sqrt{a}}\right)} \tag{4-26}$$

Substituting Equation (4-26) into the definition of the universal variable, Equation (4-24), we obtain

$$\sqrt{\mu}\,dt = a\left(1 + e\sin\frac{x + c_0}{\sqrt{a}}\right)dx$$

$$\sqrt{\mu}\,t = ax - ae\sqrt{a}\left(\cos\frac{x+c_0}{\sqrt{a}} - \cos\frac{c_0}{\sqrt{a}}\right) \tag{4-27}$$

where we assumed $x = 0$ at $t = 0$.

At this point we have developed equations for both r and t in terms of x. The constant of integration, c_0, has not been evaluated yet. Application of these equations will now be made to specific problem types.

4.4 THE PREDICTION PROBLEM

With the Kepler time-of-flight equations you can easily solve for the time of flight, $t - t_0$, if you are given a, e, ν_0 and ν. The inverse problem of finding ν when you are given a, e, ν_0 and $t - t_0$ is not so simple, as we shall see. Small,[4] in *An Account of the Astronomical Discoveries of Kepler,* relates: "This problem has, ever since the time of Kepler, continued to exercise the ingenuity of the ablest geometers; but no solution of it which is rigorously accurate has been obtained. Nor is there much reason to hope that the difficulty will ever be overcome..." This problem classically involves the solution of Kepler's equation and is often referred to as Kepler's problem.

4.4.1 Development of the Universal Variable Formulation

The prediction problem can be stated as (see Figure 4-9)

Given: $\mathbf{r}_0$, $\mathbf{v}_0$, $t_0 = 0$

Find: **r, v** at time t.

We have assumed $x = 0$ at $t = 0$. From Equation (4-26),

$$e\sin\frac{c_0}{\sqrt{a}} = \frac{r_0}{a} - 1 \tag{4-28}$$

Now differentiating Equation (4-26) with respect to time gives

$$\dot{r} = \frac{ae}{\sqrt{a}}\cos\left[\frac{x + c_0}{\sqrt{a}}\right]\frac{\sqrt{\mu}}{r} \tag{4-29}$$

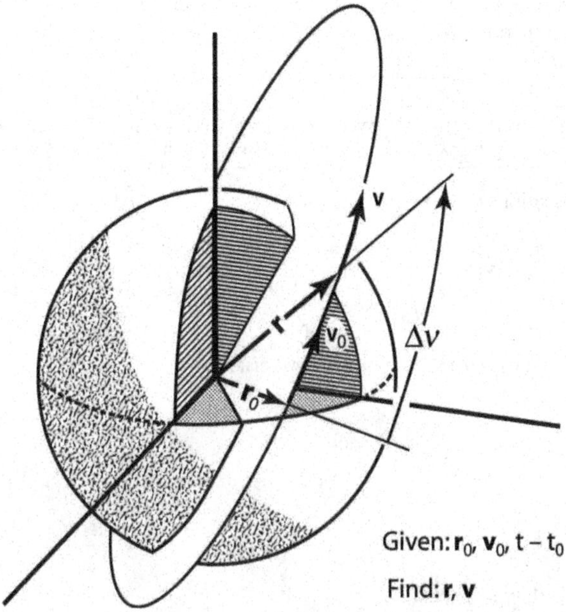

Fig. 4-9 The Kepler Problem.

Applying the initial conditions to Equation (4-29) and using the identity $\mathbf{r} \cdot \dot{\mathbf{r}} = r\dot{r}$, we get

$$e \cos \frac{c_0}{\sqrt{a}} = \frac{\mathbf{r}_0 \cdot \mathbf{v}_0}{\sqrt{\mu a}} \tag{4-30}$$

Using the trigonometric identity for the cosine of a sum we can write Equation (4-27) as

$$\sqrt{\mu}\, t = ax - ae\sqrt{a}\left(\cos\frac{x}{\sqrt{a}}\cos\frac{c_0}{\sqrt{a}} - \sin\frac{x}{\sqrt{a}}\sin\frac{c_0}{\sqrt{a}} - \cos\frac{c_0}{\sqrt{a}}\right)$$

Then substituting Equations (4-28) and (4-30) and rearranging gives

$$\sqrt{\mu}\, t = a\left(x - \sqrt{a}\sin\frac{x}{\sqrt{a}}\right) + \frac{\mathbf{r}_0 \cdot \mathbf{v}_0}{\sqrt{\mu}}\, a\left(1 - \cos\frac{x}{\sqrt{a}}\right)$$
$$+ r_0\sqrt{a}\sin\frac{x}{\sqrt{a}} \tag{4-31}$$

In a similar fashion we can use the trigonometric identity for the sine of a sum to rewrite Equation (4-26) as

$$\boxed{r = a + ae\left(\sin\frac{x}{\sqrt{a}}\cos\frac{c_0}{\sqrt{a}} + \cos\frac{x}{\sqrt{a}}\sin\frac{c_0}{\sqrt{a}}\right)} \tag{4-32}$$

Substituting Equations (4-28) and (4-30) we get:

$$r = a + a\left[\frac{\mathbf{r}_0 \bullet \mathbf{v}_0}{\sqrt{\mu a}}\sin\frac{x}{\sqrt{a}} + \left(\frac{r_0}{a} - 1\right)\cos\frac{x}{\sqrt{a}}\right] \tag{4-33}$$

At this point let us introduce another new variable

$$z = \frac{x^2}{a} \tag{4-34}$$

Then $a = \frac{x^2}{z}$

Equation (4-31) then becomes

$$\sqrt{\mu}\,t = \frac{x^2}{z}\left(x - \frac{x}{\sqrt{z}}\sin\sqrt{z} + \frac{\mathbf{r}_0 \cdot \mathbf{v}_0}{\sqrt{\mu}}\frac{x^2}{z}(1 - \cos\sqrt{z}) + r_0\frac{x}{\sqrt{z}}\sin\sqrt{z}\right)$$

which can be rearranged as

$$\sqrt{\mu}\,t = \left[\frac{\sqrt{z} - \sin\sqrt{z}}{\sqrt{z^3}}\right]x^3 + \frac{\mathbf{r}_0 \cdot \mathbf{v}_0}{\sqrt{\mu}}x^2\frac{1 - \cos\sqrt{z}}{z} + \frac{r_0 x \sin\sqrt{z}}{\sqrt{z}} \tag{4-35}$$

Similarly, Equation (4-33) becomes

$$r = \frac{x^2}{z} + \frac{\mathbf{r}_0 \cdot \mathbf{v}_0}{\sqrt{\mu}}\frac{x}{\sqrt{z}}\sin\sqrt{z} + r_0\cos\sqrt{z} - \frac{x^2}{z}\cos\sqrt{z} \tag{4-36}$$

These equations are indeterminate for $\mathbf{z} = 0$. To remedy this we will introduce two very useful functions that can be expressed as a series:

$$C(z) \equiv \frac{1 - \cos\sqrt{z}}{z} = \frac{1 - \cosh\sqrt{-z}}{z} = \frac{z^0}{2!} - \frac{z^1}{4!} + \frac{z^2}{6!} - \frac{z^3}{8!} + \ldots$$

$$= \sum_{k=0}^{\infty}\frac{(-z)^k}{(2k+2)!} \tag{4-37}$$

$$S(z) = \frac{\sqrt{z} - \sin\sqrt{z}}{\sqrt{z^3}} = \frac{\sinh\sqrt{-z} - \sqrt{-z}}{\sqrt{(-z)^3}}$$

$$= \frac{z^0}{3!} - \frac{z^1}{5!} + \frac{z^2}{7!} - \frac{z^3}{9!} + \ldots = \sum_{k=0}^{\infty} \frac{(-z)^k}{(2k+3)!} \tag{4-38}$$

The properties of these functions will be discussed in a later section. By using these functions Equations (4-35) and (4-36) become

$$\boxed{\sqrt{\mu}\, t = x^3\, S + \frac{\mathbf{r}_0 \bullet \mathbf{v}_0}{\sqrt{\mu}} x^2\, C + r_0\, x\,(1 - zS)} \tag{4-39}$$

$$\boxed{r = \sqrt{\mu}\, \frac{dt}{dx} = x^2\, C + \frac{\mathbf{r}_0 \bullet \mathbf{v}_0}{\sqrt{\mu}} x\,(1 - zS) + r_0\,(1 - zC)} \tag{4-40}$$

4.4.2 Solving for x When Time Is Known

An intermediate step to finding the radius and velocity vectors at a future time is to find x when time is known. From $\mathbf{r}_0$, $\mathbf{v}_0$ and the energy equation you can obtain the semi-major axis, a. But now we have a problem: since Equation (4-39) is transcendental in x, we cannot get it by itself on the left of the equal sign. Therefore, a trial-and-error solution is indicated.

Fortunately, the t vs. x curve is well behaved and a Newton iteration technique may be used successfully to solve for x when time of flight is given. If we let $t_0 = 0$ and choose a trial value for x—call it x_n, then

$$\sqrt{\mu}\, t_n = \frac{\mathbf{r}_0 \cdot \mathbf{v}_0}{\sqrt{\mu}} x_n^2 C + \left(1 - \frac{r_0}{a}\right) x_n^3 S + r_0 x_n \tag{4-41}$$

where t_n is the time of flight corresponding to the given $\mathbf{r}_0$, $\mathbf{v}_0$, a and the trial value of x. Equation (4-34) has been used to eliminate z. In one sense, C and S should have a subscript of n also because they are functions of the guess of x_n.

A better approximation is then obtained from the Newton iteration algorithm

$$x_{n+1} = x_n + \frac{t - t_n}{dt/dx|_{x=x_n}} \tag{4-42}$$

where t is the given time of flight, and where $dt/dx|_{x=x_n}$ is the slope of the t vs. x curve at the trial point, x_n.

An analytical expression for the slope may be obtained directly from the definition of x in Equation (4-24) as

$$\frac{dt}{dx} = \frac{1}{\dot{x}} = \frac{r}{\sqrt{\mu}} \tag{4-43}$$

Note that the slope of the t vs. x curve is directly proportional to r; it will be minimum at periapsis and maximum where r is maximum. Some typical plots of t vs. x are illustrated in Figure 4-10. Substituting for r in Equation (4-43) yields

$$\boxed{\sqrt{\mu}\frac{dt}{dx} = x^2 C + \frac{\mathbf{r}_0 \cdot \mathbf{v}_0}{\sqrt{\mu}} x(1 - zS) + r_0(1 - zC)} \tag{4-44}$$

When the difference between t and t_n becomes negligible, the iteration may be terminated.

With x known, we must then calculate the corresponding **r** and **v**. To do this we will now develop what is called the f and g expressions in terms of x and z.

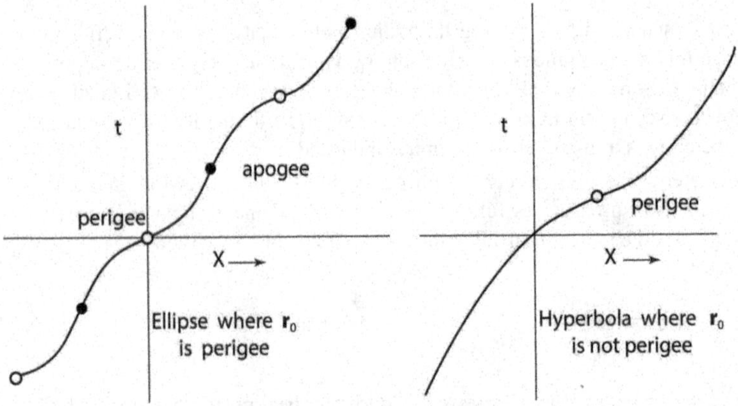

Fig. 4-10 Typical t vs. x Plots.

4.4.3 The f and g Expressions

Knowing x we now wish to calculate **r** and **v** in terms of $\mathbf{r}_0$, $\mathbf{v}_0$ and x. In determining this relationship we will make use of a fundamental theorem concerning coplanar vectors: *If* **A**, **B** *and* **C** *are coplanar vectors, and* **A** *and* **B** *are not collinear, it is possible to express* **C** *as a linear combination of* **A** *and* **B**.

Since Keplerian motion is confined to a plane, the four vectors $\mathbf{r}_0$, $\mathbf{v}_0$, **r** and **v** are all coplanar. Thus

$$\boxed{\mathbf{r} = f\ \mathbf{r}_0 + g\ \mathbf{v}_0} \tag{4-45}$$

Differentiating this expression gives

$$\boxed{\mathbf{v} = \dot{f}\ \mathbf{r}_0 + \dot{g}\ \mathbf{v}_0} \tag{4-46}$$

where $f, g, \dot{f}$ and $\dot{g}$ are time dependent scalar quantities. The main purpose of this section will be to determine expressions for these scalars in terms of the universal variable, x.

First, however, we will derive an interesting relationship among $f, g, \dot{f}$ and $\dot{g}$. Crossing Equation (4-45) into Equation (4-46) gives

$$\mathbf{r} \times \mathbf{v} = (f\ \mathbf{r}_0 + g\ \mathbf{v}_0) \times (\dot{f}\ \mathbf{r}_0 + \dot{g}\ \mathbf{v}_0)$$

$$\overbrace{\mathbf{r} \times \mathbf{v}}^{\mathbf{h}} = f\dot{f}\ \overbrace{\mathbf{r}_0 \times \mathbf{r}_0}^{0} + f\dot{g}\ \overbrace{\mathbf{r}_0 \times \mathbf{v}_0}^{\mathbf{h}} - \dot{f}g\ \overbrace{\mathbf{r}_0 \times \mathbf{v}_0}^{\mathbf{h}} + g\dot{g}\ \overbrace{\mathbf{v}_0 \times \mathbf{v}_0}^{0}$$

Equating the scalar components of $\mathbf{h}$ on both sides of the equation yields

$$\boxed{1 = f\dot{g} - \dot{f}g} \tag{4-47}$$

This equation shows that $f, g, \dot{f}$ and $\dot{g}$ are not independent; if you know any three you can determine the fourth from this useful identity.

We will now develop the f and g expressions in terms of perifocal coordinates. We can isolate the scalar, f, in Equation (4-45) by crossing the equation into $\mathbf{v}_0$:

$$\mathbf{r} \times \mathbf{v}_0 = f(\overbrace{\mathbf{r}_0 \times \mathbf{v}_0}^{h\mathbf{W}}) + g(\overbrace{\mathbf{v}_0 \times \mathbf{v}_0}^{0})$$

Since $\mathbf{r} = x_\omega \mathbf{P} + y_\omega \mathbf{Q}$ and $\mathbf{v}_0 = \dot{x}_{\omega_0} \mathbf{P} + \dot{y}_{\omega_0} \mathbf{Q}$, the left side of the equation becomes

$$\mathbf{r} \times \mathbf{v}_0 = \begin{vmatrix} \mathbf{P} & \mathbf{Q} & \mathbf{W} \\ x_\omega & y_\omega & 0 \\ \dot{x}_{\omega_0} & \dot{y}_{\omega_0} & 0 \end{vmatrix} = (x_\omega\ \dot{y}_{\omega_0} - \dot{x}_{\omega_0}\ y_\omega)\mathbf{W}$$

Equating the scalar components of $\mathbf{W}$ and solving for f,

$$\boxed{f = \frac{x_\omega\ \dot{y}_{\omega_0} - \dot{x}_{\omega_0}\ y_\omega}{h}} \tag{4-48}$$

We can isolate g in a similar manner by crossing $\mathbf{r}_0$ into Equation (4-45):

$$\mathbf{r}_0 \times \mathbf{r} = f(\mathbf{r}_0 \times \mathbf{r}_0) + g(\mathbf{r}_0 \times \mathbf{v}_0) = gh\mathbf{W}$$

$$\mathbf{r}_0 \times \mathbf{r} = \begin{vmatrix} \mathbf{P} & \mathbf{Q} & \mathbf{W} \\ x_{\omega_0} & y_{\omega_0} & 0 \\ x_\omega & y_\omega & 0 \end{vmatrix} = \left(x_{\omega_0} y_\omega - x_\omega y_{\omega_0}\right)\mathbf{W}$$

$$g = \frac{\left(x_{\omega_0} y_\omega - x_\omega y_{\omega_0}\right)}{h} \tag{4-49}$$

To obtain $\dot{f}$ and $\dot{g}$ we only need to differentiate the expressions for f and g (or we could cross $\mathbf{r}_0$ into Equation (4-46) to get $\dot{g}$ and then cross Equation (4-46) into $\mathbf{v}_0$ to get $\dot{f}$):

$$\boxed{\dot{f} = \frac{\dot{x}_\omega \dot{y}_{\omega_0} - \dot{x}_{\omega_0} \dot{y}_\omega}{h}} \tag{4-50}$$

$$\boxed{\dot{g} = \frac{x_{\omega_0} \dot{y}_\omega - \dot{x}_\omega y_{\omega_0}}{h}} \tag{4-51}$$

To get the f and g expressions in terms of x, we need to relate the perifocal coordinates to x. From the standard conic equation we obtain

$$re\cos\nu = a(1 - e^2) - r \tag{4-52}$$

Combining Equations (4-26) and (4-32) gives

$$\boxed{x_\omega = r\cos\nu = -a\left(e + \sin\frac{x + c_0}{\sqrt{a}}\right)} \tag{4-53}$$

Since $y_\omega^2 = r^2 - x_\omega^2$ we obtain

$$\boxed{y_\omega = a\sqrt{1 - e^2}\cos\frac{x + c_0}{\sqrt{a}}} \tag{4-54}$$

Now by differentiating Equations (4-53) and (4-54) and using the definition of the universal variable, Equation (4-24), we have

$$\boxed{\dot{x}_{\omega} = -\frac{\sqrt{\mu a}}{r}\cos\frac{x + c_0}{\sqrt{a}}} \tag{4-55}$$

$$\boxed{\dot{y}_{\omega} = -\frac{h}{r}\sin\frac{x + c_0}{\sqrt{a}}} \tag{4-56}$$

Substituting Equations (4-53) through (4-56) into Equation (4-48) gives

$$f = \frac{1}{h}\left[a\left(e + \sin\frac{x + c_0}{\sqrt{a}}\right)\frac{h}{r_0}\sin\frac{c_0}{\sqrt{a}} + \left(\frac{\sqrt{\mu a}}{r_0}\cos\frac{c_0}{\sqrt{a}}\right)a\sqrt{1 - e^2}\cos\frac{x + c_0}{\sqrt{a}}\right]$$

Recall that x = 0 at t = 0. Using the trigonometric identities for sine and cosine of a sum gives

$$f = \frac{a}{r_0}\left(e\sin\frac{c_0}{\sqrt{a}} + \cos\frac{x}{\sqrt{a}}\right) \tag{4-57}$$

By using the definitions of **z** and **C(z)** and Equation (4-28), Equation (4-57) becomes

$$f = 1 - \frac{a}{r_0}\left(1 - \cos\frac{x}{\sqrt{a}}\right) = 1 - \frac{x^2}{r_0}C \tag{4-58}$$

We can derive the expression for g in a similar manner as follows

$$g = \frac{1}{h}\left[-a\left(e + \sin\frac{c_0}{\sqrt{a}}\right)a\sqrt{1 - e^2}\cos\frac{x + c_0}{\sqrt{a}}\right.$$

$$\left. + a\sqrt{1 - e^2}\left(\cos\frac{c_0}{\sqrt{a}}\right)a\left(e + \sin\frac{x + c_0}{\sqrt{a}}\right)\right]$$

$$= \frac{a^2}{\sqrt{\mu a}}\left[e\left(\cos\frac{c_0}{\sqrt{a}} - \cos\frac{x}{\sqrt{a}}\cos\frac{c}{\sqrt{a}}\right.\right.$$

$$\left.\left. + \sin\frac{x}{\sqrt{a}}\sin\frac{c_0}{\sqrt{a}}\right) + \sin\frac{x}{\sqrt{a}}\right]$$

Using Equations (4-29) and (4-30) gives

$$g = \frac{a^2}{\sqrt{\mu a}}\left[\frac{\mathbf{r}_0 \bullet \mathbf{v}_0}{\sqrt{\mu a}}\left(1 - \cos\frac{x}{\sqrt{a}}\right) + \frac{r_0}{a}\sin\frac{x}{\sqrt{a}}\right]$$

Then

$$\sqrt{\mu}\,g = x^2\,\frac{\mathbf{r}_0 \bullet \mathbf{v}_0}{\sqrt{\mu}}\,C + r_0\,x\,(1 - zS) \tag{4-59}$$

Comparing this Equation (4-41) we see that

$$\sqrt{\mu}\,g = \sqrt{\mu}\,t - x^3 S \tag{4-60}$$

and

$$\boxed{g = t - \frac{x^3}{\sqrt{\mu}}\,S} \tag{4-61}$$

In a similar fashion we can show that

$$\boxed{\dot{g} = 1 - \frac{a}{r} + \frac{a}{r}\cos\frac{x}{\sqrt{a}} = 1 - \frac{x^2}{r}\,C} \tag{4-62}$$

and

$$\boxed{\dot{f} = -\frac{\sqrt{\mu a}}{r_0 r}\,\sin\frac{x}{\sqrt{a}} = \frac{\sqrt{\mu}}{r_0 r}\,x\,(zS - 1)} \tag{4-63}$$

In computation, note that Equation (4-47) can be used as a check on the accuracy of the f and g expressions. Also, in any of the equations where z appears, its definition, x^2/a, can be used. Note also that, if t_0 were not zero, the expression $(t - t_0)$ would replace t.

4.4.4 Algorithm for Solution of the Kepler Problem

1. From $\mathbf{r}_0$ and $\mathbf{v}_0$ determine r_0 and a.
2. Given $t - t_0$ (where usually t_0 is assumed to be zero), solve the universal time-of-flight equation for x using a Newton iteration scheme.
3. Evaluate f and g from Equations (4-58) and (4-61); then compute $\mathbf{r}$ and r from Equation (4-45).
4. Evaluate $\dot{f}$ and $\dot{g}$ from Equations (4-62) and (4-63) then compute $\mathbf{v}$ from Equation (4-46).

The advantages of this method over other methods are that only one set of equations is needed for all conic orbits and accuracy and convergence for nearly parabolic orbits is better.

4.5 IMPLEMENTING THE UNIVERSAL VARIABLE FORMULATION

In this section several aspects of the universal variable formulation will be presented that will increase your understanding and facilitate its computer implementation.

4.5.1 The Physical Significance of x and z

Up to now we have developed universally valid expressions for r and $t - t_0$ in terms of the auxiliary variables x and z; but we have not specified what x and z represent. Obviously, if we are going to use Equation (4-39) above we must know how x and z are related to the physical parameters of the orbit (Figure 4-11).

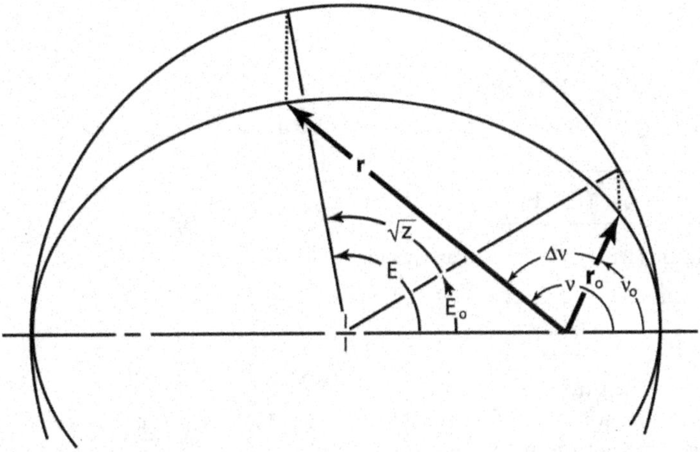

Fig. 4-11 Change in True Anomaly and Eccentric Anomaly During Time $t - t_0$.

Let's compare the expression for r in terms of x with the corresponding expression for r in terms of the eccentric anomaly:

$$r = a\left(1 + e \sin \frac{x + c_0}{\sqrt{a}}\right) \tag{4-64}$$

$$r = a\left(1 - e \cos E\right) \tag{4-65}$$

We can conclude that

$$\sin \frac{x + c_0}{\sqrt{a}} = -\cos E$$

But

$$\sin \frac{x + c_0}{\sqrt{a}} = \cos\left[\frac{\pi}{2} - \frac{x + c_0}{\sqrt{a}}\right] = -\cos\left[\frac{\pi}{2} + \frac{x + c_0}{\sqrt{a}}\right]$$

so

$$E = \frac{\pi}{2} + \frac{x + c_0}{\sqrt{a}} \tag{4-66}$$

If we compare Equations (4-26) and (4-65) at a time t_0 when $x = 0$, $r = r_0$ and $E = E_0$, we get

$$E_0 = \frac{\pi}{2} + \frac{c_0}{\sqrt{a}} \tag{4-67}$$

Subtracting Equation (4-67) from Equation (4-66) yields

$$\boxed{x = \sqrt{a}\,(E - E_0)} \tag{4-68}$$

Using the identity $F = iE$, we can conclude that

$$\boxed{x = \sqrt{-a}\,(F - F_0)} \tag{4-69}$$

whenever a is negative.

To determine what x represents on the parabolic orbit let's look at the general expression for r in terms of **x**:

$$r = x^2 C + \frac{\mathbf{r}_0 \cdot \mathbf{v}_0}{\sqrt{\mu}}\, x\,(1 - zS) + r_0\,(1 - zC) \tag{4-70}$$

For the parabolic orbit, $z = 0$ and $C = 1/2$, since $a = \infty$, so

$$r = \frac{1}{2} x^2 + \frac{\mathbf{r}_0 \cdot \mathbf{v}_0}{\sqrt{\mu}}\, x + r_0$$

We will show later in Equation (4-90) that $\mathbf{r}_0 \cdot \mathbf{v}_0 = \sqrt{\mu} D_0$ and in Equation (4-89) that $r = \frac{1}{2}(p + D^2)$ and $r_0 = \frac{1}{2}(p + D_0^2)$; therefore,

$$\frac{1}{2}(p + D^2) = \frac{1}{2} x^2 + D_0 x + \frac{1}{2}(p + D_0^2)$$

If we solve this quadratic for **x**, we get

$$x = D - D_0 \tag{4-71}$$

which is valid for parabolic orbits.

Obviously, **x** is related to the change in the eccentric anomaly that occurs between $\mathbf{r_0}$ and **r**. Since $z = x^2/a$,

$$\boxed{z = (E - E_0)^2} \tag{4-72}$$

when **z** is positive. If **z** is negative, it can only mean that E and E_0 are imaginary, so

$$\boxed{-z = (F - F_0)^2} \tag{4-73}$$

When **z** is zero, either a is infinite or the change in the eccentric anomaly is zero.

4.5.2 Some Notes on the Computer Solution of the Kepler Problem

A word of caution is in order concerning the use of the universal time-of-flight equation for solving the Kepler problem. In computing the semi-major axis, **a**, from $\mathbf{r_0}$ and $\mathbf{v_0}$ and the energy equation we get

$$a = \frac{-\mu}{v_0^2 - 2\mu/r_0}$$

If the orbit is parabolic, the denominator of this expression is zero and an error finish would result if the computation is performed on a computer. Therefore, the reciprocal of **a** should be computed and stored instead:

$$\alpha = \frac{2\mu/r_0 - v_0^2}{\mu} \tag{4-74}$$

All equations should then be modified by replacing 1/**a**, wherever it occurs, with α.

The number of iterations required to compute **x** to any desired degree of accuracy depends mainly on the initial trial value of **x**; if the initial estimate of **x** is close to the correct value, convergence will be extremely rapid.

In the case of elliptical orbits, where the given time of flight exceeds one orbital period, **t** can obviously be reduced to less than the period and the same **r** and **v** will result. Furthermore, since $x = \sqrt{a}\Delta E$, $x = 2\pi\sqrt{a}$ after one orbital period and we can make the approximation

$$\frac{x}{2\pi\sqrt{a}} \approx \frac{t - t_0}{\mathbb{P}}$$

where $\mathbb{P}$ is the period. Solving for x and letting $\mathbb{P} = 2\pi\sqrt{a^3/\mu}$, we get

$$\boxed{x \approx \frac{\sqrt{\mu}\,(t - t_0)}{a}} \tag{4-75}$$

for elliptical orbits. Use this for a first guess.

If the orbit is hyperbolic and the change in eccentric anomaly, ΔF, is large then z will be a large negative number. When z is negative the C function may be evaluated from

$$C = \frac{1 - \cosh\sqrt{-z}}{z}$$

But $\cosh\sqrt{-z} = (e^{\sqrt{-z}} + e^{-\sqrt{-z}})/2$ and if $\sqrt{-z}$ is a large positive number, then $e^{\sqrt{-z}}$ will be large compared to 1 or $e^{-\sqrt{-z}}$, so

$$C \approx \frac{-e^{\sqrt{-z}}}{2z} = \frac{-ae^{\sqrt{-z}}}{2x^2}$$

Similarly, we can say that

$$S = \frac{\sinh\sqrt{-z} - \sqrt{-z}}{\sqrt{(-z)^3}}$$

But $\sinh\sqrt{-z} = (e^{\sqrt{-z}} - e^{-\sqrt{-z}})/2$ and

$$\sqrt{(-z)^3} = \pm x^3/(-a\sqrt{-a}), \text{ so } S \approx \frac{e^{\sqrt{-z}}}{2\sqrt{(-z)^3}} = \frac{-a\sqrt{-a}\,e^{\sqrt{-z}}}{\pm 2x^3}$$

The ± sign can be resolved easily since we know (Section 4.5.3) that the S function is positive for all values of z. Therefore, if x is positive we should take the + sign and if x is negative we should take the – sign in the above expression. Anytime $t - t_0$ is positive, x will be positive and vice versa, so

$$S \approx \frac{-a\sqrt{-a}\,e^{\sqrt{-z}}}{\text{sign}\,(t - t_0)\,2x^3}$$

Substituting these approximate values for C and S into the universal time-of-flight equation and neglecting the last term, we get

$$t - t_0 \approx -a\frac{\mathbf{r}_0 \bullet \mathbf{v}_0}{2\mu}e^{\sqrt{-z}}$$

$$-\text{sign}(t - t_0)\frac{a\sqrt{-a}}{2\sqrt{\mu}}\left(1 - \frac{r_0}{a}\right)e^{\sqrt{-z}}$$

Solving for $e^{\sqrt{-z}}$ then gives

$$e^{\sqrt{-z}} \approx \frac{-2\mu(t - t_0)}{a\left[(\mathbf{r}_0 \cdot \mathbf{v}_0) + \text{sign}(t - t_0)\sqrt{-\mu a}\left(1 - \frac{r_0}{a}\right)\right]}$$

so $\sqrt{-z} = \pm\frac{x}{\sqrt{-a}}$

$$\approx \ln\left[\frac{-2\mu(t - t_0)}{a\left[(\mathbf{r}_0 \cdot \mathbf{v}_0) + \text{sign}(t - t_0)\sqrt{-\mu a}\left(1 - \frac{r_0}{a}\right)\right]}\right]$$

We can resolve the ± sign in the same way as before, recognizing that **x** will be positive when $(t - t_0)$ is positive. Thus, *for hyperbolic orbits,*

$$\boxed{x \approx \text{sign}(t - t_0)\sqrt{-a}\ \ln\left[\frac{-2\mu(t - t_0)}{a\left[(\mathbf{r}_0 \bullet \mathbf{v}_0) + \text{sign}(t - t_0)\sqrt{-\mu a}\left(1 - \frac{r_0}{a}\right)\right]}\right]}$$

(4-76)

The use of these approximations, where appropriate, for selecting the first trial value of **x** will greatly speed convergence.

4.5.3 Properties of C(z) and S(z)

The functions $C(z)$ and $S(z)$ are defined as

$$C = \frac{1 - \cos\sqrt{z}}{z} \tag{4-77}$$

$$S = \frac{\sqrt{z} - \sin\sqrt{z}}{\sqrt{z^3}} \tag{4-78}$$

We can write Equation (4-77) as

$$C = \frac{1 - \cos i\sqrt{-z}}{z}$$

where $i = \sqrt{-1}$ but $\cos i\,\theta = \cosh\theta$, so an *equivalent expression for* C is

$$\boxed{C = \frac{1 - \cosh\sqrt{-z}}{z}} \tag{4-79}$$

Similarly, we can write Equation (4-78) as

$$S = \frac{i\sqrt{-z} - \sin i\sqrt{-z}}{-i\sqrt{(-z)^3}} = \frac{-i\sin i\sqrt{-z} - \sqrt{-z}}{\sqrt{(-z)^3}}$$

But $-i \sin i\,\theta = \sinh\theta$, so an *equivalent expression for* S is

$$S = \frac{\sinh\sqrt{-z} - \sqrt{-z}}{\sqrt{(-z)^3}} \tag{4-80}$$

To evaluate C and S when z is positive, use Equations (4-37) and (4-38); if z is negative, use Equations (4-39) and (4-40). If z is near zero, the power series expansions of the functions may be used to evaluate C and S. The series expansions are easily derived from the power series for $\sin\theta$ and $\cos\theta$.

$$\cos\theta = 1 - \frac{\theta^2}{2!} + \frac{\theta^4}{4!} - \frac{\theta^6}{6!} + \ldots$$

$$\sin\theta = \theta - \frac{\theta^3}{3!} + \frac{\theta^5}{5!} - \frac{\theta^7}{7!} + \ldots$$

Substituting these series into the definitions of C and S, we get

$$C = \frac{1}{z}\left[1 - \left(1 - \frac{z}{2!} + \frac{z^2}{4!} - \frac{z^3}{6!} + \ldots\right)\right]$$

$$\boxed{C = \frac{1}{2!} - \frac{z}{4!} + \frac{z^2}{6!} - \ldots} \tag{4-81}$$

$$S = \frac{1}{\sqrt{z}^3}\left[\sqrt{z} - \left(\sqrt{z} - \frac{\sqrt{z}^3}{3!} + \frac{\sqrt{z}^5}{5!} - \frac{\sqrt{z}^7}{7!} + \ldots\right)\right]$$

$$\boxed{S = \frac{1}{3!} - \frac{z}{5!} + \frac{z^2}{7!} - \ldots} \tag{4-82}$$

Both the C and S functions approach infinity as z approaches minus infinity. The S function is 1/6 when z = 0 and decreases asymptotically to zero as z approaches plus infinity. The C function is 1/2 when z = 0 and decreases to zero at $z = (2\pi)^2, (4\pi)^2, (6\pi)^2$ etc. Figure 4-12 shows the shape of the C and S functions.

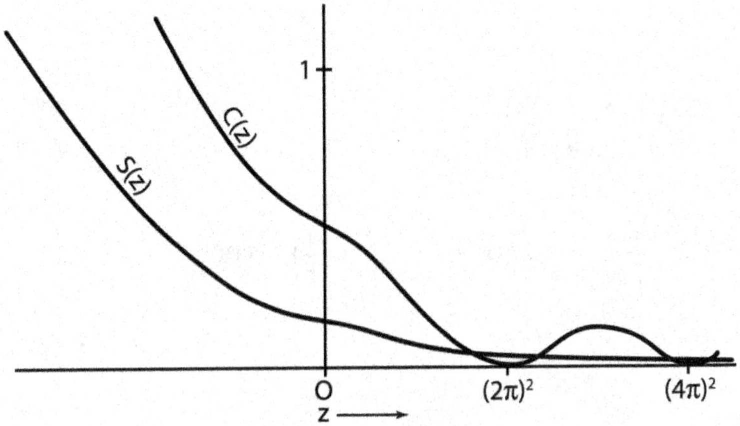

Fig. 4-12 Plot of S(z) and C(z) vs. z.

Since the C and S functions are defined by means of the cos and sin functions, it is not surprising that the derivatives, dC/dz and dS/dz, can be expressed in terms of the functions themselves. Differentiating the definition of C, we get

$$\frac{dC}{dz} = \frac{1}{2z}\left[\frac{\sin\sqrt{z}}{\sqrt{z}} - \frac{2\left(1 - \cos\sqrt{z}\right)}{z}\right]$$

$$\boxed{\frac{dC}{dz} = \frac{1}{2z}(1 - zS - 2C)} \tag{4-83}$$

Differentiating the definition of S, we get

$$\frac{dS}{dz} = \frac{1}{2z}\left[\frac{1-\cos\sqrt{z}}{z} - \frac{3\left(\sqrt{z}-\sin\sqrt{z}\right)}{\sqrt{z^3}}\right]$$

$$\boxed{\frac{dS}{dz} = \frac{1}{2z}(C - 3S)} \tag{4-84}$$

Example Problem.

Given $\mathbf{r}_0 = \mathbf{I}$ DU

$\mathbf{v}_0 = 1.1\mathbf{K}$ DU/TU

$t = 2$ TU

Find x, **r** and **v** at t = 2 TU,

$\mathbf{r}_0 = \mathbf{I},\ \mathbf{v}_0 = 1.1\mathbf{K},\ t = 2\ \mathrm{TU}_\oplus$

$\mathbf{r}_0 \cdot \mathbf{v}_0 = 0,\ v_0 > \sqrt{\mu/r_0}$

Therefore, $\mathbf{r}_0 = \mathbf{r}_p$ and

$$E = \frac{1.1^2}{2} - \frac{1}{1} = -0.395\ \mathrm{DU^2/TU^2},\ a = \frac{-1}{2E}1.266\ \mathrm{DU}$$

Since $x = \sqrt{a}\Delta E$ and $\Delta E = 2\pi$ for one complete orbit, we can make a first approximation for x:

$$\frac{x}{2\pi\sqrt{a}} \cong \frac{t-t_0}{TP} \text{ or } x \cong \frac{\sqrt{\mu}(t-t_0)}{a}$$

Therefore, $x_1 = \dfrac{2}{1.266} = 1.58$

$$z_1 = \frac{(1.58)^2}{a} = 1.973$$

Using Equations (4-41) and (4-44) to find t_1 and dt/dx_1 gives

$$\sqrt{\mu}t_1 = \frac{\cancelto{0}{\mathbf{r}_0\cdot\mathbf{v}_0}}{\sqrt{\mu}}x_1^2C_1 + \left(1-\frac{r_0}{a}\right)x_1^3S_1 + r_0x_1 = 1.705$$

$$\sqrt{\mu}\frac{dt}{dx_1} = x_1^2C_1 + \frac{\cancelto{0}{\mathbf{r}_0\cdot\mathbf{v}_0}}{\sqrt{\mu}}x_1(1-z_1S_1) + r_0(1-z_1C_1) = 1.222$$

Inserting these figures in the Newton iteration then yields

$$x_2 = x_1 + \frac{2 - 1.705}{1.222} = 1.58 + \frac{0.295}{1.222} = 1.58 + 0.241 = 1.821$$

Repeating the process, solving for t_2, dt/dx_2, x_3, etc., we can construct the following table and see how the iteration process drives our successive values for t_n toward $t = 2\text{TU}$:

x_n	t_n	dt/dx_n	x_{n+1}
1.58	1.705	1.222	1.821
1.821	2.007	1.279	1.816
1.816	2.000	1.277	1.816

After three iterations, we have found the value of x for a time of flight of 2 TU accurate to three decimal places. Using a computer, this accuracy can be improved to a desired precision.

Then from the definitions off, $\dot{f}$, g, $\dot{g}$, we have $f = -0.321$, $g = 1.124$, $\dot{f} = -0.8801$ and $\dot{g} = -0.035$. Thus

$$\mathbf{r} = f\mathbf{r}_0 + g\mathbf{v}_0 = \underline{\underline{-0.321\mathbf{I} + 1.236\mathbf{K}}}$$

$$\mathbf{v} = \dot{f}\mathbf{r}_0 + \dot{g}\mathbf{v}_0 = \underline{\underline{-0.8801\mathbf{I} - 0.039\mathbf{K}}}$$

4.6 CLASSICAL FORMULATIONS OF THE KEPLER PROBLEM

In the interest of relating to the historical development of the solution of the Kepler problem we will briefly summarize the solution using the various eccentric anomalies. But first some useful identities will be presented.

4.6.1 Some Useful Identities Involving D, E and F

We have developed time-of-flight equations for the parabola, ellipse and hyperbola that involve the auxiliary variables D, E, and F. You have already seen how these eccentric anomalies relate to the true anomaly, ν. Now let's look at the relationship between these auxiliary variables and some other physical parameters of the orbit.

Taking each of the eccentric anomalies in turn, we will derive expressions for x_ω, y_ω and r in terms of D, E or F. Then we will relate the eccentric anomalies to the dot product $\mathbf{r} \cdot \mathbf{v}$. Finally, we will examine the very interesting and fundamental relationship between E and F.

In order to simplify Barker's Equation (4-18), we introduced the parabolic eccentric anomaly, D, as

$$D = \sqrt{p}\tan\frac{\nu}{2} \tag{4-85}$$

From Figure 4-13 we can see that

$$x_\omega = r\cos\nu$$

But, for the parabola, since e = 1,

$$r = \frac{p}{1+\cos\nu}$$

so

$$x_\omega = \frac{p\cos\nu}{1+\cos\nu} \tag{4-86}$$

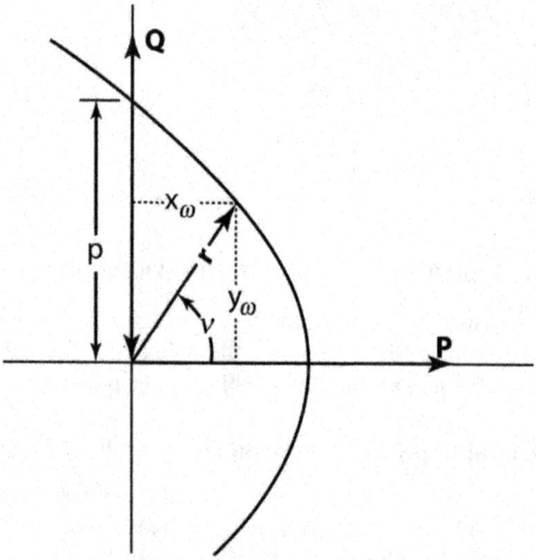

Fig. 4-13 Perifocal Components of r.

Now substituting $\cos\nu = \cos^2\frac{\nu}{2} - \sin^2\frac{\nu}{2}$ in the numerator of Equation (4-86) and substitute $\cos\nu = 2\cos^2\frac{\nu}{2} - 1$ in the denominator gives

$$x_\omega = \frac{p}{2}\left(1 - \tan^2\frac{\nu}{2}\right)$$

If we substitute from Equation (4-85), the expression for x_ω reduces to

$$\boxed{x_\omega = \frac{1}{2}\left(p - D^2\right)} \qquad (4\text{-}87)$$

The expression for y_ω is much simpler:

$$y_\omega = r\ \sin\nu = \frac{p\sin\nu}{1+\cos\nu} = p\tan\frac{\nu}{2}$$

$$\boxed{y_\omega = \sqrt{p}\ D} \qquad (4\text{-}88)$$

Since x_ω and y_ω are the rectangular components of the vector $\mathbf{r}$ in the perifocal coordinate system,

$$r^2 = x_\omega^2 + y_\omega^2$$

$$\boxed{r = \frac{1}{2}\left(p + D^2\right)} \qquad (4\text{-}89)$$

Now let's find an expression for the dot product $\mathbf{r}\cdot\mathbf{v}$. Using $\mathbf{r}\cdot\dot{\mathbf{r}} = r\dot{r}$ and Equation (2-18) and setting $e = 1$ in this equation and substituting for r from the polar equation of a parabola yields

$$\boxed{\mathbf{r}\bullet\mathbf{v} = \frac{p}{1+\cos\nu}\sqrt{\frac{\mu}{p}}\ \sin\nu = \sqrt{\mu p}\ \tan\frac{\nu}{2}}$$

If we now substitute from Equation (4-85), we get

$$\boxed{D = \frac{\mathbf{r}\bullet\mathbf{v}}{\sqrt{\mu}}} \qquad (4\text{-}90)$$

which is useful for evaluating D when $\mathbf{r}$ and $\mathbf{v}$ are known.

For the eccentric anomaly, in Figure 4-14 we have drawn an ellipse with its auxiliary circle in the perifocal coordinate system. The origin of the perifocal system is at the focus of the ellipse and the distance between the focus and the center (0) is just $c = ae$ as shown in Figure 4-14.

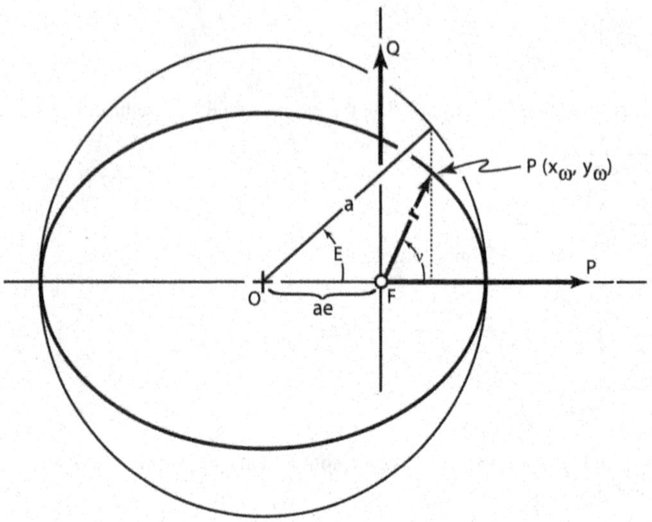

Fig. 4-14 Perifocal Components of r.

From Figure 4-14, we see that

$$x_\omega = a \cos E - ae$$

or

$$\boxed{x_\omega = a(\cos E - e)} \quad (4\text{-}91)$$

From geometry and the relationship between the y_ω-ordinates of the ellipse and circle, we get

$$y_\omega = \frac{b}{a}(a \sin E)$$

But, since $b^2 = a^2 - c^2$ for the ellipse and $c = ae$,

$$y_\omega = a\sqrt{1 - e^2} \sin E \quad (4\text{-}92)$$

Just as we did for the parabola, we can say that

$$r^2 = x_\omega^2 + y_\omega^2$$

Substituting from Equations (4-91) and (4-92) and simplifying, we obtain

$$r = a(1 - e \cos E) \quad (4\text{-}93)$$

If we solve this equation for e cos E we get another useful expression:

$$\boxed{e \cos E = 1 - \frac{r}{a}} \tag{4-94}$$

We will find it convenient to have an expression for e sin E also. To get it we need to differentiate Equation (4-93), which gives us

$$\dot{r} = (ae \sin E)\dot{E} \tag{4-95}$$

To find ($\dot{E}$) we differentiate the Kepler time-of-flight equation

$$t - T = \sqrt{\frac{a^3}{\mu}}\,(E - e \sin E) \tag{4-96}$$

to get

$$1 = \sqrt{\frac{a^3}{\mu}}\,(1 - e \cos E)\,\dot{E}$$

Solving this equation for $\dot{E}$ and substituting for $e \cos E$ from Equation (4-94), we get

$$\dot{E} = \frac{1}{r}\sqrt{\frac{\mu}{a}} \tag{4-97}$$

and so

$$\dot{r} = \sqrt{\mu a}\,\frac{e \sin E}{r}$$

Finally, solving this equation for e sin E and again noting that $r\dot{r} = \mathbf{r} \cdot \mathbf{v}$, we obtain

$$e \sin E = \frac{\mathbf{r} \cdot \mathbf{v}}{\sqrt{\mu a}} \tag{4-98}$$

which is particularly useful since the term e sin E appears in the Kepler time-of-flight equation.

For the hyperbolic eccentric anomaly we will exploit the relationship between E and F to arrive at a set of identities involving F that are analogous to the ones involving E.

The relationship between the eccentric anomalies and the true anomaly, ν, is given by

$$\cos E = \frac{e + \cos \nu}{1 + e\cos \nu} \tag{4-99}$$

$$\cosh F = \frac{e + \cos \nu}{1 + e\cos \nu} \tag{4-100}$$

from which we may conclude that

$$\cosh F = \cos E$$

Using the identity $\cosh \theta = \cos i\theta$ we see that apparently

$$\boxed{E = \pm iF} \tag{4-101}$$

In other words, when E is a real number, F is imaginary; when F is a real number, E is imaginary. The $\pm$ sign is a result of defining E in the range of 0 to 2π while F is defined from minus infinity to plus infinity. The proper sign can always be determined from physical reasoning.

From Equation (4-91) we can write

$$\boxed{\begin{aligned} x_\omega &= a(\cos E - e) \\ &= a(\cos iF - e) \end{aligned}}$$

But $\cos iF = \cosh F$, so

$$\boxed{x_\omega = a(\cosh F - e)} \tag{4-102}$$

Similarly,

$$\begin{aligned} y_\omega &= a\sqrt{1-e^2}\,\sin E \\ &= ai\sqrt{e^2-1}\,\sin iF \end{aligned}$$

But $i \sin iF = -\sinh F$, so

$$y_\omega = -a\sqrt{e^2-1}\,\sinh F \tag{4-103}$$

The following identities are obtained in analogous fashion:

$$\boxed{r = a(1 - e\cosh F)} \tag{4-104}$$

$$\boxed{e\cosh F = 1 - \frac{r}{a}} \tag{4-105}$$

$$\boxed{e\sinh F = \frac{\mathbf{r} \cdot \mathbf{v}}{\sqrt{-\mu a}}} \tag{4-106}$$

This last identity is particularly useful since the expression e sinh F appears in the hyperbolic time-of-flight equation.

4.6.2 The f and g Expressions in Terms of $\Delta\nu$

In Figure (4-15) we have drawn an orbit in the perifocal system. Although an ellipse is shown we need to make no assumption concerning the type of conic.

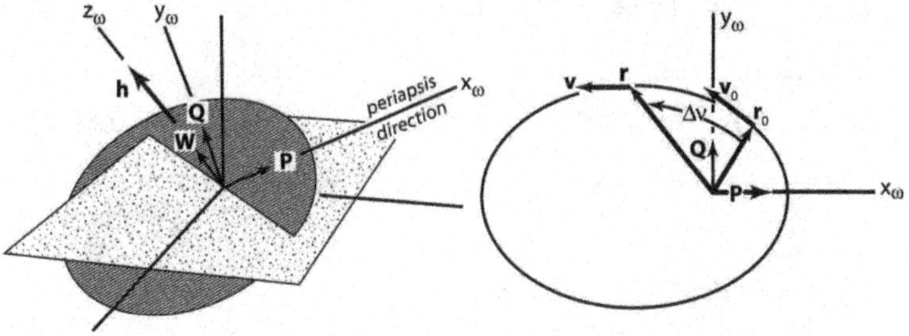

Fig. 4-15 Perifocal Components of Position and Velocity.

The rectangular components of a general position vector, **r**, may be written as

$$x_\omega = r\cos\nu \tag{4-107}$$

$$y_\omega = r\sin\nu \tag{4-108}$$

From Equation (2-20), the rectangular components of the velocity vector, **v**, are

$$\dot{x}_\omega = -\sqrt{\frac{\mu}{p}}\sin\nu \tag{4-109}$$

$$\dot{y}_\omega = \sqrt{\frac{\mu}{p}}(e + \cos\nu) \tag{4-110}$$

If we substitute these expressions into Equation (4-48), and note that $h = \sqrt{\mu p}$, we get

$$f = \frac{(r\cos\nu)\sqrt{\frac{\mu}{p}}(e + \cos\nu_0) + \sqrt{\frac{\mu}{p}}\sin\nu_0\,(r\sin\nu)}{\sqrt{\mu p}}$$

But $\cos\nu\cos\nu_0 + \sin\nu\sin\nu_0 = \cos\Delta\nu$, so

$$\boxed{f = 1 - \frac{r}{p}(1 - \cos\Delta\nu)} \tag{4-111}$$

where $\Delta\nu = \nu - \nu_0$.

Similarly, we obtain

$$\boxed{g = \frac{rr_0\sin\Delta\nu}{\sqrt{\mu p}}} \tag{4-112}$$

$$\boxed{\dot{g} = 1 - \frac{r_0}{p}(1 - \cos\Delta\nu)} \tag{4-113}$$

and

$$\boxed{\dot{f} = \sqrt{\frac{\mu}{p}}\tan\frac{\Delta\nu}{2}\left(\frac{1-\cos\Delta\nu}{p} - \frac{1}{r} - \frac{1}{r_0}\right)} \tag{4-114}$$

4.6.3 The f and g Expressions in Terms of the Eccentric Anomaly

From Equations (4-91) and (4-92), the rectangular components of velocity may be obtained directly by differentiation. Thus

$$\dot{x}_\omega = -a\dot{E}\sin E$$

But, using Equation (4-97) gives

$$\dot{x}_\omega = -\frac{1}{r}\sqrt{\mu a}\sin E \tag{4-115}$$

Differentiating the expression for $\mathbf{y}_\omega$ above yields

$$\dot{y}_\omega = a\sqrt{1-e^2}\,\dot{E}\cos E$$

Therefore,

$$\dot{y}_\omega = \frac{1}{r}\sqrt{\mu a(1-e^2)}\,\cos E \tag{4-116}$$

If we now substitute into Equation (4-48) and recognize that $h = \sqrt{\mu a(1-e^2)}$ we get

$$f = \frac{a\,(\cos E - e)\sqrt{\mu a(1-e^2)}\,\cos E_0}{r_0\sqrt{\mu a(1-e^2)}} + \frac{\sqrt{\mu a}\,\sin E_0\; a\sqrt{(1-e^2)}\,\sin E}{r_0\sqrt{\mu a(1-e^2)}}$$

$$= \frac{a}{r_0}(\cos E\cos E_0 + \sin E\sin E_0 - e\cos E_0)$$

But $\cos E \cos E_0 + \sin E \sin E_0 = \cos \Delta E$ and using Equation (4-94), we have

$$f = 1 - \frac{a}{r_0}(1 - \cos\Delta E) \tag{4-117}$$

Similarly,

$$\boxed{g = (t - t_0) - \sqrt{\frac{a^3}{\mu}}(\Delta E - \sin\Delta E)} \tag{4-118}$$

$$\boxed{\dot{f} = -\frac{\sqrt{\mu a}\sin\Delta E}{r\ r_0}} \tag{4-119}$$

and

$$\boxed{\dot{g} = 1 - \frac{a}{r}(1 - \cos\Delta E)} \tag{4-120}$$

As before, we will use the relationship that $\Delta E = i\Delta F$ and the identities relating the circular and hyperbolic functions to derive the hyperbolic expressions directly from the ones involving ΔE.

From Equation (4-117) we get

$$f = 1 - \frac{a}{r_0}(1 - \cos i\Delta F)$$

But $\cos i\Delta F = \cosh \Delta F$, so

$$\boxed{f = 1 - \frac{a}{r_0}(1 - \cosh \Delta F)} \tag{4-121}$$

In the same manner, we can obtain

$$\boxed{g = (t - t_0) - \sqrt{\frac{(-a)^3}{\mu}}(\sinh\Delta F - \Delta F)} \tag{4-122}$$

$$\boxed{\dot{f} = \frac{-\sqrt{-\mu a}\sinh\Delta F}{r\, r_0}} \tag{4-123}$$

and

$$\boxed{\dot{g} = 1 - \frac{a}{r}(1 - \cosh\Delta F)} \tag{4-124}$$

4.6.4 Kepler Problem Algorithm

In classical terms, the Kepler problem is basically the solution of the equation

$$M = E - e \sin E \tag{4-125}$$

where M is known as $n(t - T)$. M can be obtained from

$$M = \sqrt{\frac{\mu}{a^3}}(t - t_0) - 2k\pi + M_0$$

Even though Equation (4-125) is one equation in one unknown, it is transcendental in E; there is no way of getting E by itself on the left of the equal sign. Kepler himself realized this, of course, and Small[4] tells us: "But, with respect to the direct solution of the problem—from the mean anomaly given to find the true—[Kepler] tells us that he found it impracticable, and that he did not believe there was any geometrical or rigorous method of attaining to it."

The first approximate solution for E was quite naturally made by Kepler himself. The next was by Newton in the *Principia*, from a graphical construction involving the cycloid he was able to find an approximate solution for the eccentric anomaly. A very large number of analytical and graphical solutions have since been discovered because nearly every prominent mathematician since Newton has given some attention to the problem. We will resort again to the Newton iteration method.

It would, of course, be possible to graph Kepler's equation as we have done in Figure 4-16 and then determine what value of E corresponds to a known value of M; but this is not very accurate. Since we can derive an analytical expression for the slope of the M vs. E curve, we can formulate a Newton iteration scheme as follows: first select a trial value for E—call it E_n. Next compute the mean anomaly, M_n, that results from this trial value:

$$M_n = E_n - e \sin E_n \tag{4-126}$$

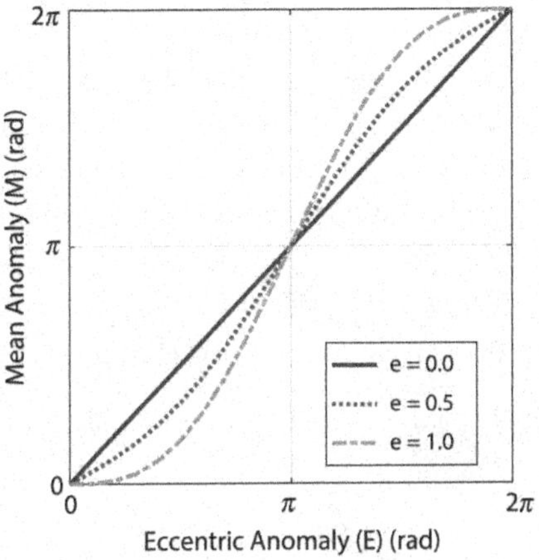

Fig. 4-16 M vs. E Plot.

Now, select a new trial value, E_{n+1}, from

$$E_{n+1} = E_n + \frac{M - M_n}{dM/dE\big|_{E=E_n}} \tag{4-127}$$

where $dM / dE|_{E=E_n}$ is the slope of the M vs. E curve at the trial value, E_n.

The slope expression is obtained by differentiating Kepler's equation:

$$\frac{dM}{dE} = 1 - e \cos E \tag{4-128}$$

Therefore, Equation (4-127) may be written as

$$\boxed{E_{n+1} = E_n + \frac{M - M_n}{1 - e \cos E_n}} \tag{4-129}$$

When the difference $M - M_n$ becomes acceptably small we can quit iterating. Since the slope of the M vs. E curve approaches zero at E = 0 or 2π when e is nearly 1, we can anticipate convergence difficulties for the near-parabolic orbits. Picking a first trial value of $E_1 = \pi$ should guarantee convergence, however, even when e is nearly 1.

Once E is determined by any method, the true anomaly may be found from Equation (4-12). Exactly analogous methods may be used to solve for ν on a hyperbolic orbit when a, e, ν_0 and the time of flight, $t - t_0$, are given.

We may now state the algorithm for solving the Kepler problem that can be used for $\Delta\nu$ or ΔE:

1. From $\mathbf{r}_0$ and $\mathbf{v}_0$ determine r_0, a, e, p and ν_0.
2. Given $t - t_0$, solve the appropriate Kepler time-of-flight equation for E or F using a trial-and-error method such as the Newton iteration. Solve for ν if needed.
3. Solve for r from the polar equation of a conic or Equation (4-14) or the similar expression for the hyperbola.
4. Evaluate the f and g expressions above using r, r_0, p and $\Delta\nu$ (or ΔE or ΔF).
5. Determine $\mathbf{r}$ and $\mathbf{v}$ from Equations (4-45) and (4-46).

The algorithm using ΔE (or ΔF) is shorter than that using $\Delta\nu$ since neither p or ν needed to be calculated.

Exercises

4.1 The equation of a body in Earth orbit is

$$r = \frac{1.5}{1 + 0.5\cos\nu}\,\text{DU}$$

Calculate the time of flight from one end of the minor axis out to apogee.

(Answer: TOF = 5.85 TU)

4.2 In deriving TOF on an ellipse, it was stated that the area beneath the ellipse was to the area beneath the auxiliary circle as b/a, i.e.,

$$\frac{\text{area PSV}}{\text{area QSV}} = \frac{b}{a}$$

as a result of the fact that

$$\frac{\text{Y ellipse}}{\text{Y circle}} = \frac{b}{a}$$

Explain why this must be so.

4.3 Given that $\mathbf{r}_0 = \mathbf{I} + \mathbf{J}$ DU and $\mathbf{v}_0 = 2\mathbf{K}$ DU/TU, find $\mathbf{r}$ and $\mathbf{v}$ for $\Delta\nu = 60°$.

(Partial answer: $\mathbf{v} = -0.348\mathbf{I} - 0.348\mathbf{J} + 1.5\mathbf{K}$ DU/TU)

4.4 If, in a computer solution for position and velocity on an ellipse, given $\mathbf{r}_0$, $\mathbf{v}_0$, and t, one modifies t by subtracting an integer number of periods to make t < TP, how should the area of search for x be limited, to reduce iterations to a minimum?

4.5 A radar ship at 150° W on the equator picks up an object directly overhead. Returns indicate a position and velocity of

$$\mathbf{r}_0 = 1.2\mathbf{I}\ \mathrm{DU}_\oplus$$

$$\mathbf{v}_0 = 0.1\mathbf{I} + \mathbf{J}\ \mathrm{DU}_\oplus/\mathrm{TU}_\oplus$$

Four hours later another ship at 120° W on the equator spots the same object directly overhead. Find the values of f, g, $\dot{f}$ and $\dot{g}$ that could be used to calculate position and velocity at the second sighting.

(Partial Answer: $\dot{f} = -0.625$)

4.6 For the data given in Problem 4.3, find the universal variable x, corresponding to $\Delta\nu$ of 60°.

4.7 The text, in Equations (4-75) and (4-76), gives analytic expressions to use as a first guess for x in an iterative solution for either the elliptical or hyperbolic trajectory. Develop, and give your analytic reasoning behind, an expression for a first guess for x for the parabolic trajectory.

4.8 Why is the slope of the t vs. x curve a minimum at a point corresponding to periapsis? If the slope equals zero at that point, what type of conic section does the curve represent? Draw the family of t vs. x curves for ellipses with the same period but different eccentricities (show e = 0, e = 0.5, and e = 0.99).

4.9 At burnout, a space probe has the following position and velocity:

$$\mathbf{r}_{bo} = 1.1\mathbf{J}\ \mathrm{DU}$$

$$\mathbf{v}_{bo} = \sqrt{2}\mathbf{I}\ \mathrm{DU/TU}$$

How long will it take for the probe to cross the x-axis?

(Answer: TOF = 2.22 TU)

4.10 A satellite is in a polar orbit, with a perigee above the north pole. $r_p = 1.5$ DU, $r_a = 2.5$ DU. Find the time required to go from a point above 30° N latitude to a point above 30° S latitude.

(Answer: 2.73 TU)

4.11 For Problem 4.9 compare the calculations using classical and universal variable methods. Do the same for Problem 4.10. Which method would be most convenient to program on a computer?

4.12 Construct a flow chart for an algorithm that will read in values for $\mathbf{r}_0$, $\mathbf{v}_0$, and Δv and will solve for f, g, $\dot{f}$ and $\dot{g}$.

4.13 Construct a flowchart for an algorithm that will read in values for $\mathbf{r}_0$, $\mathbf{v}_0$ and $t - t_0$ and solve for **r** and **v**.

4.14 Any continuous time-varying function can be expressed as a Taylor series expansion about a starting value, i.e., if x = x(t), then

$$x = x_0 + \frac{(t-t_0)}{1!}\dot{x}_0 + \frac{(t-t_0)^2}{2!}\ddot{x}_0 + \frac{(t-t_0)^3}{3!}\dddot{x}_0 + \ldots$$

where $\dot{x}_0 = \left.\frac{dx}{dt}\right|_{x = x_0}$

By defining $U \triangleq -\frac{\mu}{r^3}$ and using it in our equation of motion

$$\ddot{\mathbf{r}} = -\frac{\mu}{r^3}\mathbf{r} = U\mathbf{r}$$

expand **r** and **v** in a Taylor series and substitute the equation of motion to derive series expressions for f, g, $\dot{f}$ and $\dot{g}$ in terms of $(t - t_0)$ and derivatives of U_0. Find the first three terms of each expression.

4.15 Derive analytically the expression for time of flight on a parabola, Equation (4-19).

4.16 Verify the results expressed in Equations (4-62) and (4-63).

4.17 Derive the expression for g, $\dot{f}$ and $\dot{g}$ in terms of the eccentric anomaly, ΔE. See Equations (4-118) through (4-120).

4.18 *A lunar probe is given just escape speed at a burnout altitude of $0.2 DU_\oplus$ and a flight-path angle of 45°. How long will it take to get to the vicinity of the Moon (r_2 = 60 DU) disregarding the Moon's gravity?

(Answer: TOF = 219.6 TU)

List of References

1. Einstein, Albert. Introduction to *Johannes Kepler: Life and Letters.* New York, NY, Philosophical Library, 1951.

2. Koestler, Arthur. *The Sleepwalkers*. New York, NY, Macmillan, 1959.

3. Battin, Richard H. *Astronautical Guidance*. New York, NY, McGraw-Hill Book Company, 1964.

4. Small Robert. *An Account of the Astronomical Discoveries of Kepler.* A reprinting of the 1804 text with a foreword by William D. Stahlman. Madison, Wisconsin, University of Wisconsin Press, 1963.

5. Moulton, Forest Ray. *An Introduction to Celestial Mechanics*. New York, NY, The MacMillan Company, 1914.

6. Sundman, K. F. "Memoire sur le probleme des trois corps," *Acta Mathmatia*. Vol. 36, pp 105–179, 1912.

7. Goodyear, W. H. "Completely General Closed Form Solution for Coordinates and Partial Derivatives of the Two-body Problem," *Astron. J.* Vol. 70, pp 189–192, 1965.

8. Lemmon, W. W. and J. E. Brooks. *A Universal Formulation for Conic Trajectories—Basic Variables and Relationships.* Report 3400-6019-TU000 TRW/Systems, Redondo Beach, California, 1965.

9. Herrick, S. H. "Universal Variables," *Astron, J.* Vol. 70, pp 309–315, 1965.

10. Stumpff, K. "Neue Formeln and Hilfstafeln zur Ephemeridenrechnung," *Astron. Nach.* Vol. 275, pp 108–128, 1947.

11. Sperling, H. "Computation of Keplerian Conic Sections," *ARS J.* Vol. 31, pp 660–661, 1961.

12. Bate, Roger R. Department of Astronautics and Computer Science, United States Air Force Academy. Unpublished notes.

Chapter 5

Orbit Determination from Two Positions and Time

Probably all mathematicians today regret that Gauss was deflected from his march through darkness by "a couple of clods of dirt which we call planets"—his own words—which shone out unexpectedly in the night sky and led him astray. Lesser mathematicians than Gauss—Laplace for instance—might have done all that Gauss did in computing the orbits of Ceres and Pallas, even if the problem of orbit determination was of a sort which Newton said belonged to the most difficult in mathematical astronomy. But the brilliant success of Gauss in these matters brought him instant recognition as the first mathematician in Europe and thereby won him a comfortable position where he could work in comparative peace; so perhaps those wretched lumps of dirt were after all his lucky stars.

-Eric Temple Bell[1]

5.1 HISTORICAL BACKGROUND

The most brilliant chapter in the history of orbit determination was written by Carl Fredrich Gauss, a 24 year-old German mathematician, in the first year of the 19th century. Ever since Sir William Herschel had discovered the seventh planet, Uranus, in 1781, astronomers had been looking for further members of the solar system—especially since Bode's law predicted the existence of a planet between the orbits of Mars and Jupiter. A plan was formed dividing the sky into several areas that were to be searched for evidence of a new planet. But, before the search operation could begin, one of the prospective participants, Giuseppe Piazzi of Palermo, on New Year's Day of 1801, observed what he first mistook for a small comet approaching the Sun. The object turned out to be Ceres, the first of the swarm of asteroids or minor planets circling the Sun between Mars and Jupiter.

It is ironic that the discovery of Ceres coincided with the publication by the famous philosopher Hegel of a vitriolic attack on astronomers for wasting their time in search for an eighth planet. If they paid some attention to philosophy,

Hegel asserted, they would see immediately that there can be precisely seven planets, no more, no less. This slight lapse on Hegel's part no doubt has been explained by his disciples even if they cannot explain away the hundreds of minor planets that mock his philosophic ban.

To understand why computing the orbit of Ceres was such a triumph for Gauss, you must appreciate the meager data that were available in the case of sighting a new object in the sky. Without radar or any other means of measuring the distance or velocity of the object, the only information astronomers had to work with was the line-of-sight direction at each sighting. To compound the difficulty in the case of Ceres, Piazzi was only able to observe the asteroid for about one month before it was lost in the glare of the Sun. The challenge of rediscovering the insignificant clod of dirt when it reappeared from behind the Sun seduced the intellect of Gauss and he calculated as he had never calculated before. Ceres was rediscovered on New Year's Day in 1802, exactly one year later, precisely where the ingenious and detailed calculations of the young Gauss had predicted it must be found.[1]

The method that young Gauss had used is just as pertinent today as it was in 1801, but for a different reason. The data that Gauss used to determine the orbit of Ceres consisted of the right ascension and declination at three observation times. His method is much simplified if the original data consists of two position vectors and the time of flight between them. The technique of determining an orbit from two positions and time is of considerable interest to modern astrodynamics since it has direct application in the solution of intercept and rendezvous or ballistic missile targeting problems. Because of its importance, and for convenience in referring to it later, we will formally define the problem of orbit determination from two positions and time and give it a name—"the Gauss problem."

5.2 THE GAUSS PROBLEM—GENERAL METHODS OF SOLUTION

We may define the Gauss problem as follows: Given $\mathbf{r}_1$, $\mathbf{r}_2$, the time of flight from $\mathbf{r}_1$ to $\mathbf{r}_2$, which we will call t, and the direction of motion, find $\mathbf{v}_1$ and $\mathbf{v}_2$.

By "direction of motion" we mean whether the satellite is to go from $\mathbf{r}_1$ to $\mathbf{r}_2$ the "short way," through an angular change $\Delta\nu$ of less than π radians, or the "long way," through an angular change greater than π.

Obviously, there are an infinite number of orbits passing through $\mathbf{r}_1$ and $\mathbf{r}_2$, but only two that have the specified time of flight—one for each possible direction of motion.

One thing is immediately obvious from Figure 5-1: the two vectors $\mathbf{r}_1$ and $\mathbf{r}_2$ uniquely define the plane of the orbit. If the vectors $\mathbf{r}_1$ and $\mathbf{r}_2$ are collinear and in opposite directions, ($\delta\nu = \pi$), the plane of the orbit is not determined and a unique solution for $\mathbf{v}_1$ and $\mathbf{v}_2$ is not possible. If the two position vectors are collinear and in the same direction, ($\delta\nu = 0$ or 2π), the orbit is a degenerate conic, but a unique solution is possible for $\mathbf{v}_1$ and $\mathbf{v}_2$. In the latter case, the method of solution may have to be modified as there may be a mathematical singularity in the equations used, particularly if the parameter appears in the denominator of any expression.

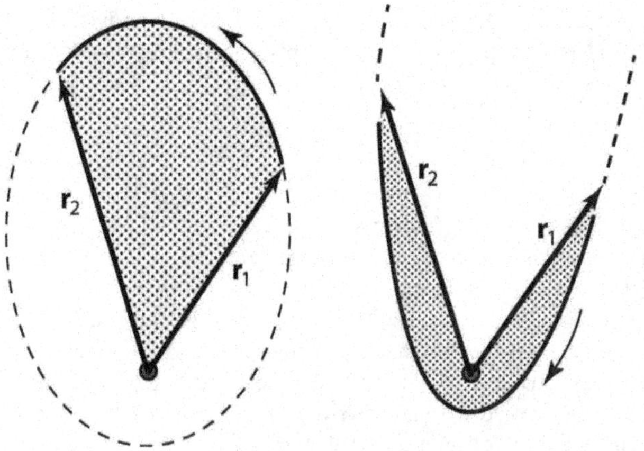

Fig. 5-1 Short-way and Long-way Trajectories with the Same Time of Flight.

The relationship among the four vectors $\mathbf{r}_1$, $\mathbf{r}_2$, $\mathbf{v}_1$ and $\mathbf{v}_2$ is contained in the f and g expressions, which were developed in Chapter 4. It is not surprising, therefore, that nearly every known method for solving the Gauss problem may be derived from the f and g relations. We will rewrite them as follows making an obvious change in notation to be consistent with the definition of the Gauss problem:

$$\mathbf{r}_2 = f\mathbf{r}_1 + g\mathbf{v}_1 \tag{5-1}$$

$$\mathbf{v}_2 = \dot{f}\mathbf{r}_1 + \dot{g}\mathbf{v}_1 \tag{5-2}$$

where

$$f = 1 - \frac{r_2}{p}(1 - \cos \Delta\nu) = 1 - \frac{a}{r_1}(1 - \cos \Delta E) \tag{5-3}$$

$$g = \frac{r_1 r_2 \sin \Delta\nu}{\sqrt{\mu p}} = t - \sqrt{\frac{a^3}{\mu}}(\Delta E - \sin \Delta E) \tag{5-4}$$

$$\dot{f} = \sqrt{\frac{\mu}{p}} \tan \frac{\Delta\nu}{2}\left(\frac{1 - \cos \Delta\nu}{p} - \frac{1}{r_1} - \frac{1}{r_2}\right) = \frac{-\sqrt{\mu a}}{r_1 r_2} \sin \Delta E \tag{5-5}$$

$$\dot{g} = 1 - \frac{r_1}{p}(1 - \cos \Delta \nu) = 1 - \frac{a}{r_2}(1 - \cos \Delta E) \tag{5-6}$$

Actually, this last equation is not needed since we have already shown that only three of the f and g expressions are truly independent. From Equation (5-1) we see that

$$\mathbf{v}_1 = \frac{\mathbf{r}_2 - f\mathbf{r}_1}{g} \tag{5-7}$$

Since Equations (5-7) and (5-2) express the vectors $\mathbf{v}_1$ and $\mathbf{v}_2$ in terms of f, g, $\dot{f}$, $\dot{g}$ and the two known vectors $\mathbf{r}_1$ and $\mathbf{r}_2$, the solution to the Gauss problem is reduced to evaluating the scalars f, g, $\dot{f}$ and $\dot{g}$.

Consider Equations (5-3), (5-4) and (5-5). There are seven variables—r_1, r_2, $\Delta\nu$, t, p, a and ΔE; but the first four are known, so what we have is three equations in three unknowns. The only trouble is that the equations are transcendental in nature so a trial-and-error solution is necessary. We may outline the general method of solution as follows:

1. Guess a trial value for one of the three unknowns, p, a or ΔE, directly or indirectly by guessing some other parameter of the transfer orbit (which in turn establishes p, a or ΔE).
2. Use Equations (5-3) and (5-5) to compute the remaining two unknowns.
3. Test the result by solving Equation (5-4) for t and check it against the given value of time of flight.
4. If the computed value of t does not agree with the given value, adjust the trial value of the iteration variable and repeat the procedure until it does agree.

This last step is perhaps the most important of all, since the method used to adjust the trial value of the iteration variable is what determines how quickly the procedure converges to a solution. This point is frequently overlooked by authors who suggest a method of accomplishing the first three steps but give no guidance on how to adjust the iterative variable.

Several methods for solving the Gauss problem will be discussed in this chapter including the original method suggested by Gauss. In each case a scheme for adjusting the trial value of the iterative variable will be suggested and the relative advantages of one method over another will be discussed. What we are referring to as the Gauss problem can also be stated in terms of the Lambert theorem. A solution using the Lambert theorem will not be treated here because the universal variable method avoids much of the awkwardness of special cases that must be treated when applying Lambert's theorem.

5.3 SOLUTIONS OF THE GAUSS PROBLEM VIA UNIVERSAL VARIABLES

In discussing general methods for solving the Gauss problem earlier in this chapter, we indicated that the f and g expressions provide us with three independent equations in three unknowns, p, a and ΔE or ΔF. Later we will show that a trial-and-error solution based on guessing a value of p can be formulated. A direct iteration on the variable a is more difficult since picking a trial value for a does not determine a unique value for p or ΔE. A solution based on guessing a trial value of ΔE or ΔF would work, however, and would enable us to use the universal variables x and z introduced in Chapter 4, since $z = \Delta E^2$ and $-z = \Delta F^2$.

To see how such a scheme might work, let's write the expressions for f, g, $\dot{f}$, and $\dot{g}$ in terms of the universal variables:

$$f = 1 - \frac{r_2}{p}(1 - \cos\Delta\nu) = 1 - \frac{x^2}{r_1}C \tag{5-8}$$

$$g = \frac{r_1 r_2 \sin\Delta\nu}{\sqrt{\mu p}} = t - \frac{x^3}{\sqrt{\mu}}S \tag{5-9}$$

$$\dot{f} = \sqrt{\frac{\mu}{p}}\frac{(1 - \cos\Delta\nu)}{\sin\Delta\nu}\left(\frac{1 - \cos\Delta\nu}{p} - \frac{1}{r_1} - \frac{1}{r_2}\right) = \frac{-\sqrt{\mu}}{r_1 r_2}x(1 - zS) \tag{5-10}$$

$$\dot{g} = 1 - \frac{r_1}{p}(1 - \cos\Delta\nu) = 1 - \frac{x^2}{r_2}C \tag{5-11}$$

Solving for x in Equation (5-8), we get

$$x = \sqrt{\frac{r_1 r_2(1 - \cos\Delta\nu)}{pC}} \tag{5-12}$$

Substitution for x in Equation (5-10) and canceling $\sqrt{\mu/p}$ from both sides, yields

$$\frac{1 - \cos\Delta\nu}{\sin\Delta\nu}\left(\frac{1 - \cos\Delta\nu}{p} - \frac{1}{r_1} - \frac{1}{r_2}\right) = -\sqrt{\frac{1 - \cos\Delta\nu}{r_1 r_2}}\frac{(1 - zS)}{\sqrt{C}} \tag{5-13}$$

If we multiply both sides by $r_1\, r_2$ and rearrange, we obtain

$$r_1 r_2\frac{(1 - \cos\Delta\nu)}{p} = r_1 + r_2 - \frac{\sqrt{r_1 r_2}\sin\Delta\nu}{\sqrt{1 - \cos\Delta\nu}}\frac{(1 - zS)}{\sqrt{C}} \tag{5-14}$$

We can write this equation more compactly if we define a constant, A, as

$$\boxed{A = \frac{\sqrt{r_1 r_2}\,\sin\Delta\nu}{\sqrt{1 - \cos\Delta\nu}}} \tag{5-15}$$

We will also find it convenient to define another auxiliary variable, y, such that

$$y = \frac{r_1 r_2 (1 - \cos \Delta\nu)}{p} \tag{5-16}$$

By using these definitions of A and y, Equation (5-14) may be written more compactly as

$$\boxed{y = r_1 + r_2 - A \frac{(1 - zS)}{\sqrt{C}}} \tag{5-17}$$

We can also express x in Equation (5-12) more concisely as

$$\boxed{x = \sqrt{\frac{y}{C}}} \tag{5-18}$$

If we now solve for t from Equation (5-9), we get

$$\sqrt{\mu}\, t = x^3 S + \frac{r_1 r_2 \sin \Delta v}{\sqrt{p}} \tag{5-19}$$

But, by using Equations (5-15) and (5-16), the last term of this expression may be simplified, so that

$$\boxed{\sqrt{\mu}\, t = x^3 S + A\sqrt{y}} \tag{5-20}$$

The simplification of the equations resulting from the introduction of the constant and the auxiliary variable can be extended to f and g expressions themselves. From Equations (5-8), (5-9) and (5-11), we can obtain the following simplified expressions:

$$\boxed{f = 1 - \frac{y}{r_1}} \tag{5-21}$$

$$\boxed{g = A\sqrt{\frac{y}{\mu}}} \tag{5-22}$$

$$\boxed{\dot{g} = 1 - \frac{y}{r_2}} \tag{5-23}$$

Since $\mathbf{r}_2 = f\,\mathbf{r}_1 + g\,\mathbf{v}_1$, we can compute $\mathbf{v}_1$ from

$$\boxed{\mathbf{v}_1 = \frac{\mathbf{r}_2 - f\mathbf{r}_1}{g}} \tag{5-24}$$

The velocity, $\mathbf{v}_2$, may be expressed as

$$\mathbf{v}_2 = \dot{f}\mathbf{r}_1 + \dot{g}\mathbf{v}_1$$

By substituting for $\mathbf{v}_1$ from Equation (5-24) and using the identity $f\dot{g} - \dot{f}g = 1$ from Equation (4-47), this last expression simplifies to

$$\boxed{\mathbf{v}_2 = \frac{\dot{g}\mathbf{r}_2 - \mathbf{r}_1}{g}} \tag{5-25}$$

A simple algorithm for solving the Gauss problem via universal variables may now be stated as follows:

1. From $\mathbf{r}_1$, $\mathbf{r}_2$ and the "direction of motion," evaluate the constant using Equation (5-15).
2. Pick a trial value for z. Since $z = \Delta E^2$ and $-z = \Delta F^2$, this amounts to guessing the change in eccentric anomaly. The usual range for a scalar is from minus values to $(2\pi)^2$. Values of $z > (2\pi)^2$ correspond to changes in the eccentric anomaly of more than 2π and can occur only if the satellite passes back through $\mathbf{r}_1$ en route to $\mathbf{r}_2$.
3. Evaluate the functions S and C for the selected trial value of z using Equations (4-37) and (4-38).
4. Determine the auxiliary variable y from Equation (5-17).
5. Determine x from Equation (5-18).
6. Check the trial value of z by computing t from Equation (5-20) and compare it with the desired time of flight. If it is not nearly the same, adjust the trial value of z and repeat the procedure until the desired value of t is obtained. A Newton iteration scheme for adjusting z will be discussed in the next section.
7. When the method has converged to a solution, evaluate f, g and $\dot{g}$ from Equations (5-21), (5-22) and (5-23), then compute $\mathbf{v}_1$ and $\mathbf{v}_2$ from Equations (5-24) and (5-25).

5.3.1 Selecting a New Trial Value of z

Although any iterative scheme, such as a Bolzano bisection technique or linear interpolation, may be used successfully to pick a better trial value for z, a Newton iteration, which converges more rapidly, may be used if we can determine the slope of the t vs. z curve at the last trial point.

The derivative, dt/dz, necessary for a Newton iteration can be determined by differentiating Equation (5-20) for t:

$$\sqrt{\mu}t = x^3S + A\sqrt{y} \tag{5-26}$$

$$\sqrt{\mu}\frac{dt}{dz} = 3x^2\frac{dx}{dz}S + x^3\frac{dS}{dz} + \frac{A}{2\sqrt{y}}\frac{dy}{dz} \tag{5-27}$$

Differentiating Equation (5-18) for x yields

$$\frac{dx}{dz} = \frac{1}{2xC}\left(\frac{dy}{dz} - x^2\frac{dC}{dz}\right) \tag{5-28}$$

Differentiating Equation (5-17) for y, we get

$$\frac{dy}{dz} = -\frac{A}{C}\left[\sqrt{C}\left(-S - z\frac{dS}{dz}\right) - \frac{(1-zS)}{2\sqrt{C}}\frac{dC}{dz}\right] \tag{5-29}$$

But, in Section 4.5.3, we showed that

$$\frac{dS}{dz} = \frac{1}{2z}(C - 3S) \tag{5-30}$$

$$\frac{dC}{dz} = \frac{1}{2z}(1 - zS - 2C) \tag{5-31}$$

so, the expression for dy/dz reduces to

$$\frac{dy}{dz} = \frac{A}{4}\sqrt{C} \tag{5-32}$$

If we now substitute Equations (5-28) and (5-32) into Equation (5-27), we get

$$\sqrt{\mu}\frac{dt}{dz} = x^3\left(S' - \frac{3SC'}{2C}\right) + \frac{A}{8}\left(\frac{3S\sqrt{y}}{C} + \frac{A}{x}\right) \tag{5-33}$$

where S' and C' are the derivatives of S and C with respect to z. These derivatives may be evaluated from Equations (5-30) and (5-31) except when z is nearly zero (near-parabolic orbit). If we differentiate the power series expansion of C and S, Equations (4-37) and (4-38), we get

$$C' = -\frac{1}{4!} + \frac{2z}{6!} - \frac{3z^2}{8!} + \frac{4z^3}{10!} - \ldots \tag{5-34}$$

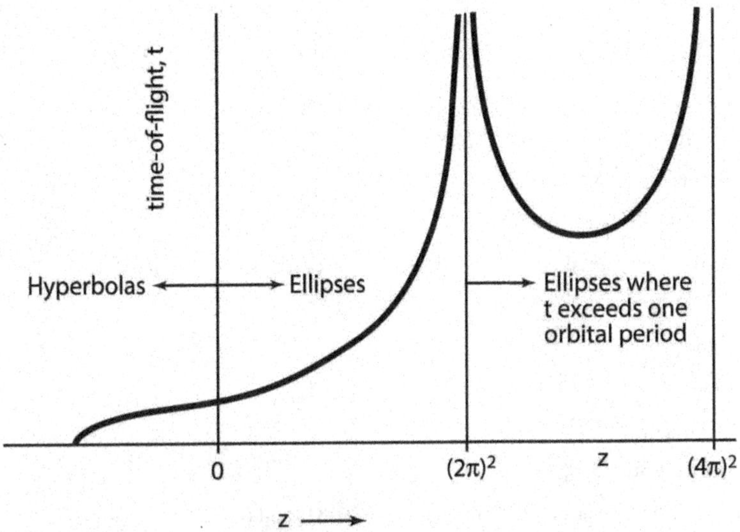

Fig. 5-2 Typical t vs. z Plot for a Fixed r_1 and r_2.

$$S' = -\frac{1}{5!} + \frac{2z}{7!} - \frac{3z^2}{9!} + \frac{4z^3}{11!} - \ldots \tag{5-35}$$

which may be used to evaluate the derivatives when z is near zero.

Example Problem. From radar measurements you know the position vector of a space object at 0432:00Z to be: $\mathbf{r}_1 = 0.5\mathbf{I} + 0.6\mathbf{J} + 0.7\mathbf{K}$ DU. At 0445:00Z the position of the object was: $\mathbf{r}_2 = 0.0\mathbf{I} + 1.0\mathbf{J} + 0.0\mathbf{K}$ DU. Using universal variables, determine on what two paths the object could have moved from position one to position two. Assume the object occupies each position only once during this time period.

There is only one path for each direction of motion. Thus the two paths sought are the "short way" and "long way." We first find $\mathbf{v}_1$ and $\mathbf{v}_2$ for each direction by solving the Gauss problem and then use $\mathbf{r}_1$, $\mathbf{r}_2$ and the method of Chapter 2 to solve for the orbital elements.

To facilitate our solution let us define an integer quantity DM, "Direction of Motion":

$$\text{DM} \triangleq \text{sign}(\pi - \Delta\nu) \tag{5-36}$$

For DM = –1 we have the "long way" $(\Delta\nu > \pi)$; for DM = +1 we have the "short way" $(\Delta\nu < \pi)$.

For ease of numerical solutions, a convenient form of Equation (5-15) is

$$A = DM\sqrt{r_1 r_2(1 + \cos\Delta\nu)} \tag{5-37}$$

which does not present the problem that Equation (5-15) does when $\Delta\nu$ is small.

Now follow the algorithm of Section 5.3:

Step 1: $\mathbf{r}_1 = 1.0488088482$ DU

$\mathbf{r}_2 = 1.0000000000$ DU

$\Delta\nu$ short way = the smallest angle between $\mathbf{r}_1$ and $\mathbf{r}_2 = 0.962$ radians
$\Delta\nu$ long way = $2\pi - \Delta\nu$ short way = 5.321 radians

Using Equation (5-37) gives

$$A_{\text{long way}} = -1.28406$$

$$A_{\text{short way}} = 1.28406$$

Step 2 and Step 3:

Let the first estimate be Z = 0; from Equations (4-81) and (4-82)

$$S(Z)\big|_{Z=0} = \frac{1}{6},\quad C(Z)\big|_{Z=0} = \frac{1}{2}$$

Step 4:

Using Equation (5-17) gives

$$y_{\text{long way}} = 1.0488088482 + 1.0000 + 1.28406(1/\sqrt{1/2}) = 3.86474323$$

$$y_{\text{short way}} = 1.0488088482 + 1.0000 - 1.28406(1/\sqrt{1/2}) = 0.23287446$$

Here we must ensure that we heed the caution concerning the sign of **y** for the short way.

Step 5:

Using Equation (5-18) gives

$$x_{\text{long way}} = \sqrt{\frac{y_{\text{long}}}{C(0)}} = \sqrt{\frac{3.86474323}{1/2}} = 2.78019540$$

$$x_{\text{short way}} = \sqrt{\frac{y_{\text{short}}}{C(0)}} = \sqrt{\frac{0.23287446}{1/2}} = 0.68245800$$

Step 6:

Now using Equation (5-20), we solve for t_{long} to get

$$t_{\text{long}} = 1/\sqrt{\mu}\, x_{\text{long}}^3 S + A\sqrt{y_{\text{long}}}$$

$$= 21.489482706\left(\frac{1}{6}\right) - 1.28406\sqrt{3.86474323}$$

$$= 1.0575424 \text{ TU}$$

$$t_{short} = 0.317854077\left(\frac{1}{6}\right) + 1.28406\sqrt{0.23287446}$$

$$= 0.67262516 \text{ TU}$$

Now, from the given data, the desired time of flight is

$$\Delta t = 0445{:}00 \text{ Zulu} - 0432{:}00 \text{ Zulu} = \frac{13 \text{ min}}{13.44685 \text{ min /TU}}$$

$$= 0.9667663 \text{ TU}$$

We see that our computed values of t using $z=0$ are too far off, so we adjust the value of z using a Newton iteration and repeat steps 2 through 6. Care must be taken to ensure that the net trial value of z does not cause y to be negative. (The iterations *must* be performed separately for the long way and the short way.)

Convergence criteria should be chosen with the size of Δt taken into consideration. Since Δt for this problem is less than 1.0, we do not normalize.

When $\Delta t - t \leq 10^{-4}$ (within 0.1 ms) we consider the problem solved.

We have

Long way (after 2 iterations)	Short way (after 2 iterations)
$z = -3.61856634$	$z = 0.83236253$
$x = 2.66224213$	$x = 0.94746230$
$A = -1.28406$	$A = 1.28406$
$y = 4.74994739$	$y = 0.41856019$
$t = 0.96681012$	$t = 0.96670788$

Step 7:

Using Equations (5-21), (5-22), (5-23), (5-24) and (5-25) we get

Long way:

$f = -3.52889714$ $\quad v_1 = 1.554824$ DU/TU

$g = -2.79852734$ $\quad v_2 = 1.584516$ DU/TU

$\dot{g} = -3.74994739$

$\mathbf{v}_1 = -0.6304918096\mathbf{I} - 1.1139209665\mathbf{J} - 0.8826885334\mathbf{K}$

$\mathbf{v}_2 = 0.1786653974\mathbf{I} + 1.5544139777\mathbf{J} - 0.250135563\mathbf{K}$

Short way:

$$f = 0.60091852 \qquad v_1 = 0.989794 \text{ DU/TU}$$
$$g = 0.83073807 \qquad v_2 = 1.035746 \text{ DU/TU}$$
$$\dot{g} = 0.58143981$$
$$\mathbf{v}_1 = -0.36167749\mathbf{I} + 0.76973587\mathbf{J} - 0.50634848\mathbf{K}$$
$$\mathbf{v}_2 = -0.60187442\mathbf{I} - 0.02234181\mathbf{J} - 0.84262419\mathbf{K}$$

Then using the given **r** vectors and these **v** vectors we have specified the two paths.

Notice that the long way trajectory is a hyperbola:

energy	$= \underline{\underline{0.2553 \text{ DU}^2/\text{TU}^2}}$
eccentricity	$= \underline{\underline{3.96}}$
perigee radius	$= \underline{\underline{0.0191 \text{ DU}}}$

while the short way is an ellipse:

energy	$= \underline{\underline{-4.636 \text{ DU}^2/\text{TU}^2}}$
eccentricity	$= \underline{\underline{0.076832}}$
perigee radius	$= \underline{\underline{0.9958 \text{ DU}}}$

Since the hyperbola passes through Earth between the two positions, and the ellipse intersects Earth (but only after it passes the second position), the ellipse is the only trajectory a real object could travel on. This result illustrates the beauty of the universal variables approach to this problem: with one set of equations we solved problems involving two different types of conics.

There is one pitfall in the solution of the Gauss problem via universal variables that you should be aware of. For "short way" trajectories, where $\Delta\nu$ is less than π, the t vs. z curve crosses the t=0 axis at some negative lower limit for permissible values of z when $\Delta\nu<\pi$. The reason for this may be seen by examining Equations (5-16) and (5-17).

From Equation (5-16) it is obvious that y cannot be negative, yet Equation (5-17) will result in a negative value for y if $\Delta\nu<\pi$ and z is too large a negative number. This is apparent from the fact that A is positive whenever $\Delta\nu<\pi$ and negative whenever $\pi<\Delta\nu<2\pi$. Because both C and S become large positive numbers when z is large and negative (see Figure 4-12), the expression

$$\frac{A(1 - zS)}{\sqrt{C}}$$

can become a large positive number if A is positive. Whenever

$$\frac{A(-zS)}{\sqrt{C}} > r_1 + r_2$$

the value of y be negative and x will be imaginary in Equation (5-18).

Any computational algorithm for solving the Gauss problem via universal variables should include a check to see whether y is negative prior to evaluating x. This check is only necessary when A is positive. For "long way" trajectories y is positive for all values of z and the t vs. z curve approaches zero asymptotically as z approaches minus infinity.

5.4 THE p-ITERATION METHOD

The next method of solving the Gauss problem that we will look at could be called a direct p-iteration technique. It differs from the p-iteration method first proposed by Herrick and Liu in 1959[6] since it does not directly involve eccentricity. The method consists of guessing a trial value of p from which we can compute the other two unknowns a and ΔE. The trial values are checked by solving for t and comparing it with the given time of flight.

The p-iteration method presented below is unusual in that it will enable us to develop an analytical expression for the slope of the t vs. p curve; hence, a Newton iteration scheme is possible for adjusting the trial value of p.

5.4.1 Expressing p as a Function of ΔE

From Equation (5-5), if we cancel $\sqrt{\mu}$ from both sides and write $\tan(\Delta\nu/2)$ as $(1-\cos\Delta\nu)/(\sin\Delta\nu)$, we obtain

$$\frac{1-\cos\Delta\nu}{\sqrt{p}\,\sin\Delta\nu}\left(\frac{1-\cos\Delta\nu}{p}-\frac{1}{r_2}-\frac{1}{r_2}\right)=\frac{-\sqrt{a}\,\sin\Delta E}{r_1\,r_2} \tag{5-38}$$

From Equation (5-3), we can solve for a and get

$$a=\frac{r_1 r_2(1-\cos\Delta\nu)}{p(1-\cos\Delta E)} \tag{5-39}$$

Substituting this expression for a into Equation (5-38) and rearranging yields

$$\frac{1-\cos\Delta\nu}{p}-\frac{1}{r_1}-\frac{1}{r_2}=\frac{-1}{\sqrt{r_1\,r_2}}\frac{\sin\Delta\nu}{\sqrt{1-\cos\Delta\nu}}\frac{\sin\Delta E}{\sqrt{1-\cos\Delta E}} \tag{5-40}$$

Using the trigonometric identity, $(\sin x)/(\sqrt{1-\cos x}) = \sqrt{2}\cos(x/2)$ and solving for p, we get

$$\boxed{p=\frac{r_1\,r_2(1-\cos\Delta\nu)}{r_1+r_2-2\sqrt{r_1\,r_2}\,\cos\dfrac{\Delta\nu}{2}\cos\dfrac{\Delta E}{2}}} \tag{5-41}$$

5.4.2 Expressing a as a Function of p

The first step in the solution is to find an expression for a as a function of p and the given information. We will find it convenient to define three constants that may be determined from the given information:

$$\boxed{\begin{aligned} k &= r_1 r_2 (1 - \cos \Delta\nu) \\ \ell &= r_1 + r_2 \\ m &= r_1 r_2 (1 + \cos \Delta\nu) \end{aligned}} \tag{5-42}$$

By using these definitions, a may be written as

$$a = \frac{k}{p(1 - \cos\Delta E)} = \frac{k}{2p\sin^2 \frac{\Delta E}{2}} = \frac{k}{2p\left(1 - \cos^2 \frac{\Delta E}{2}\right)} \tag{5-43}$$

Using these same definitions, and noting that $\sqrt{2r_1 r_2}\cos\frac{\Delta\nu}{2} = \pm\sqrt{r_1 r_2(1 + \cos\Delta\nu)}$, we can rewrite Equation (5-41) as

$$p = \frac{k}{\ell \pm \sqrt{2m}\cos\frac{\Delta E}{2}} \tag{5-44}$$

Solving for $\cos(\Delta E/2)$, we get

$$\cos\frac{\Delta E}{2} = \frac{k - \ell p}{\pm\sqrt{2m}p}$$

$$\text{and so } \cos^2\frac{\Delta E}{2} = \frac{(k - \ell p)^2}{2mp^2} \tag{5-45}$$

If we substitute this last expression into Equation (5-43) and simplify, we obtain

$$\boxed{a = \frac{mkp}{(2m - \ell^2)p^2 + 2k\ell p - k^2}} \tag{5-46}$$

where k, ℓ and m are all constants that can be determined from $\mathbf{r_1}$ and $\mathbf{r_2}$. It is clear from this equation that once p is specified a unique value of a is determined.

In Figure 5-3 we have drawn a typical plot for a vs. p for a fixed $\mathbf{r}_1$, $\mathbf{r}_2$ and $\Delta\nu$ less than π. Notice that, for those orbits where a is positive (ellipses), a may not be smaller than some minimum value a_m. The point where a is a minimum corresponds to the "minimum energy ellipse" joining $\mathbf{r}_1$ and $\mathbf{r}_2$.

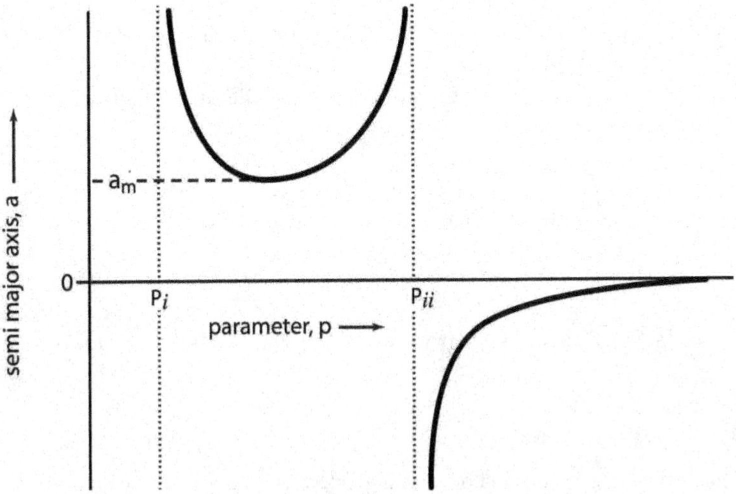

Fig. 5-3 Typical Plot of a vs. p for a Fixed $\mathbf{r}_1$ and $\mathbf{r}_2$.

The two points where a approaches infinity correspond to parabolic orbits joining $\mathbf{r}_1$ and $\mathbf{r}_2$. The values of p that specify the parabolic orbits are labeled p_i and p_{ii} and will be important to us later.

Those orbits where a is negative are hyperbolic. The limiting case where p approaches infinity and a approaches zero is the straight-line orbit connecting $\mathbf{r}_1$ and $\mathbf{r}_2$. It would require infinity energy and have a time of flight of zero.

5.4.3 Checking the Trial Value of p

Once we have selected a trial value of p and computed a from Equation (5-46), we are ready to solve for t and check it against the given time of flight. First, however, we need to determine ΔE (or ΔF in case a is negative).

From the trial value of p and the known information, we can compute f, g and $\dot{f}$ from Equations (5-3), (5-4) and (5-5). If a is positive, we can determine ΔE from Equations (5-3) and (5-5):

$$\cos\Delta E = 1 - \frac{r_1}{a}(1 - f) \tag{5-47}$$

$$\sin\Delta E = \frac{-r_1 r_2 \dot{f}}{\sqrt{\mu a}} \tag{5-48}$$

If a is negative, the corresponding f and g expressions involving ΔF yield

$$\cosh \Delta F = 1 - \frac{r_1}{a}(1 - f) \tag{5-49}$$

Since we always assume ΔF is positive there is no ambiguity in determining ΔF from this one equation.

The time of flight may now be determined from Equation (5-4) or the corresponding equation involving ΔF:

$$t = g + \sqrt{\frac{a^3}{\mu}}(\Delta E - \sin \Delta E) \tag{5-50}$$

$$t = g + \sqrt{\frac{-a^3}{\mu}}(\sinh \Delta F - \Delta F) \tag{5-51}$$

5.4.4 The t vs. p Curve

Before discussing the method of selecting a new trial value of p, we need to understand what the t vs. p curve looks like. To get a feeling for the problem, let's look at the family of orbits that can be drawn between a given $\mathbf{r}_1$ and $\mathbf{r}_2$. In Figure 5-4 we have drawn $\mathbf{r}_1$ and $\mathbf{r}_2$ to be of equal length. The conclusion we will reach from examining this illustration also applies to the more general case where $\mathbf{r}_1 \neq \mathbf{r}_2$.

First, let's consider the orbits that permit traveling from $\mathbf{r}_1$ to $\mathbf{r}_2$ the "short way." The quickest way to get from $\mathbf{r}_1$ to $\mathbf{r}_2$ is obviously along the straight line from $\mathbf{r}_1$ to $\mathbf{r}_2$. This is the limiting case where p is infinite and t is zero. As we take longer to make the journey from $\mathbf{r}_1$ to $\mathbf{r}_2$, the trajectories become more "lofted" until we approach the limiting case where we try to go through the open end of the parabolas joining $\mathbf{r}_1$ and $\mathbf{r}_2$. Notice that, although t approaches infinity for this orbit, p approaches a finite minimum value, which we will call p_i.

If we look at "long way" trajectories, we see that the limiting case of zero time of flight is achieved on the degenerate hyperbola that goes straight down $\mathbf{r}_1$ to the focus and then straight up $\mathbf{r}_2$ and has p equal to zero. As we choose trajectories with longer time of flight, p increases until we reach the limiting case where we try to travel through the open end of the other parabola that joins $\mathbf{r}_1$ and $\mathbf{r}_2$. For this case, t approaches infinity as p approaches a finite limiting value, which we will call p_{ii}.

From the discussion above, we can construct a typical plot of t vs. p for fixed $\mathbf{r}_1$ and $\mathbf{r}_2$. In Figure 5-5 the solid line represents "short way" trajectories and the dashed line represents "long way" trajectories.

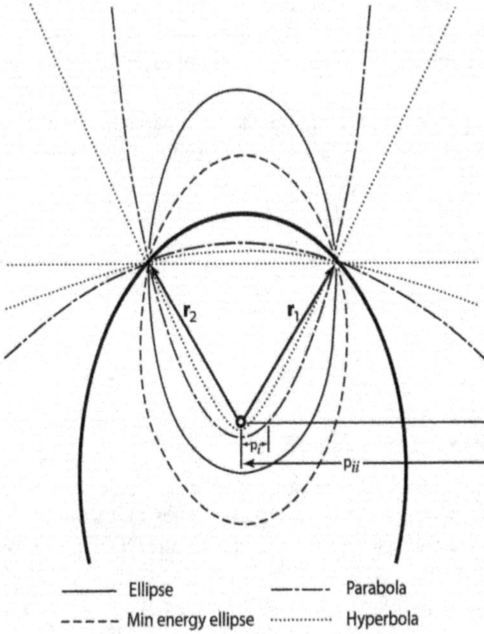

Fig. 5-4 Family of Possible Transfer Orbits Connecting r_1 and r_2.

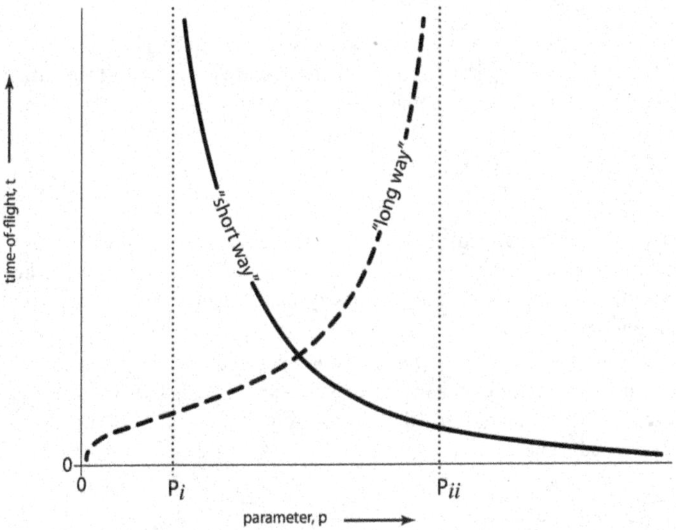

Fig. 5-5 Typical t and p Plot for Fixed r_1 and r_2.

For $\Delta\nu$ less than π, p must lie between p_i and infinity; for $\Delta\nu$ greater than π, p must lie between 0 and p_{ii}. Because it is important that the first trial value, as well as all subsequent guesses for p, lie within the prescribed limits, we should first compute p_i or p_{ii}.

The limiting values of p correspond to the two parabolic orbits passing through $\mathbf{r_1}$ and $\mathbf{r_2}$. Since $\Delta E = 0$ for all parabolic orbits, we can obtain, from Equation (5-44),

$$p_i = \frac{k}{\ell + \sqrt{2m}} \tag{5-52}$$

$$p_{ii} = \frac{k}{\ell - \sqrt{2m}} \tag{5-53}$$

5.4.5 Selecting a New Trial Value of p

The method used to adjust the trial value of p to give the desired time of flight is crucial in determining how rapidly p converges to a solution. Several simple methods may be used successfully, such as the "Bolzano bisection" technique or "linear interpolation" (regula falsi).

In the *bisection method* we must find two trial values of p: one that gives too small a value for t, and one that gives too large a value. The solution is then bracketed and, by choosing our next trial value half way between the first two, we can keep it bracketed while reducing the interval of uncertainty to some arbitrarily small value.

In the linear interpolation method we choose two trial values of p—call them p_{n-1} and p_n. If t_{n-1} and t_n are the times of flight corresponding to these trial values of p, then we select a new value from

$$p_{n+1} = p_n + \frac{(t - t_n)(p_n - p_{n-1})}{(t_n - t_{n-1})} \tag{5-54}$$

This scheme can be repeated, always retaining the latest two trial values of p and their corresponding times of flight for use in computing a still better trial value from Equation (5-54). It is not necessary that the initial two trial values bracket the answer.

An even faster method is a *Newton iteration*, but it requires that we compute the slope of the t vs. p curve at the last trial value of p. If this last trial value is called p_n and t_n is the time of flight that results from it, then a better estimate of p may be obtained from

$$p_{n+1} = p_n + \frac{t - t_n}{dt/dp|_{p = p_n}} \tag{5-55}$$

where t is the desired time of flight and $dt/dp|_{p=p_n}$ is the slope at the last trial point. We must now obtain an expression for the derivative in Equation (5-55).

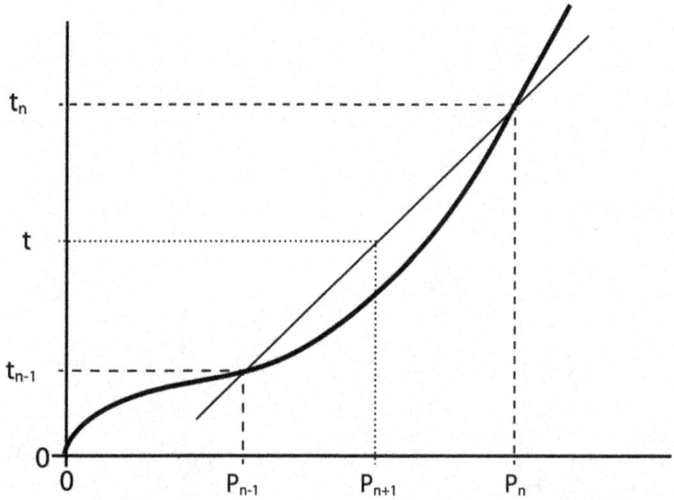

Fig. 5-6 Selecting a New Trial Value of p by Linear Interpolation.

From the f and g expressions we can write

$$t = g + \sqrt{\frac{a^3}{\mu}}(\Delta E - \sin \Delta E) \tag{5-56}$$

Differentiating this expression with respect to p, we get

$$\frac{dt}{dp} = \frac{dg}{dp} + \frac{3}{2}\sqrt{\frac{a}{\mu}}(\Delta E - \sin \Delta E)\frac{da}{dp} + \sqrt{\frac{a^3}{\mu}}(1 - \cos \Delta E)\frac{d\Delta E}{dp} \tag{5-57}$$

The expression for g is

$$g = \frac{r_1 \, r_2 \, \sin \Delta \nu}{\sqrt{\mu p}} \tag{5-58}$$

Differentiating with respect to p yields

$$\frac{dg}{dp} = \frac{-r_1 \, r_2 \, \sin \Delta \nu}{2p\sqrt{\mu p}} = \frac{-g}{2p} \tag{5-59}$$

The derivative da/dp comes directly from differentiating the expression for a:

$$a = \frac{mkp}{(2m-\ell^2)p^2 + 2k\ell p - k^2} \tag{5-60}$$

Therefore, $\frac{da}{dp} = \frac{mk(2m-\ell^2)p^2 + 2mk^2\ell p - mk^3 - 2mk(2m-\ell^2)p^2 - 2mk^2\ell p}{[(2m-\ell^2)p^2 + 2k\ell p - k^2]^2}$

which simplifies to

$$\frac{da}{dp} = \frac{-a^2[k^2 + (2m-\ell^2)p^2]}{mkp^2} \tag{5-61}$$

From Equation (5-45), we can write

$$\cos^2\frac{\Delta E}{2} = \frac{(k-\ell p)^2}{2mp^2} \tag{5-62}$$

Differentiating both sides of this equation yields

$$\left(-\frac{1}{2}\right)2\cos\frac{\Delta E}{2}\sin\frac{\Delta E}{2}\frac{d\Delta E}{dp} = \frac{-4mp^2(k-\ell p)\ell - (k-\ell p)^2 4mp}{4m^2p^4}$$

which simplifies to

$$\frac{1}{2}\sin\frac{\Delta E}{2}\frac{d\Delta E}{dp} = \frac{k(k-\ell p)}{mp^3} \tag{5-63}$$

Solving for dΔE/dp and noting, from Equation (5-45), that $mp^2 = (k-\ell p)^2/(1 + \cos\Delta E)$, we obtain

$$\frac{d\Delta E}{dp} = \frac{2k(1+\cos\Delta E)}{p(k-\ell p)\sin\Delta E} \tag{5-64}$$

We are now ready to substitute into Equation (5-57) from (5-59), (5-61) and (5-64) to get

$$\frac{dt}{dp} = \frac{-g}{2p} - \frac{3a}{2}\sqrt{\frac{a^3}{\mu}}(\Delta E - \sin\Delta E)\left[\frac{k^2 + (2m-l^2)p^2}{mkp^2}\right]$$
$$+ \sqrt{\frac{a^3}{\mu}}\frac{(1-\cos\Delta E)(1+\cos\Delta E)2k}{\sin\Delta E p(k-lp)}$$

which simplifies to

$$\boxed{\frac{dt}{dp}=\frac{-g}{2p}-\frac{3}{2}a(t-g)\left(\frac{k^2+(2m-\ell^2)p^2}{mkp^2}\right)+\sqrt{\frac{a^3}{\mu}}\frac{2k\sin\Delta E}{p(k-\ell p)}} \tag{5-65}$$

which is valid for the elliptical portion of the t vs. p curve.

By an analogous derivation starting with the equation

$$t=g+\sqrt{\frac{(-a)^3}{\mu}}(\sinh\Delta F-\Delta F)$$

we can arrive at the following slope expression that is valid for the hyperbolic portion of the t vs. p curve:

$$\frac{dt}{dp}=\frac{-g}{2p}-\frac{3}{2}a(t-g)\left(\frac{k^2+(2m-\ell^2)p^2}{mkp^2}\right)-\sqrt{\frac{(-a)^3}{\mu}}\frac{2k\sinh\Delta F}{p(k-\ell p)} \tag{5-66}$$

To evaluate the slope at p_n, the values of g, a, t and ΔE or ΔF obtained from the trial value p_n, should be used in Equation (5-65) or (5-66).

We can summarize the steps involved in solving the Gauss problem via the p-iteration technique as follows:

1. Evaluate the constants k, ℓ and m from $\mathbf{r}_1$, $\mathbf{r}_2$ and $\Delta\nu$ using Equation (5-42).
2. Determine the limits on the possible values of p by evaluating p_i and p_{ii} from Equations (5-52) and (5-54).
3. Pick a trial value of p within the appropriate limits.
4. Using the trial value of p, solve for a from Equation (5-46). The type of conic orbit will be known from the value of a.
5. Solve for f, g and $\dot{f}$ from Equations (5-3), (5-4) and (5-5).
6. Solve for ΔE or ΔF, as appropriate, using Equations (5-47) and (5-48) or Equation (5-49).
7. Solve for t from Equation (5-50) or (5-51) and compare it with the desired time of flight.
8. Adjust the trial value of p using one of the iteration methods discussed above until the desired time of flight is obtained.
9. Evaluate $\dot{g}$ from Equation (5-11) and then solve for $\mathbf{v}_1$ and $\mathbf{v}_2$ using Equations (5-7) and (5-2).

The p-iteration method converges in all cases except when $\mathbf{r}_1$ and $\mathbf{r}_2$ are collinear. Its main disadvantage is that separate equations are used for the ellipse and hyperbola. This defect may be overcome by using the universal variables x and z introduced in Chapter 4 and discussed earlier in this chapter.

5.5 THE GAUSS PROBLEM USING THE f AND g SERIES

In this section we will develop another method for solving the Gauss problem. Instead of using the f and g expressions. We will develop and use the f and g *series*. As stated earlier, the motion of a body in a Keplerian orbit is in a plane which contains the radius vector from the center of force to the body and the velocity vector. If we know the position $\mathbf{r}_0$ and the velocity $\mathbf{v}_0$ at some time t_0 then we know that the position vector at any time t can be expressed as a linear combination of $\mathbf{r}_0$ and $\mathbf{v}_0$ because it always lies in the same plane as $\mathbf{r}_0$ and $\mathbf{v}_0$. The coefficients of $\mathbf{r}_0$ and $\mathbf{v}_0$ in this linear combination will be functions of time and will depend upon the vectors $\mathbf{r}_0$ and $\mathbf{v}_0$:

$$\boxed{\mathbf{r} = f(\mathbf{r}_0, \mathbf{v}_0, t)\mathbf{r}_0 + g(\mathbf{r}_0, \mathbf{v}_0, t)\mathbf{v}_0} \tag{5-67}$$

We can determine the functions f and g by expanding $\mathbf{r}$ in a Taylor series expansion for around $t = t_0$,

$$\mathbf{r} = \sum_{n=0}^{\infty} \frac{(t-t_0)^n}{n!}\mathbf{r}_0^{(n)} \tag{5-68}$$

where

$$\mathbf{r}_0^{(n)} \equiv \left.\frac{d^n\mathbf{r}}{dt^n}\right|_{t=t_0} \tag{5-69}$$

Since the motion is in a plane, all the time derivatives of $\mathbf{r}$ must line in the plane of $\mathbf{r}$ and $\mathbf{v}$. Therefore, we can write in general

$$\mathbf{r}^{(n)} = F_n\mathbf{r} + G_n\mathbf{v} \tag{5-70}$$

Differentiating with respect to time gives

$$\mathbf{r}^{(n+1)} = \dot{F}_n\mathbf{r} + F_n\mathbf{v} + \dot{G}_n\mathbf{v} + G_n\ddot{\mathbf{r}} \tag{5-71}$$

but $\ddot{\mathbf{r}} = -\frac{\mu}{r^3}\mathbf{r}$ and if we define $u \equiv \frac{\mu}{r^3}$ we can write

$$\mathbf{r}^{(n+1)} = (\dot{F}_n - uG_n)\mathbf{r} + (F_n + \dot{G}_n)\mathbf{v} \tag{5-72}$$

Comparing with Equation (5-70), we have the following recursion formulas:

$$F_{n+1} = \dot{F}_n - uG_n \tag{5-73}$$

$$G_{n+1} = F_n + \dot{G}_n \tag{5-74}$$

To determine F_0 and G_0 so that we can start the recursion, we write Equation (5-70) for $n = 0$:

$$\mathbf{r}^{(0)} = \mathbf{r} = F_0\mathbf{r} + G_0\dot{\mathbf{r}} \tag{5-75}$$

from which it is obvious that $F_0 = 1$ and $G_0 = 0$.

5.5.1 Development of the Series Coefficients

Before continuing with the development of the functions F_n and G_n it is convenient to digress at this point to introduce and discuss three quantities u, p and q that will be useful later. We define them as follows:

$$u \equiv \frac{\mu}{r^3} \tag{5-76}$$

$$p \equiv \frac{1}{r^2}(\mathbf{r}\bullet\mathbf{v}) \text{ (not the semi-latus rectum)} \tag{5-77}$$

$$q \equiv \frac{1}{r^2}(v^2) - u \tag{5-78}$$

These quantities can all be determined if the position $\mathbf{r}$ and the velocity $\mathbf{v}$ are known. These quantities are useful because their time derivatives can be expressed in terms of the quantities u, p and q, as we shall demonstrate. The time derivative of u is

$$\dot{u} = \frac{d}{dt}\left(\frac{\mu}{r^3}\right) = -\frac{3\mu}{r^4}\dot{r} \tag{5-79}$$

By using the relationship $\mathbf{r} \cdot \mathbf{v} = r\dot{r}$ and the definition of p, this can be written:

$$\dot{u} = -\left(\frac{3\mu}{r^5}\right)(\mathbf{r}\bullet\dot{\mathbf{r}}) = -3up \tag{5-80}$$

$$\dot{p} = \frac{d}{dt}\left[\frac{1}{r^2}(\mathbf{r}\bullet\dot{\mathbf{r}})\right] = \frac{1}{r^2}(v^2) + \frac{1}{r^2}(\mathbf{r}\bullet\ddot{\mathbf{r}}) - \frac{2}{r^3}\dot{r}(\mathbf{r}\bullet\mathbf{v}) \tag{5-81}$$

By using the relationship $\mathbf{r} \cdot \mathbf{v} = r\dot{r}$ and the equation of motion $\ddot{\mathbf{r}} = -u\mathbf{r}$ this can be written as

$$\dot{p} = \frac{1}{r^2}(v)^2 - u - \frac{2}{r^4}(\mathbf{r}\bullet\mathbf{v})^2 \tag{5-82}$$

Using the definitions of q and p, we have

$$\dot{p} = q - 2p^2 \tag{5-83}$$

$$\dot{q} = \frac{d}{dt}\left[\frac{1}{r^2}(v)^2 - u\right] = \frac{2}{r^2}(\mathbf{v}\bullet\ddot{\mathbf{r}}) - \frac{2}{r^3}\dot{r}(v)^2 - \dot{u} \tag{5-84}$$

Similarly,

$$\dot{q} = -\frac{2}{r^2}u(\mathbf{r}\bullet\dot{\mathbf{r}}) - \frac{2}{r^4}(\mathbf{r}\bullet\dot{\mathbf{r}})(v)^2 - \dot{u} \tag{5-85}$$

Using the definitions of p and q and the expression for $\dot{u}$, we have

$$\dot{q} = -2p(u+q+u)+3up \tag{5-86}$$

$$\dot{q} = -p(u+2q) \tag{5-87}$$

In summary,

$$\boxed{\begin{array}{ll} u \equiv \dfrac{\mu}{r^3}, & \dot{u} = -3up \\ p \equiv \dfrac{1}{r^2}(\mathbf{r}\bullet\mathbf{v}), & \dot{p} = q - 2p^2 \\ q \equiv \dfrac{1}{r^2}(v)^2 - u, & \dot{q} = -p(u+2q) \end{array}} \tag{5-88}$$

After this digression we are now in a position to carry out the recursion of F_n and G_n. We have already seen that

$F_0 = 1$ and $G_0 = 0$

Applying the recursion formulas (5-73) and (5-74), we obtain

$$\begin{aligned} F_1 &= \dot{F}_0 - uG_0 = 0 \\ G_1 &= F_0 + \dot{G}_0 = 1 \\ F_2 &= \dot{F}_1 - uG_1 = -u \\ G_2 &= F_1 + \dot{G}_1 = 0 \end{aligned}$$

It may be appropriate to stop at this point and check that these results make sense. Writing Equations (5-70) for $n = 0, 1, 2$, we have

$$\begin{aligned} \mathbf{r}^{(0)} &= \mathbf{r} = F_0\mathbf{r} + G_0\dot{\mathbf{r}} = \mathbf{r} \\ \mathbf{r}^{(1)} &= \dot{\mathbf{r}} = F_1\mathbf{r} + G_1\dot{\mathbf{r}} = \dot{\mathbf{r}} \\ \mathbf{r}^{(2)} &= \ddot{\mathbf{r}} = F_2\mathbf{r} + G_2\mathbf{r} = -u\mathbf{r} = -\frac{\mu}{r^3}\mathbf{r} \end{aligned}$$

Since everything seems to be working, we shall continue and write

$$F_3 = \dot{F}_2 - uG_2 = -\dot{u} = 3up$$
$$G_3 = F_2 + \dot{G}_2 = -u$$

In the next step we shall see the values of u, p and q:

$$\begin{aligned} F_4 &= \dot{F}_3 - uG_3 = 3\dot{u}p + 3u\dot{p} + u^2 \\ &= 3p(-3up) + 3u(q - 2p^2) + u^2 \\ &= u(u - 15p^2 + 3q) \\ G_4 &= F_3 + \dot{G}_3 = 3up + 3up = 6up \end{aligned}$$

Continuing, we have

$$\begin{aligned} F_5 &= \dot{F}_4 - uG_4 = \dot{u}(u - 15p^2 + 3q) + u(\dot{u} - 30p\dot{p} + 3\dot{q}) - 6u^2p \\ &= -3up(u - 15p^2 + 3q) + u\left[-3up - 30p(q - 2p^2) - 3p(u + 2q)\right] - 6u^2p \\ &= -15up(u - 7p^2 + 3q) \\ G_5 &= F_4 + \dot{G}_4 = u(u - 15p^2 + 3q) + 6u\dot{p} + 6\dot{u}p \\ &= u(u - 15p^2 + 3q) + 6p(q - 2p^2) + 6p(-3up) \\ &= u(u - 45p^2 + 9q) \end{aligned}$$

To continue much further is obviously going to become tedious and laborious; therefore, the algebra was programmed on a computer to get further terms (which are indicated after Equation (5-92)), but we shall use our results so far to determine the functions f and g to terms in τ^5. Referring to Equations (5-68) and (5-70), we have

$$\begin{aligned} \mathbf{r} &= \sum_{n=0}^{\infty} \frac{(t - t_0)^n}{n!} \mathbf{r}_0^{(n)} \\ &= \sum_{n=0}^{\infty} \frac{(t - t_0)^n}{n!} \left[(F_n \mathbf{r} + G_n \mathbf{v})\right]_{t=t_0} \\ &= \left(\sum_{n=0}^{\infty} \frac{\tau^n}{n!} F_n\right)\mathbf{r}_0 + \left(\sum_{n=0}^{\infty} \frac{\tau^n}{n!} G_n\right)\mathbf{v}_0 \end{aligned} \tag{5-89}$$

where $\tau \equiv t - t_0$. Comparing with Equation (5-67), we have

$$f(\mathbf{r}_0, \mathbf{v}_0, t) = \sum_{n=0}^{\infty} \frac{\tau^n}{n!} [F_n]_{t=t_0} \tag{5-90}$$

$$g(\mathbf{r}_0, v_0, t) = \sum_{n=0}^{\infty} \frac{\tau^n}{n!} [G_n]_{t=t_0} \tag{5-91}$$

Using the results for F_n and G_n we have previously derived then gives us

$$\boxed{\begin{aligned} f = 1 - \frac{1}{2}u_0\tau^2 + \frac{1}{2}u_0 p_0 \tau^3 + \frac{1}{24}u_0(u_0 - 15p_0^2 + 3q_0)\tau^4 \\ + \frac{1}{8}u_0 p_0(7p_0^2 - u_0 - 3q_0)\tau^5 + \ldots \end{aligned}} \tag{5-92}$$

$$F_0 = 1,\ F_1 = 0,\ F_2 = -u,\ F_3 = 3up$$

$$F_4 = u(-15p^2 + 3q + u),\ F_5 = 15up(7p^2 - 3q - u)$$

$$F_6 = 105up^2(-9p^2 + 6q + 2u) - u(45q^2 + 24up + u^2)$$

$$F_7 = 315up^3(33p^2 - 30q - 10u) + 63up(25q^2 + 14up + u^2)$$

$$\begin{aligned} F_8 &= 10395up^4(-13p^5 + 15q + 5u) - 315up^2(15q + 7u)(9q + u) \\ &\quad + u(1575q^3 + 1107uq^2 + 117u^2q + u^3) \end{aligned}$$

$$\begin{aligned} F_9 &= 135135up^5(15p^2 - 21q - 7u) + 3465up^3(315q^2 + 186uq + 19u^2) \\ &\quad - 15up(6615q^3 + 4959uq^2 + 729u^2q + 17u^3) \end{aligned}$$

$$\begin{aligned} F_{10} &= 675675up^6(-15p^2 + 84q + 28u) - 1891890up^4(15q^2 + 9uq + u^2) \\ &\quad + 660up^2(6615q^3 + 5184uq^2 + 909u^2q + 32u^3) \\ &\quad - u(99225q^4 + 85410uq^3 + 15066u^2q^2 + 498u^3q + u^4) \end{aligned}$$

$$G_0 = 0,\ G_1 = 1,\ G_2 = 0,\ G_3 = -u,\ G_4 = 6up$$

$$G_5 = u(-45p^2 + 9q + u),\ G_6 = 30up(14p^2 - 6q - u)$$

$$G_7 = 315up^2(-15p^2 + 10q + 2u) - u(225q^2 + 54uq + u^2)$$

$$G_8 = 630up^3(99p^2 - 90q - 20u) + 126up(75q^2 + 24uq + u^2)$$

$$\begin{aligned} G_9 &= 10395up^4(-91p^2 + 105q + 25u) - 945up^2(315q^2 + 118up + 7u^2) \\ &\quad + u(11025q^3 + 4131uq^2 + 243u^2q + u^3) \end{aligned}$$

$$\begin{aligned} G_{10} &= 810810up^5(20p^2 - 28q - 7u) + 13860up^3(630q^2 + 261uq + 19u^2) \\ &\quad - 30up(26460q^3 + 12393uq^2 + 1170u^2q + 17u^3) \end{aligned}$$

$$\boxed{g = \tau - \frac{1}{6}u_0\tau^3 + \frac{1}{4}u_0 p_0 \tau^4 + \frac{1}{120}u_0(u_0 - 45p_0^2 + 9q_0)\tau^5 + \ldots} \tag{5-93}$$

where u_0, p_0 and q_0 are the values of u, p and q at $t = t_0$.

The f and g series will be used in a later section to determine an orbit from sighting directions only.

5.5.2 Solutions of the Gauss Problem

The f and g series may be used to solve the Gauss problem if the time interval between the two measurements is not too large. This method has the advantage of not having a quadrant ambiguity as some other methods. We assume that we are given the positions $\mathbf{r_1}$ and $\mathbf{r_2}$ at time t_1 and t_2 and we wish to find $\mathbf{v_1}$. From Equation (5-67), we have

$$\mathbf{r}_2 = f(\mathbf{r}_1, \mathbf{v}_1, t_2 - t_1)\mathbf{r}_1 + g(\mathbf{r}_1, \mathbf{v}_1, t_2 - t_1)\mathbf{v}_1 \tag{5-94}$$

from which we find

$$\mathbf{v}_1 = \frac{\mathbf{r}_2 - f(\mathbf{r}_1, \dot{\mathbf{r}}_1, \tau)\mathbf{r}_1}{g(\mathbf{r}_1, \dot{\mathbf{r}}_1, \tau)} \tag{5-95}$$

If we guess a value of $\mathbf{v_1}$ we can compute f and g and then use Equation (5-95) to compute a new value of $\mathbf{v_1}$. This method of successive approximations can be continued until $\mathbf{v_1}$ is determined with sufficient accuracy. This method converges very rapidly if τ is not too large.

5.6 THE ORIGINAL GAUSS METHOD

In the interest of its historic and illustrative value we will examine the method that was originally proposed by Gauss in 1809.[4] Although we will assume that the transfer orbit connecting $\mathbf{r_1}$ and $\mathbf{r_2}$ is an ellipse, the extension of the method to cover hyperbolic orbits will be obvious. The derivation of the necessary equations "from scratch" is long and tedious and may be found in Escobal[3] or Moulton.[5] Since all of the relationships we need are contained in the f and g expressions, we will present a very compact and concise development of the Gauss method using only Equations (5-3), (5-4) and (5-5).

5.6.1 Ratio of Sector to Triangle

In going from $\mathbf{r_1}$ to $\mathbf{r_2}$ the radius vector sweeps out the shaded area shown in Figure 5-7. In Chapter 1 we showed that the area is swept out at a constant rate:

$$dt = \frac{2}{h}dA \tag{5-96}$$

Since $h = \sqrt{\mu p}$ the area of the shaded sector, A_S, becomes

$$A_S = \frac{1}{2}(\sqrt{\mu p})t$$

where t is the time of flight from $\mathbf{r_1}$ to $\mathbf{r_2}$.

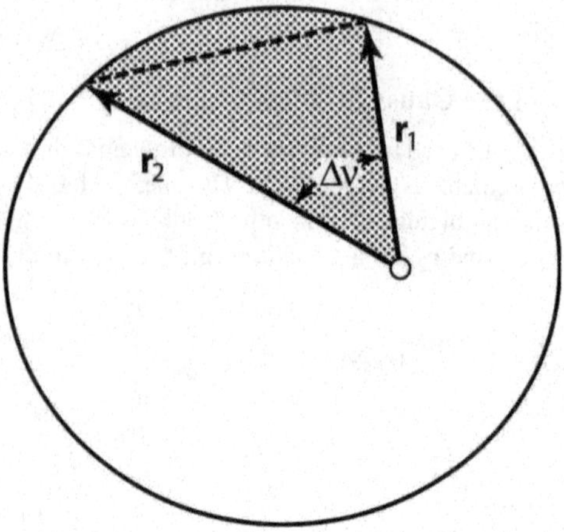

Fig. 5-7 Sector and Triangle Area.

The area of the triangle formed by the two radii and the subtended chord is just one half the base times the altitude, so

$$A_t = \frac{1}{2} r_1 r_2 \sin \Delta \nu$$

Gauss called the *ratio of sector to triangle area y*, thus:

$$\boxed{y = \frac{(\sqrt{\mu p})t}{r_1 r_2 \sin \Delta \nu}} \tag{5-97}$$

The Gauss method is based on obtaining two independent equations relating y and the change in eccentric anomaly, ΔE. A trial value of y (usually $y \approx 1$) is selected and the first equation is solved for ΔE. This value of ΔE is then used in the second equation to compute a better trial value of y. This technique of successive approximations will converge rapidly if y is nearly one, but it fails completely if the radius vector spread is large.

The first equation of Gauss will be obtained by substituting for p in Equation (5-97) an expression that contains ΔE as the only unknown.

5.6.2 The First Equation of Gauss

If we square the expression for the sector-to-triangle ratio we obtain

$$y^2 = \frac{\mu p t^2}{(r_1 r_2 \sin \Delta \nu)^2}$$

By substituting for p from Equation (5-41) and using the identity $(\sin^2 x)/(1 - \cos x) = 2\cos^2(x/2)$, this expression becomes

$$y^2 = \frac{\mu t^2}{2 r_1 r_2 \cos^2 \frac{\Delta \nu}{2} \left(r_1 + r_2 - 2\sqrt{r_1 r_2} \cos \frac{\Delta \nu}{2} \cos \frac{\Delta E}{2} \right)}$$

To simplify this expression, let

$$\boxed{s = \frac{r_1 + r_2}{4\sqrt{r_1 r_2} \cos \frac{\Delta \nu}{2}} - \frac{1}{2}} \tag{5-98}$$

$$\boxed{w = \frac{\mu t^2}{\left(2\sqrt{r_1 r_2} \cos \frac{\Delta \nu}{2} \right)^3}} \tag{5-99}$$

Note that s and w are constants that may be evaluated from the given information.

A little trigonometric manipulation will prove that y^2 may be expressed compactly as

$$\boxed{y^2 = \frac{w}{s + \frac{1}{2}\left(1 - \cos \frac{\Delta E}{2} \right)}} \tag{5-100}$$

which is known as "the first equation of Gauss."

5.6.3 The Second Equation of Gauss

Another completely independent expression for y involving ΔE as the only unknown may be derived from Equations (5-4) and (5-97). From the first of these equation, we see that

$$\frac{r_1 r_2 \sin \Delta \nu}{\sqrt{\mu p}} = t - \sqrt{\frac{a^3}{\mu}} (\Delta E - \sin \Delta E)$$

But $r_1 r_2 \sin \Delta \nu / \sqrt{\mu p} = t / y$, so

$$1 - \frac{1}{y} = \frac{1}{t} \sqrt{\frac{a^3}{\mu}} (\Delta E - \sin \Delta E) \tag{5-101}$$

We still need to eliminate a from this expression. By using the identity, $\sin \Delta \nu = 2 \sin(\Delta \nu / 2) \cos(\Delta \nu / 2)$, Equation (5-96) becomes

$$y = \frac{\sqrt{\mu p}\, t}{2 r_1 r_2 \sin \frac{\Delta \nu}{2} \cos \frac{\Delta \nu}{2}} \tag{5-102}$$

From Equation (5-3) we can write

$$1-\cos\Delta\nu=\frac{ap}{r_1 r_2}(1-\cos\Delta E)$$

$$2\sin^2\frac{\Delta\nu}{2}=2\frac{ap}{r_1 r_2}\sin^2\frac{\Delta E}{2}$$

$$\sin\frac{\Delta\nu}{2}=\sqrt{\frac{ap}{r_1 r_2}}\sin\frac{\Delta E}{2}$$

Substituting this last expression into Equation (5-102) eliminates $\sqrt{p}$ in favor of $\sqrt{a}$:

$$y=\frac{\sqrt{\mu}\,t}{2\sqrt{ar_1 r_2}\sin\frac{\Delta E}{2}\cos\frac{\Delta\nu}{2}} \tag{5-103}$$

If we now cube this equation and multiply it by Equation (5-101), a will be eliminated and we end up with

$$y^3\left(1-\frac{1}{y}\right)=\frac{\mu t^2}{\left(2\sqrt{r_1 r_2}\cos\frac{\Delta\nu}{2}\right)^3}\frac{(\Delta E-\sin\Delta E)}{\sin^3\frac{\Delta E}{2}}$$

Recognizing the first factor as w, we may write, more compactly,

$$y^2(y-1)=w\frac{(\Delta E-\sin\Delta E)}{\sin^3\frac{\Delta E}{2}}$$

Substituting for y^2 from Equation (5-100) and solving for y, we get

$$\boxed{y=1+\left(\frac{\Delta E-\sin\Delta E}{\sin^3\frac{\Delta E}{2}}\right)\left(s+\frac{1-\cos\frac{\Delta E}{2}}{2}\right)} \tag{5-104}$$

which is known as "the second equation of Gauss."

5.6.4 Solution of the Equations

To review what we have done so far, recall that we started with three Equations, (5-3), (5-4) and (5-5), in three unknowns, p, a and ΔE. We then added another independent Equation, (5-97), and one more unknown y. By a process of eliminating p and a between these four equations we now have reduced the set to two equations in two unknowns, y and ΔE. Unfortunately, Equations (5-100) and (5-104) are transcendental, so a trial-and-error solution is necessary.

The first step is to evaluate the constants s and w from $\mathbf{r}_1$, $\mathbf{r}_2$, $\Delta\nu$ and t. Next, pick a trial value for y; since this method only works well if $\Delta\nu$ is less than about 90°, a good first guess is $y \approx 1$.

We can now solve Gauss's first equation for ΔE, using the trial value of y:

$$\cos\frac{\Delta E}{2} = 1 - 2\left(\frac{w}{y^2} - s\right) \tag{5-105}$$

If we assume that ΔE is less than 2π (which will always be the case unless the satellite passes back through $\mathbf{r}_1$ en route to $\mathbf{r}_2$), there is no problem determining the correct quadrant for ΔE.

We are now ready to use this approximate value for ΔE to compute a better approximation for y from Gauss' second equation. This better value of y is then used in Equation (5-105) to compute a still better value of ΔE, and so on, until two successive approximations for y are nearly identical.

When convergence has occurred the parameter p may be computed from Equation (5-41) and the f and g expressions evaluated. The determination of $\mathbf{v}_1$ and $\mathbf{v}_2$ from Equations (5-7) and (5-2) completes the solution.

Since the equations above involve ΔE, they are valid only if the transfer orbit from $\mathbf{r}_1$ to $\mathbf{r}_2$ is elliptical. The extension of Gauss' method to include hyperbolic and parabolic orbits is the subject of the next section.

5.6.5 Extension of Gauss' Method to Any Type of Conic Orbit

If the given time of flight is short, the right-hand side of Equation (5-105) may become greater than one, indicating that ΔE is imaginary. Since we already know that ΔE is imaginary and ΔF is real, we can conclude that the transfer orbit is hyperbolic when this occurs. By noting that $\Delta E = i\Delta F$ and $\cos(i\Delta F) = \cosh(\Delta F)$, Equation (5-105) may also be written as

$$\cosh\frac{\Delta F}{2} = 1 - 2\left(\frac{w}{y^2} - s\right) \tag{5-106}$$

whenever the right side is greater than one.

By using the identity $-i\sin(i\Delta F) = \sinh(\Delta F)$, Equation (5-104) becomes

$$y = 1 + \left(\frac{\sinh\Delta F - \Delta F}{\sinh^3\frac{\Delta F}{2}}\right)\left(s + \frac{1 - \cosh\frac{\Delta F}{2}}{2}\right) \tag{5-107}$$

These equations may be used exactly as Equations (5-104) and (5-105) to determine y.

If the transfer orbit being sought happens to be parabolic, then ΔE and ΔF will be zero and both Equations (5-104) and (5-107) become indeterminate. For this reason, difficulties may be anticipated any time ΔE or ΔF are close to zero.

Gauss solved this problem by defining two auxiliary variables, x (not to be confused with the universal variable of Chapter 4) and X as follows:

$$x = \frac{1}{2}\left(1 - \cos\frac{\Delta E}{2}\right)$$

$$X = \frac{\Delta E - \sin\Delta E}{\sin^3\frac{\Delta E}{2}}$$

The first equation of Gauss may then be written, as

$$y^2 = \frac{w}{s+x}$$

$$\boxed{x = \frac{w}{y^2} - s} \qquad (5\text{-}108)$$

The second equation of Gauss may be written as

$$\boxed{y = 1 + X(s+x)} \qquad (5\text{-}109)$$

Now, it is possible to expand the function X as a power series in x. This may be accomplished by first writing the power series expansion for X in terms of ΔE, and then expressing ΔE as a power series in x. The result, which is developed by Moulton, is

$$X = \frac{4}{3}\left(1 + \frac{6}{5}x + \frac{(6)\cdot(8)}{(5)\cdot(7)}x^2 + \frac{(6)\cdot(8)\cdot(10)}{(5)\cdot(7)\cdot(9)}x^3 \ldots\right) \qquad (5\text{-}110)$$

We may now reformulate the algorithm for *solving the Gauss problem via the Gauss method* as follows:

1. Compute the constants, s and w, from $\mathbf{r}_1$, $\mathbf{r}_2$, $\Delta\nu$ and t using Equations (5-98) and (5-99).
2. Assume $y \approx 1$ and compute x from Equation (5-108).
3. Determine X from Equation (5-110) and use it to compute a better approximation to y from Equation (5-109). Repeat this cycle until y converges to a solution.
4. The type of conic orbit is determined at this point of the orbit as being an ellipse, parabola, or hyperbola according to whether x is positive, zero, or negative. Depending on the type of conic, determine ΔE or ΔF from Equation (5-105) or (5-106).
5. Determine p from Equation (5-41), replacing $\cos((\Delta E)/2)$ with $\cosh(\Delta F)/2$ in the case of the hyperbolic orbit.
6. Evaluate f, g, $\dot{f}$ and $\dot{g}$ from Equations (5-3), (5-4), (5-5) and (5-6).
7. Solve $\mathbf{v}_1$ and $\mathbf{v}_2$ from Equations (5-7) and (5-2).

The method outlined above is perhaps the most accurate and rapid technique known for solving the Gauss problem when $\Delta\nu$ is less than 90°; the iteration to determine y fails to converge shortly beyond this point.

5.7 PRACTICAL APPLICATIONS OF THE GAUSS PROBLEM— INTERCEPT AND RENDEZVOUS

A fundamental problem of astrodynamics is that of getting from one point in space to another in a predetermined time. Usually, we would like to know what velocity is required at the first point in order to coast along a conic orbit and arrive at the destination at a prescribed time.

Applications of the Gauss problem are almost limitless and include interplanetary transfers, satellite intercept and rendezvous, ballistic missile targeting, and ballistic missile interception, as illustrated in Figure 5-8. The subject of ballistic missile targeting is covered in Chapter 6 and interplanetary trajectories are covered in Chapter 8. The Gauss problem is also applicable to lunar trajectories, which is the subject of Chapter 7.

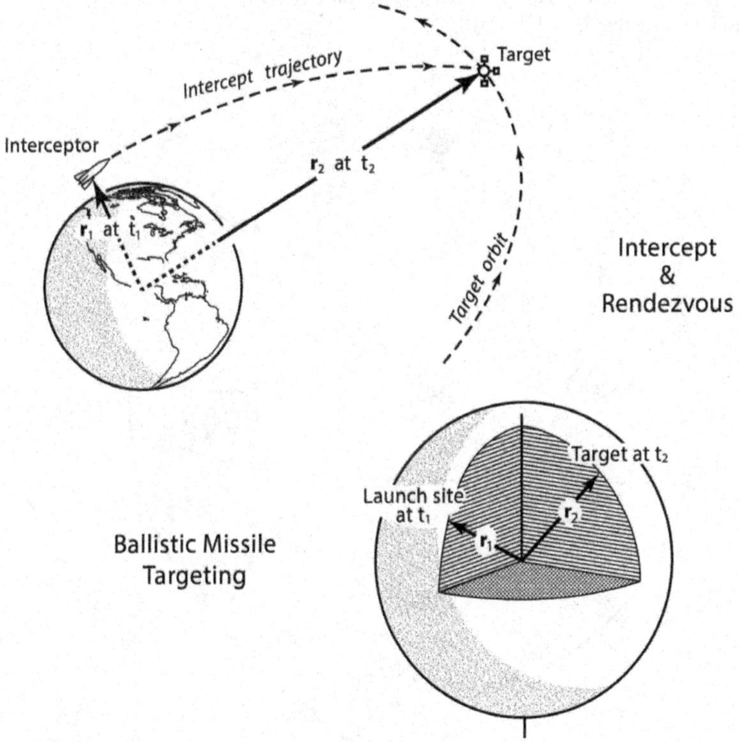

Fig. 5-8 Important Practical Applications of the Gauss Problem.

Orbit determination from two positions and time is usually part of an even larger problem that we may call "mission planning." Mission planning includes determining the optimum timing and sequence of maneuvers for a particular mission and the cost of the mission in terms of Δv.

In all but the simplest cases, the problem of determining the optimum sequencing of velocity changes to give the minimum total Δv defies analytical solution, and we must rely on a computer analysis to establish suitable "launch windows" for a particular mission. To avoid generalities let's define a hypothetical mission and show how such an analysis is performed.

Let's assume that we have the position and velocity of a target satellite at some time t_0 and we wish to intercept this target satellite from a ground launch site. We will assume that a single impulse is added to the launch vehicle to give it its launch velocity. The problem is to establish the optimum launch time and time of flight for the interceptor.

To make the problem even more specific, we will assume that the target satellite is in a nearly circular orbit inclined 65° to the equator and, at time t_0, it is over the Aleutian Islands heading southeastward. Our launch site will be at Johnston Island in the Pacific. The situation at time t_0 is illustrated in Figure 5-9. We will assume that t_0 is 1200 GMT.

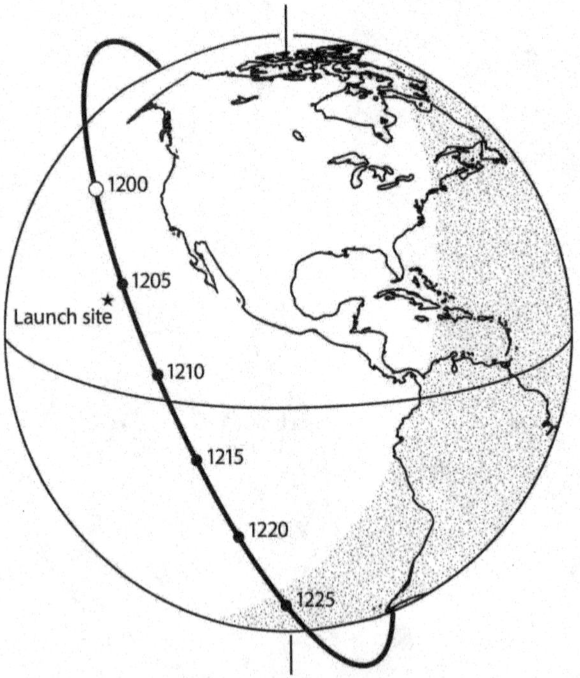

Fig. 5-9 Example Intercept Problem.

The Δv required to intercept the target depends on two parameters, which we are free to choose arbitrarily, the launch time and the time of flight of the interceptor. Suppose we pick a launch time of 1205 and a time of flight of 5 minutes. The first step in determining Δv is to find the position and velocity of the interceptor at 1205. Since it is stationary on the launch pad at Johnston Island, we need to find the position vector from the center of Earth to the launch site at 1205. We have already discussed this problem in Chapter 2. The velocity of the interceptor is due solely to Earth rotation and is in the eastward direction at the site.

The next step is to determine where the target will be at 1210, which is when the intercept will occur. Since we know the position and velocity of the target at 1200, we can update **r** and **v** to 1210 by solving the Kepler problem, which we discussed in Chapter 4.

We now have two position vectors and the time of flight between them, so we can solve the Gauss problem to find what velocity is required at the launch point and what velocity the interceptor will have at the intercept point.

As shown in Figure 5-10 the difference between the required launch velocity and the velocity that the interceptor already has by virtue of Earth rotation is the Δv_1 that the booster rocket must provide to put the interceptor on a collision course with the target. The difference between the velocity of the target and interceptor at the intercept point is the $\Delta \mathbf{v}_2$ that would have to be added if the rendezvous with the target is desired. Since the interceptor must be put on a collision course with the target for a rendezvous mission, the total cost of intercept and rendezvous is the scalar sum of $\Delta \mathbf{v}_1 + \Delta \mathbf{v}_2$.

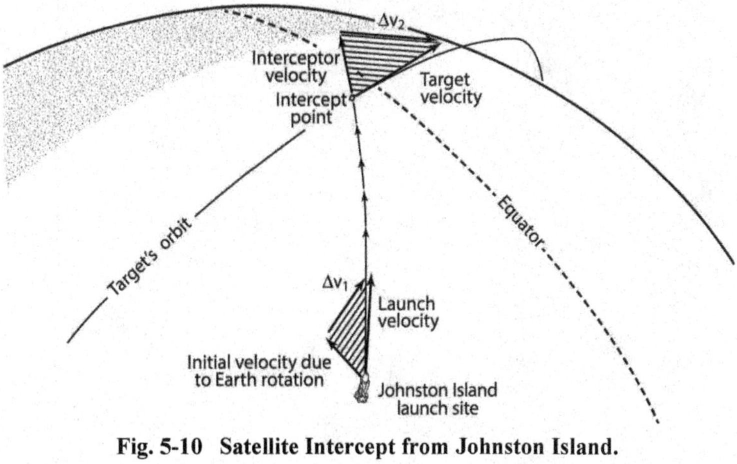

Fig. 5-10 Satellite Intercept from Johnston Island.

For our notational example we take a specific set of orbital parameters and go through the sequence of calculations to get $\Delta v_1 = 4.121$ km/s and $\Delta v_1 + \Delta v_2 = 11.238$ km/s. But how do we know that some other launch time and time of flight might not be cheaper in terms of Δv_1? To find out, we need to repeat the calculations for various combinations of launch time and time of flight. The results may be displayed in tabular form or as we have done in Figure 5-11, where lines of constant Δv indicate the regions of interest.

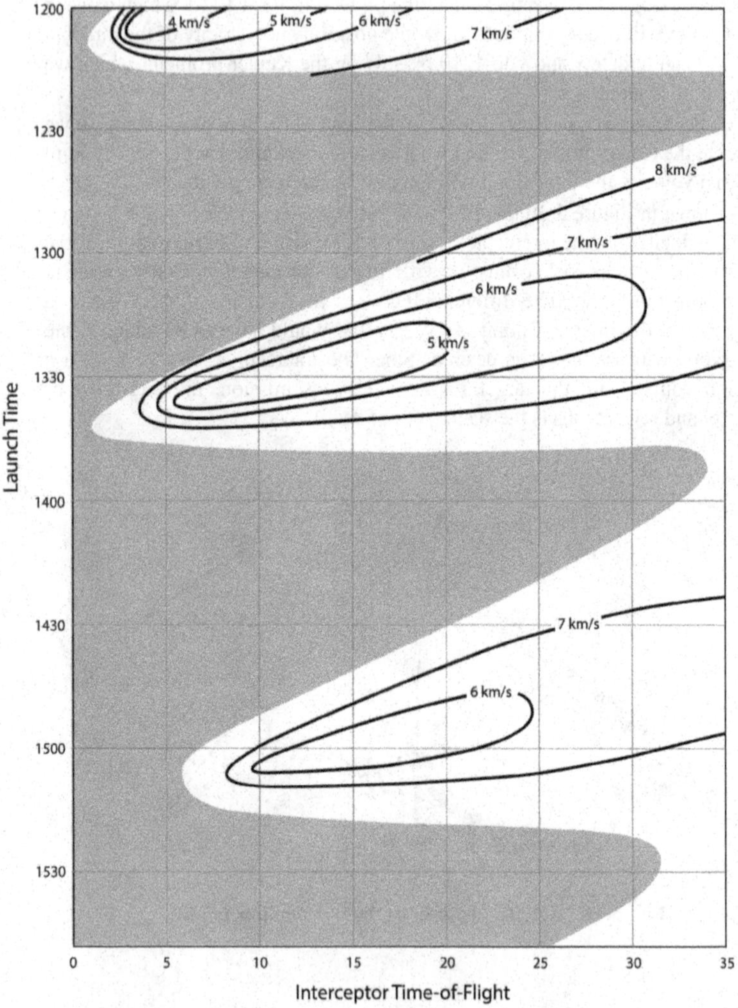

Fig. 5-11 Δv_1 Plot for Example Intercept Problem.

5.7.1 Interpretation of the Δv Plot

A great deal may be learned about a particular mission just by studying a Δv plot. In Figure 5-11 the shaded area represents those conditions that result in the interceptor striking Earth en route to the target. This is most likely to occur when the intercept point is located far from the launch site and time of flight is short. The lowest Δv's for intercept are likely to occur when the intercept point is close to the launch site. From Figure 5-9 we can see that this will first occur when the intercept time is about 1210 (launch at 1205 plus 5 min time of flight). Keep in mind that Earth is rotating and, although the target satellite returns to the same position in space after one orbit period, the launch site will have moved eastward.

The target satellite in our example will again pass close to Johnston Island at about 1340 GMT and another good launch opportunity will present itself. Because of Earth rotation, the next pass at 1510 will not bring the target satellite nearly so close to the launch site. After 12 hours the launch site will again pass through the plane of the target's orbit and another good series of launch windows should occur.

In summary, we can say that the most important single factor in determining the Δv required for intercept is the choice of where the intercept is made relative to the launch site; the closer the intercept point is to the launch site, the better.

A similar Δv plot for the intercept plus rendezvous would show that the lowest Δv's occur when the transfer orbit is coplanar with the target's orbit. This can only occur if the launch takes place at the exact time the launch site is passing through the target's orbit plane and this only occurs twice every 24 hours at most.

5.7.2 Definition of "Optimum Launch Conditions"

It would be a mistake to assume that those launch conditions that minimize Δv are the optimum for a particular mission. We have to know more about the mission before we can properly interpret a Δv plot.

Suppose the mission is to intercept the target as quickly as possible with an interceptor that has a fixed Δv capability. In this case "optimum" launch conditions are those that result in the earliest intercept time without exceeding the Δv limitations of the interceptor.

There are two common instances where we would be trying to minimize Δv. One is where we have a fixed payload and are trying to minimize the size of the launch vehicle required to accomplish the mission. The other is where we have fixed the launch vehicle and are trying to maximize the payload it can carry.

Although we have examined a very specific example, the technique of producing and interpreting a Δv plot as an aid in mission planning is generally applicable to all types of mission analysis.

5.8 DETERMINATION OF ORBIT FROM SIGHTING DIRECTIONS AT STATION

Though one method of orbit determination from optical sightings was presented in Chapter 2, now that we have developed the f and g series, another method will be developed.

Let us assume that we can measure the right ascension and declination of an Earth satellite from some station on Earth at three times t_1, t_2 and t_3. The unit vector $\mathbf{L_i}$ pointing in the direction of the satellite from the station is

$$\mathbf{L}_i = \mathbf{I}\cos\delta_i \cos\alpha_i + \mathbf{J}\cos\delta_i \sin\alpha_i + \mathbf{K}\sin\delta_i \tag{5-111}$$

The vector $\mathbf{r}$ from the center of Earth is

$$\mathbf{r} = \mathbf{R} + \rho\mathbf{L} \tag{5-112}$$

We expand the vector $\mathbf{r}$ in terms of the f and g series evaluated at t_2 so that

$$\mathbf{r} = f(\mathbf{r}_2, \mathbf{v}_2, t - t_2)\mathbf{r}_2 + g(\mathbf{r}_2, \mathbf{v}_2, t - t_2)\mathbf{v}_2 \tag{5-113}$$

then letting $f_i = f(\mathbf{r_2}, \mathbf{v_2}, t_i - t_2)$ and $g_i = g(\mathbf{r_2}, \mathbf{v_2}, t_i - t_2)$ gives

$$f_1\mathbf{r}_2 + g_1\mathbf{v}_2 = \mathbf{R}_1 + \rho_1\mathbf{L}_1 \tag{5-114}$$

$$\mathbf{r}_2 = \mathbf{R}_2 + \rho_2\mathbf{L}_2 \tag{5-115}$$

$$f_3\mathbf{r}_2 + g_3\mathbf{v}_2 = \mathbf{R}_3 + \rho_3\mathbf{L}_3 \tag{5-116}$$

This is a set of nine equations in the nine unknown $\mathbf{r_2}$, $\mathbf{v_2}$, ρ_1, ρ_2 and ρ_3, i.e., the three components or $\mathbf{r_2}$, the three components of $\mathbf{v_2}$ and the three quantities of ρ_1, ρ_2 and ρ_3.

We can eliminate ρ_1 by cross-multiplying the ith equation by $\mathbf{L_i}$ to obtain

$$f_1\mathbf{L}_1 \times \mathbf{r}_2 + g_1\mathbf{L}_1 \times \mathbf{v}_2 = \mathbf{L}_1 \times \mathbf{R}_1 \tag{5-117}$$

$$\mathbf{L}_2 \times \mathbf{r}_2 = \mathbf{L}_2 \times \mathbf{R}_2 \tag{5-118}$$

$$f_3\mathbf{L}_3 \times \mathbf{r}_2 + g_3\mathbf{L}_3 \times \mathbf{v}_2 = \mathbf{L}_3 \times \mathbf{R}_3 \tag{5-119}$$

Although it appears that there are now nine equations in six unknowns, i.e., the components of $\mathbf{r_2}$ and $\mathbf{v_2}$, in fact only six of the equations are independent. By letting the Cartesian components of $\mathbf{r_2}$ be x, y and z, letting the components of $\mathbf{v_2}$ be $\dot{x}$, $\dot{y}$ and $\dot{z}$, and eliminating the $\mathbf{K}$ components, these equations are

$$\boxed{\begin{array}{ll} f_1L_{1z}x - f_1L_{1x}z + g_1L_{1z}\dot{x} - g_1L_{1x}\dot{z} & = R_{1x}L_{1z} - R_{1z}L_{1x} \\ f_1L_{1z}y - f_1L_{1y}z + g_1L_{1z}\dot{y} - g_1L_{1y}\dot{z} & = R_{1y}L_{1z} - R_{1z}L_{1y} \\ L_{2z}x - L_{2x}z & = R_{2x}L_{2z} - R_{2z}L_{2x} \\ L_{2z}y - L_{2y}z & = R_{2y}L_{2z} - R_{2z}L_{2y} \\ f_3L_{3z}x - f_3L_{3x}z + g_3L_{3z}\dot{x} - g_3L_{3x}\dot{z} & = R_{3x}L_{3z} - R_{3z}L_{3x} \\ f_3L_{3z}y - f_3L_{3y}z + g_3L_{3z}\dot{y} - g_3L_{3y}\dot{z} & = R_{3y}L_{3z} - R_{3z}L_{3y} \end{array}} \tag{5-120}$$

A procedure that may be used to solve this set of equations is as follows:

a. Estimate the magnitude of $\mathbf{r}_2$.

b. Using this estimate compute $u_2 = \frac{\mu}{r_2^3}$.

c. Compute the values of f_1, g_1, f_3 and g_3 using the terms of Equations (5-92) and (5-93) that are independent of p_2 and q_2.

d. Substitute these values of f_1, g_1, f_3 and g_3 and the known values of the components of $\mathbf{L}_1$, $\mathbf{L}_2$, $\mathbf{L}_3$, $\mathbf{R}_1$, $\mathbf{R}_2$ and $\mathbf{R}_3$ into Equations (5-120) and solve the resulting linear algebraic equations for the six unknowns $x, y, z, \dot{x}, \dot{y}$ and $\dot{z}$, which are the components of $\mathbf{r}_2$ and $\mathbf{v}_2$.

e. Compute new values of u_2, p_2 and q_2 from these $\mathbf{r}_2$ and $\mathbf{v}_2$ using their definitions in Equations (5-88). Then compute new values of f_1, g_1, f_3 and g_3, from Equations (5-92) and (5-93), using as many terms as are necessary to obtain required accuracy.

f. Repeat steps d and e until the process converges to correct values of $\mathbf{r}_2$ and $\mathbf{v}_2$.

This process converges very rapidly if the time intervals $t_3 - t_2$ and $t_2 - t_1$ are not too large, but it is probably too tedious for hand computation because of the time required to solve the system of six linear algebraic equations several times. The process is well suited to solution on a digital computer.

Exercises

5.1 Several methods for solving the Gauss problem have been developed in this chapter. Discuss and rank each method for each of the following criteria:

a. Limitations

b. Ease of computation

c. Accuracy

5.2 As a mission planner you could calculate Δv's for various combinations of reaction time and time of flight to the target for a given interceptor location. What relative orientation between launch site and target would minimize the Δv required for intercept?

5.3 Verify Equations (5-21) through (5-23) by developing them from Equations (5-8), (5-9) and (5-17).

5.4 Verify the development of Equation (5-46).

5.5 Make a plot similar to Figure 5-3 for specific values of $\mathbf{r}_1$ and $\mathbf{r}_2$ (such as $\mathbf{r}_1 = 2\mathbf{I}$, $\mathbf{r}_2 = \mathbf{I} + \mathbf{J}$ DU). Using specific numbers, verify and amplify the descriptive statements used in discussing Figure 5-3.

5.6 Derive Equation (5-66) for hyperbolic orbits.

5.7 Given two position vectors $\mathbf{r}_1$ and $\mathbf{r}_2$ and the distance between them, d. Find an expression for the semi-major axis of the minimum energy ellipse, which will contain both position vectors, as a function of $\mathbf{r}_1$, $\mathbf{r}_2$ and d.

Note: The remaining exercises are well suited for computer use, although a few iterations can be made by hand.

5.8 An Earth satellite is observed at two times t_1 and t_2 to have the following positions:

$\mathbf{r}_1 = \mathbf{I}$ DU

$\mathbf{r}_2 = \mathbf{I} + \mathbf{J} + \mathbf{K}$ DU

Find p, **e** and **h** for $t_2 - t_1 = 1.0922$ TU. Use the p-iteration technique with p = 2 as a starting value.

(Answer: $\mathbf{h} = 1.009\,(-\mathbf{J} + \mathbf{K})$)

5.9 Repeat Problem 5.8 using the universal variable.

5.10 An Earth satellite's positions at two times t_1 and t_2 are measured by radar to be

$\mathbf{r}_1 = \mathbf{I}$ DU

$$\mathbf{r}_2 = \mathbf{I} + \frac{1}{8}\mathbf{J} + \frac{1}{8}\mathbf{K} \text{ DU}$$

and $t_2 - t_1 = \frac{1}{8}$TU

Find the velocity $\mathbf{v}_1$ at t using the f and g series.

(Answer: $\mathbf{v}_1 = 0.061858026\mathbf{I} + 1.00256\,(\mathbf{J} + \mathbf{K})$).

5.11 For the following data sets for $\mathbf{r}_1$, $\mathbf{r}_2$ and $t_2 - t_1$, determine $\mathbf{v}_2$ using

a. The universal variable method.

b. The p-iteration technique.

c. The original Gauss method.

Compare the accuracy and speed of convergence.

a. $\mathbf{r}_1 = 0.5\mathbf{I} + 06\mathbf{J} + 0.7\mathbf{K}$ DU, $t_2 - t_1 = 20$ TU

$\mathbf{r}_2 = -\mathbf{J}$ DU (Use "long way" trajectory)

(Answer: $\mathbf{v}_2 = 0.66986992\mathbf{I} + 0.48048471\mathbf{J} + 0.93781789\mathbf{K}$ DU/TU)

b. $\mathbf{r}_1 = 1.2\mathbf{I}$ DU, $t_2 - t_1 = 10$ TU

$\mathbf{r}_2 = 2\mathbf{J}$ DU (Use "short way" trajectory)

c. $\mathbf{r}_1 = \mathbf{I}$, $t_2 - t_1 = 0.0001$ TU

$\mathbf{r}_2 = \mathbf{J}$ (Use "short way" trajectory)

d. $\mathbf{r}_1 = 4\mathbf{I}$, $t_2 - t_1 = 10$ TU

$\mathbf{r}_2 = -2\mathbf{I}$ (Use "short way" trajectory. Why is this data set insoluable?)

e. $\mathbf{r}_1 = 2\mathbf{I}$, $t_2 - t_1 = 20$ TU

$\mathbf{r}_2 = -2\mathbf{I} - 0.2\mathbf{J}$ (Use "long way" trajectory)

List of References

1. Bell, Eric Temple. "The Prince of Mathematicians," in *The World of Mathematics*. Vol. 1, New York, NY, Simon and Schuster, 1956.

2. Battin, Richard H. *Astronautical Guidance*. New York, NY, McGraw-Hill Book Company, 1964.

3. Escobal, Pedro Ramon. *Methods of Orbit Determination*. New York, NY, John Wiley & Sons, Inc., 1965.

4. Gauss, Carl Friedrich. *Theory of the Motion of the Heavenly Bodies Revolving about the Sun in Conic Sections*, a translation of *Theoria Motus* (1857) by Charles H. Davis, New York, NY, Dover Publications, Inc., 1963.

5. Moulton, Forest Ray. *An Introduction to Celestial Mechanics*. New York, NY, Macmillan, 1914.

6. Herrick, S. and A. Liu. *Two Body Orbit Determination from Two Positions and Time of Flight*. Appendix A, Aeronutronic Publ. No. C-365, 1959.

Chapter 6

Ballistic Missile Trajectories

'Tis a principle of war that when you can use the lightning 'tis better than cannon.

-Napoleon I

6.1 Historical Background

While the purist might insist that the history of ballistic missiles stretches back to the first use of crude rockets in warfare by the Chinese, long-range ballistic missiles, which concern us in this chapter, have a relatively short history.

The impetus for developing the long-range rocket as a weapon of war, was, ironically, the Treaty of Versailles, which ended World War I. It forbade the Germans to develop long-range artillery. As a result, the German High Command was more receptive to suggestions for rocket development than were military commands in other countries. The result is well known. Efforts began in 1932 under the direction of then Captain Walter Dornberger, culminating in the first successful launch of an A-4 ballistic missile (commonly known as the V-2) on 3 October 1942. During 1943 and 1944 over 280 test missiles were fired from Peenemünde. The first two operational missiles were fired against Paris on 6 September 1944 and an attack on London followed 2 days later. By the end of the war in May 1945 over 3,000 V-2's had been fired in anger.[1]

General Dornberger summed up the use of the long-range missile in World War II as "too late."[2] He might have added that it was not a particularly effective weapon as used since it had a dispersion at the target of 10 miles over a range of 200 miles.[3] Nevertheless, it was more than just an extension of long-range artillery—it was the first ballistic missile as we know it today.

After the war, there was a mad scramble to "capture" the German scientists of Peenemünde who were responsible for this technological miracle. The United States Army obtained the services of most of the key scientists and technicians including Dornberger and Wernher von Braun. The Soviets were able to assemble at Khimki a staff of about 80 men under the former propulsion expert Werner Baum. They were assigned the task of designing a rocket motor with a thrust of 120tons-force (metric) and later one of 230 tons-force (metric).

At an intelligence briefing at Wright-Patterson Air Force Base in August 1952, these disquieting facts were revealed by Dornberger who had interviewed many of his former colleagues on a return trip to Germany. Even more disquieting should have been the report that the Soviets had built a separate factory building adjacent to that occupied by the German workers. No German was permitted to enter this separate building.

The experts displayed no particular sense of immediacy. After all, the atomic bomb was still too heavy to be carried by a rocket. According to Dr. Darol Froman of the Los Alamos Scientific Laboratory the key question in the 1950's was "when could the Atomic Energy Commission come up with a warhead light enough to make missiles practical?"

The ballistic missile program in the U.S.A. was still essentially in abeyance when a limited contract for the intercontinental ballistic missile (ICBM) called "Atlas" was awarded. But in November 1952 at Eniwetok the thermonuclear "Mike" shot ended all doubts and paved the way for the "Shrimp" shot of March 1954, which revolutionized the program.[4]

Accordingly, in June 1953, Trevor Gardner, Assistant Secretary of the Air Force for Research and Development, convened a special group of the nation's leading scientists known as the Teapot Committee. The group was led by Professor John von Neumann of the Institute for Advanced Studies at Princeton. They met in a vacant church in Inglewood, California, and the result of their study was a recommendation to the Air Force that the ballistic missile program be reactivated with top priority.

In October 1953 a study contract was placed with the Ramo-Wooldridge Corporation and by May 1954 the new ICBM program had highest Air Force Priority. In July 1954, Brigadier General Bernard A. Schriever was given the monumental task of directing the accelerated ICBM program and began handpicking a staff of military assistants. When he reported to the West Coast he had with him a nucleus of four officers—among them Lieutenant Colonel Benjamin P. Blasingame, later to become the first Head of the Department of Astronautics at the United States Air Force Academy.

Within a year of its beginning in a converted parochial school building in Inglewood the program had passed from top Air Force priority to top national priority. From two main contractors at the beginning, the program had, by mid-1959, 30 main contractors and more than 80,000 people participating directly.

Progress was rapid. After three unsuccessful attempts, the first successful flight of a Series A Atlas took place on 17 December 1957. Only four months after the Soviet Union had announced that it had an intercontinental ballistic missile, Atlas was a reality.

The trajectory of a missile differs from a satellite orbit in only one respect—it intersects the surface of Earth. Otherwise, it follows a conic orbit during the free-flight portion of its trajectory and we can analyze its behavior according to principles that were covered earlier.

Ballistic missile targeting is just a special application of the Gauss problem that we treated rigorously in Chapter 5. In this chapter we will present a somewhat simplified scalar analysis of the problem, so that you may gain some fresh insight into the nature of ballistic trajectories. To compute precise missile trajectories requires the full complexity of perturbation theory. In this chapter we are concerned mainly with concepts.

6.2 THE GENERAL BALLISTIC MISSILE PROBLEM

A ballistic missile trajectory is composed of three parts—the *powered flight* portion, which lasts from launch to thrust cutoff or burnout, the *free-flight* portion, which constitutes most of the trajectory, and the *reentry* portion, which begins at some ill-defined point where atmospheric drag becomes a significant force in determining the missile's path and lasts until impact.

Since mechanical energy is continuously being added to the missile during powered flight, we cannot use two-body mechanics to determine its path from launch to burnout. The path of the missile during this critical part of the flight is determined by the guidance and navigation system. This is the topic of an entire course and will not be covered here.

During free flight, the trajectory is part of a conic orbit—almost always an ellipse—which we can analyze using the principles learned in Chapter 1.

Reentry involves the dissipation of energy by friction with the atmosphere. It will not be discussed in this text.

We will begin by assuming that Earth does not rotate and that the altitude at which reentry starts is the same as the burnout altitude. This latter assumption ensures that the *free-flight trajectory is symmetrical* and will allow us to derive a fairly simple expression for the free-flight range of a missile in terms of its burnout conditions.

We will then answer a more practical question—"Given r_{bo}, v_{bo} and a desired free-flight range, what flight-path angle at burnout is required?"

Following a discussion of maximum range trajectories, we will determine the time of flight for the free-flight portion of the trajectory.

6.2.1 Geometry of the Trajectory

The book already covered the terminology of orbital mechanics, so such terms as "height at burnout," "height of apogee," "flight-path angle at burnout," etc., need not be redefined. There are, however, a few new and unfamiliar terms that must be understood before we embark on any derivations. Figure 6-1 defines these new quantities.

6.2.2 The Nondimensional Parameter, Q

We will find it very convenient to define a nondimensional parameter called Q such that

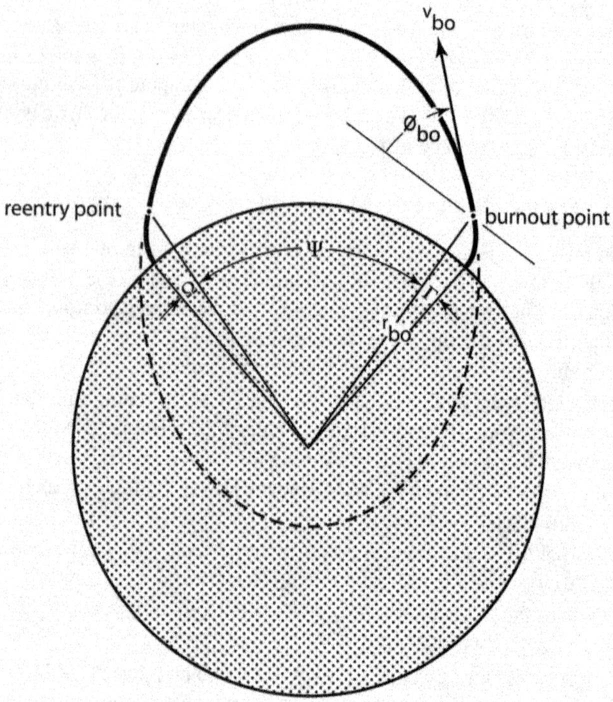

Γ - powered flight range angle

Ψ - free-flight range angle

Ω - reentry range angle

Λ - total range angle

R_p - ground range of powered flight

R_{ff} - ground range of free flight

R_{re} - ground range of reentry

R_t - total ground range

$$\Lambda = \Gamma + \Psi + \Omega$$

$$R_t = R_p + R_{ff} + R_{re}$$

Fig. 6-1 Geometry of the Ballistic Missile Trajectory.

$$Q \equiv \frac{v^2 r}{\mu} \tag{6-1}$$

Q can be evaluated at any point in an orbit and may be thought of as the squared ratio of the speed of the missile to circular satellite speed at that point. Since $v_{cs} = \sqrt{\mu/r}$,

$$Q = \left(\frac{v}{v_{cs}}\right)^2 = \frac{v^2 r}{\mu} \qquad (6\text{-}2)$$

The value of Q is not constant for a missile but varies from point to point in the orbit. When Q is equal to 1, the missile has exactly local circular satellite speed. This condition (Q = 1) exists at every point in a circular orbit and at the end of the minor axis of *every* elliptical orbit.

If Q = 2 it means that the missile has exactly escape speed and is on a parabolic orbit. If Q is greater than 2, the missile is on a hyperbolic orbit. It would be rare to find a ballistic missile with a Q equal to or greater than 2.

We can take the familiar energy equation of Chapter 1,

$$E = \frac{v^2}{2} - \frac{\mu}{r} = -\frac{\mu}{2a}$$

and substitute for v^2 the expression $(\mu Q)/r$ from Equation (6-1). This yields both of the following useful relationships:

$$\boxed{a = \frac{r}{2-Q}} \qquad (6\text{-}3)$$

or

$$\boxed{Q = 2 - \frac{r}{a}} \qquad (6\text{-}4)$$

6.2.3 The Free-Flight Range Equation

Since the free-flight trajectory of a missile is a conic section, the general equation of a conic can be applied to the burnout point,

$$r_{bo} = \frac{p}{1 + e\cos\nu_{bo}} \qquad (6\text{-}5)$$

Solving for $\cos \nu_{bo}$, we get

$$\cos\nu_{bo} = \frac{p - r_{bo}}{e r_{bo}} \qquad (6\text{-}6)$$

Since the free-flight trajectory is assumed to be symmetrical ($h_{bo} = h_{re}$), half the free-flight range angle, Ψ, lies on each side of the major axis, and

$$\cos\frac{\Psi}{2} = -\cos\nu_{bo} \qquad (6\text{-}7)$$

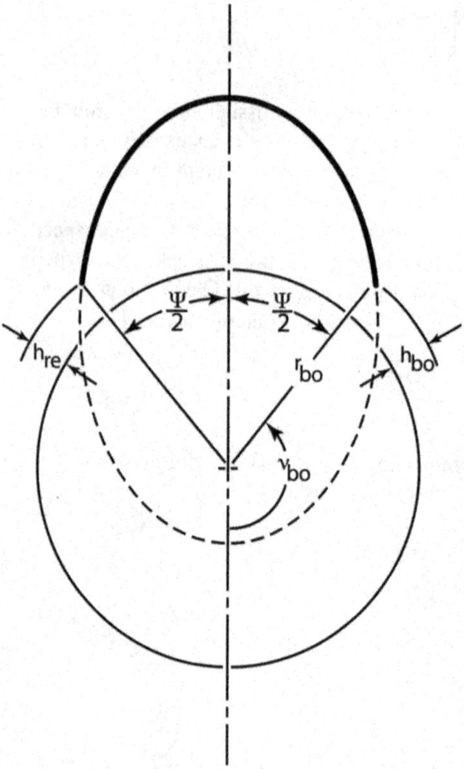

Fig. 6-2 Symmetrical Trajectory.

Equation (6-6) above can, therefore, be written as

$$\cos\frac{\Psi}{2} = \frac{r_{bo} - p}{e r_{bo}} \tag{6-8}$$

We now have an expression for the free-flight range angle in terms of p, e, and r_{bo}. Since $p = h^2/\mu$ and $h = rv\cos\phi$, we can use the definition of Q to obtain

$$p = \frac{r^2 v^2 \cos^2\phi}{\mu} = r\,Q\cos^2\phi \tag{6-9}$$

Now, since $p = a(1 - e^2)$,

$$e^2 = 1 - \frac{p}{a} \tag{6-10}$$

Substituting $p = rQ\cos^2\phi$ and $a = r/(2 - Q)$ gives

$$e^2 = 1 + Q(Q - 2)\cos^2\phi \tag{6-11}$$

If we now substitute Equations (6-9) and (6-11) into Equation (6-8) we have one form of the *free-flight range equation*:

$$\cos\frac{\Psi}{2} = \frac{1 - Q_{bo}\cos^2\phi_{bo}}{\sqrt{1 + Q_{bo}(Q_{bo} - 2)\cos^2\phi_{bo}}} \tag{6-12}$$

From this equation we can calculate the free-flight range angle resulting from any given combination of burnout conditions r_{bo}, v_{bo} and ϕ_{bo}.

While this will prove to be a very valuable equation, it is not particularly useful in solving the typical ballistic missile problem, which can be stated as follows: Given a particular launch point and target, the total range angle, Λ, can be calculated, as we shall see later in this chapter. If we know how far the missile will travel during powered flight and reentry, the required free-flight range angle, Ψ, also becomes known. If we now specify r_{bo} and v_{bo} for the missile, what should the flight-path angle, ϕ_{bo}, for the missile to hit the target?

In other words, it would be nice to have an equation for ϕ_{bo} in terms of r_{bo}, v_{bo} and Ψ.

We could, with a little algebra, choose the straightforward way of solving the range equation for $\cos\phi_{bo}$. Instead, we will derive an expression for ϕ_{bo} geometrically because it demonstrates some rather interesting geometrical properties of the ellipse.

6.2.4 The Flight-Path Angle Equation

In Figure 6-3 we have drawn the local horizontal at the burnout point and also the tangent and normal at the burnout point. The line from the burnout point to the secondary focus F′ is called r'_{bo}.

The angle between the local horizontal and the tangent (in the direction of v_{bo}) is the flight-path angle, ϕ_{bo}. Since r_{bo} is perpendicular to the local horizontal, and the normal is perpendicular to the tangent, the angle between r_{bo} and the normal is also ϕ_{bo}.

Now, it can be proven (although we won't do it) that the angle between r_{bo} and r'_{bo} is bisected by the normal. This fact gives rise to many interesting applications for the ellipse. It means, for example, that if the ellipse represented the surface of a mirror, light emanating from one focus would be reflected to the other focus since the angle of reflection equals the angle of incidence. If the ceiling of a room were made in the shape of an ellipsoid, a person standing at a particular point in the room corresponding to one focus could be heard clearly by a person standing at the other focus even if that person were whispering. This is, in fact, the basis for the so-called "whispering gallery."

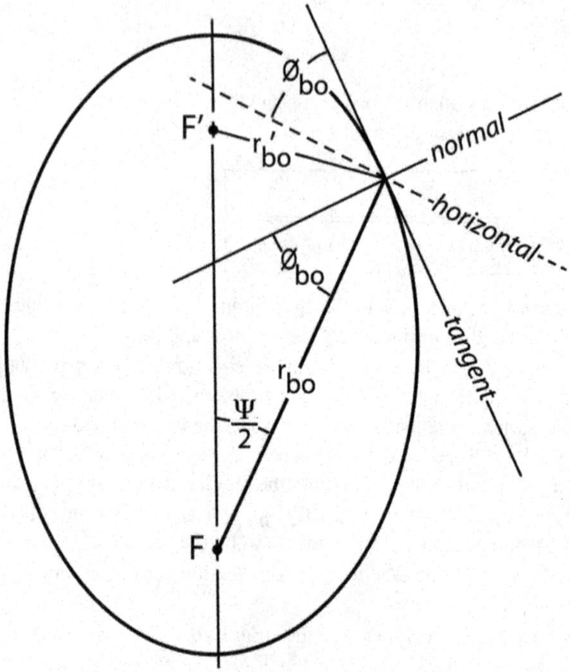

Fig. 6-3 Ellipse Geometry.

What all this means for our derivation is simply that the angle between r_{bo} and r'_{bo} is $2\phi_{bo}$.

Let us concentrate on the triangle formed by F, F′ and the burnout point. We know two of the angles in this triangle and the third can be determined from the fact that the angles of a triangle sum to 180°. If we divide the triangle into two right triangles by the dashed line, d, shown in Figure 6-4, we can express d as

$$d = r_{bo} \sin \frac{\Psi}{2} \tag{6-13}$$

and also as

$$d = r'_{bo} \sin\left[180° - \left(2\phi_{bo} + \frac{\Psi}{2}\right)\right] \tag{6-14}$$

Combining these two equations and noting that sin (180° – x) = sin x, we get

$$\sin\left(2\phi_{bo} + \frac{\Psi}{2}\right) = \frac{r_{bo}}{r'_{bo}} \sin \frac{\Psi}{2} \tag{6-15}$$

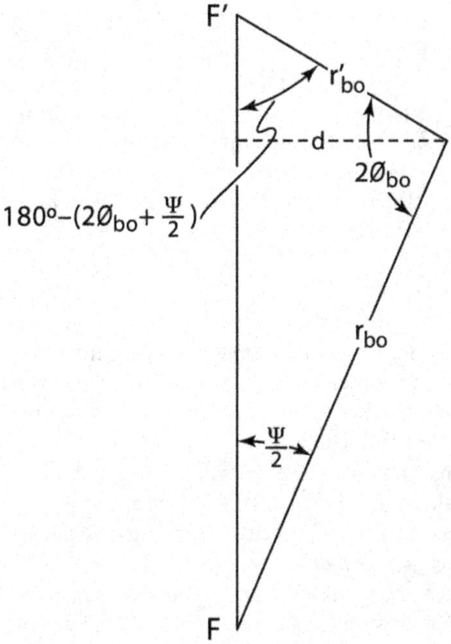

Fig. 6-4 Ellipse Geometry.

Since $r_{bo} = a(2 - Q_{bo})$ from Equation (6-3) and $r'_{bo} + r_{bo} = 2a$,

$$\boxed{\sin\left(2\phi_{bo} + \frac{\Psi}{2}\right) = \frac{2 - Q_{bo}}{Q_{bo}} \sin\frac{\Psi}{2}} \tag{6-16}$$

This is called the *flight-path angle equation* and it points out some interesting and important facts about ballistic missile trajectories.

Suppose we want a missile to travel a free-flight range of 90° and it has $Q_{bo} =$ 0.9. Substituting these values into Equation (6-16) gives us

$$\sin\left(2\phi_{bo} + 45°\right) = \frac{2 - 0.9}{0.9} \sin 45° = 0.866 \tag{6-17}$$

But there are *two angles* whose sine equals 0.866, so

$$2\phi_{bo} + 45° = 60° \text{ and } 120°$$

and

$$\phi_{bo} = 7.5° \text{ and } 37.5°$$

There are two trajectories to the target that result from the same values of r_{bo} and v_{bo}. The trajectory corresponding to the larger value of flight-path angle is call the *high trajectory*; the trajectory associated with the smaller flight-path angle is the *low trajectory*.

The fact that there are two trajectories to the target should not surprise you since even very short-range ballistic trajectories exhibit this property. A familiar illustration of this result is the behavior of water discharged from a garden hose. With constant water pressure and nozzle setting, the speed of the water leaving the nozzle is fixed. If a target well within the maximum range of the hose is selected, the target can be hit by a flat or lofted trajectory.

The nature of the high and low trajectory depends primarily on the value of Q_{bo}. If Q_{bo} is less than 1 there will be a limit to how large Ψ may be such that the value of the right side of Equation (6-16) does not exceed 1. This implies that *there is a maximum range for a missile with Q_{bo} less than 1*. This maximum range will always be less than 180° for Q_{bo} less than 1. Provided that Ψ is attainable, there will be *both a high and a low trajectory to the target.*

If Q_{bo} is exactly 1, one of the trajectories to the target will be the circular orbit connecting the burnout and reentry points. This would not be a very practical missile trajectory, but it does represent the borderline case where ranges of 180° and more are just attainable.

If Q_{bo} is greater than 1, Equation (6-16) will always yield one positive value and one negative value for ϕ_{bo}, regardless of range. A negative ϕ_{bo} is not practical since the trajectory would penetrate Earth, so *only the high trajectory can be realized for Q_{bo} greater than 1.*

The real significance of Q_{bo} greater than 1 is that ranges in excess of 180° are possible. An illustration of such a trajectory would be a missile directed at the North American continent from Asia via the south pole. While such a trajectory would avoid detection by the northern radar "fences," it would be costly in terms of payload delivered and accuracy attainable. Nevertheless, the shock value of such a surprise attack in terms of what it might do toward creating chaos among defensive forces would not be overlooked by military planners.

Since both the high and low trajectories result from the same r_{bo} and v_{bo}, they both have the same energy. Because $a = -\mu/2E$, the major axes of the high and low trajectories are the same length.

Table 6-1 shows which trajectories are possible for various combinations of Q_{bo} and Ψ. Figure 6-5 should be helpful in visualizing each case.

Table 6-1: Significance of Q_{bo}.

	$\Psi < 180°$	$\Psi > 180°$
$Q_{bo} < 1$	Both high and low if $\Psi <$ maximum range	Impossible
$Q_{bo} = 1$	Both high and low ($\phi_{bo} = 0$ for low)	High has $\phi_{bo} = 0$ Low trajectory skims Earth
$Q_{bo} > 1$	Both high and low if $\Psi <$ maximum range	High only—low trajectory hits Earth

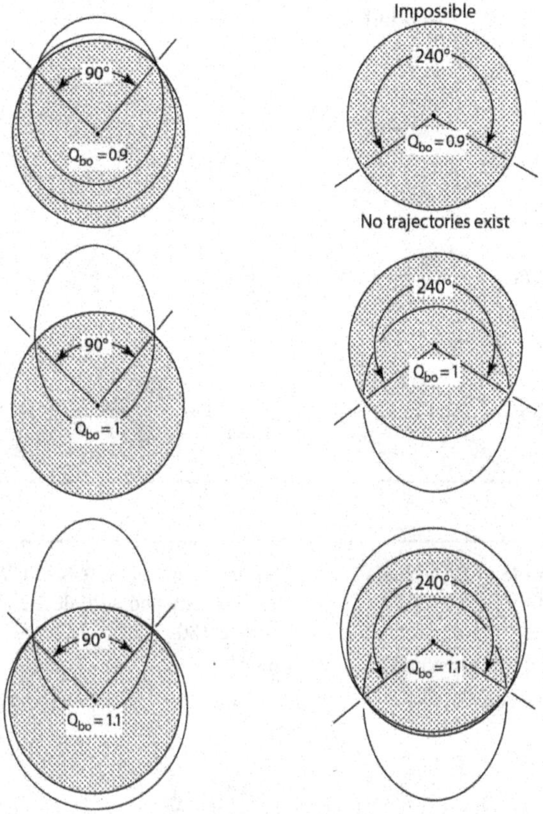

Fig. 6-5 Example Ballistic Missile Trajectories.

Example Problem. During the test firing of a ballistic missile, the following measurements were made: h_{bo} = 1,275.6 km, v_{bo} = 5.2702 km/s and h_{apogee} = 3,189.1 km. Assuming a symmetrical trajectory, what was the free-flight range of the missile during this test in kilometers?

Before we can use the free-flight range equation to find Ψ, we must find ϕ_{bo} and Q_{bo}. First, we calculate the energy and velocity:

$$E = \frac{v_{bo}^2}{2} - \frac{\mu}{r_{bo}} = -38.1914 \text{ km}^2/\text{s}^2$$

$$v_a = \sqrt{2\left(\frac{\mu}{r_a} + E\right)} = 2.6351 \text{ km/s}$$

Then $h = r_a v_a = 25{,}211\ \mathrm{km^2/s}$

but $h = r_{bo}\, v_{bo} \cos\phi_{bo}$

so $\cos\phi_{bo} = 0.625$

and $Q_{bo} = \dfrac{v_{bo}^2 r_{bo}}{\mu} = 0.533$

Using Equation (6-12), $\cos\dfrac{\Psi}{2} = 0.95$, so

$\Psi = 36°24'$ or $36.4°$

$$R_{ff} = (36.4\ \mathrm{deg})\left(111.12\frac{\mathrm{km}}{\mathrm{deg}}\right) = \underline{\underline{4{,}043.6\ \mathrm{km}}}$$

Example Problem. A missile's coordinates at burnout are: 30°N, 30°W. Reentry is planned for 30°S, 120°W. The burnout velocity and altitude are 8.5512 km/s and 159.4534 km, respectively. Ψ is less than 180°.

What must the flight-path angle be at burnout?

Before we can use the flight-path angle equation to find ϕ_{bo}, we must find Q_{bo} and Ψ.

$$Q_{bo} = \frac{v_{bo}^2 r_{bo}}{\mu} = 1.1933$$

From spherical trigonometry,

$$\cos\Psi = \cos 60° \cos 120° + \sin 60° \sin 120° \cos 90° = -0.250$$

so $\Psi = 104.48°$

From the flight-path angle equation,

$$\sin\left(2\phi_{bo} + \frac{104.48°}{2}\right) = \frac{2-1.2}{1.2}\sin\left(\frac{104.48°}{2}\right) = 0.5277$$

Therefore, $2\phi_{bo} + 52.24° = 142.10°$

or $\phi_{bo} = \underline{\underline{47.9776°}}$

6.2.5 The Maximum Range Trajectory

Suppose we plot the free-flight range, Ψ, versus the flight-path angle, ϕ_{bo}, for a fixed value of Q_{bo} less than 1. We get a curve like that shown in Figure 6-6. As the flight path angle is varied from 0° to 90° the range first increases then reaches a maximum and decreases to zero again. Notice that for every range except the maximum there are two values of ϕ_{bo} corresponding to a high and a low trajectory. *At maximum range there is only one path to the target.*

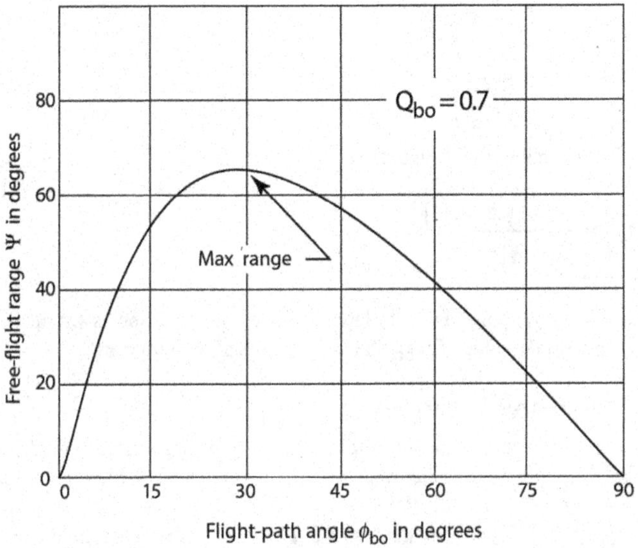

Fig. 6-6 Range vs. ϕ_{bo}.

There are at least two ways that we could derive expressions for the maximum range conditions. One way is to derive an expression for $\delta\Psi/\delta\phi$ and set it equal to zero. A simpler method is to see under what conditions the flight-path angle equation yields a single solution.

If the right side of Equation (6-16) equals exactly 1, we get only a single answer for ϕ_{bo}. This must, then, be the maximum range condition:

$$\sin\left(2\phi_{bo} + \frac{\Psi}{2}\right) = \frac{2 - Q_{bo}}{Q_{bo}} \sin\frac{\Psi}{2} = 1 \tag{6-18}$$

from which $2\phi_{bo} + \frac{\Psi}{2} = 90°$

and

$$\boxed{\phi_{bo} = \frac{1}{4}\left(180° - \Psi\right)} \tag{6-19}$$

for maximum range conditions only.

We can easily find the maximum range angle attainable with a given Q_{bo}. From Equation (6-18),

$$\boxed{\sin\frac{\Psi}{2} = \frac{Q_{bo}}{2 - Q_{bo}}} \tag{6-20}$$

for maximum range conditions.

If we solve this equation for Q_{bo}, we get

$$\boxed{Q_{bo} = \frac{2\sin(\Psi/2)}{1 + \sin(\Psi/2)}} \tag{6-21}$$

for maximum range conditions. This latter form of the equation is useful for determining the lowest value of Q_{bo} that will attain a given range angle.

6.2.6 Time of Free Flight

The time-of-flight methods developed in Chapter 4 are applicable to the free-flight portion of a ballistic missile trajectory, but owing to the symmetry of the case where $h_{re} = h_{bo}$, the equations are considerably simplified. From the symmetry of Figure 6-7 you can see that the time of flight from burnout to re-entry is just twice the time of flight from burnout (point 1) to apogee (point 2).

By inspection, the eccentric anomaly of point 2 is π radians or 180°. The value of E_1 can be computed from Equation (4-8), by noting that $\nu_1 = 180° - \Psi/2$:

$$\cos E_1 = \frac{e - \cos\frac{\Psi}{2}}{1 - e\cos\frac{\Psi}{2}} \tag{6-22}$$

If we substitute into Equation (4-9), we get the time of free flight,

$$t_{ff} = 2\sqrt{\frac{a^3}{\mu}}\left(\pi - E_1 + e\sin E_1\right) \tag{6-23}$$

The semi-major axis, a, and the eccentricity, e, can be obtained from Equations (6-3) and (6-11).

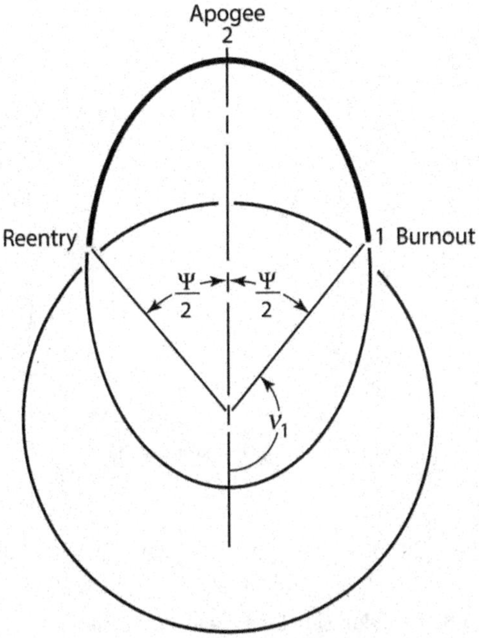

Fig. 6-7 Free Flight Symmetry.

Figure 6-9 is an excellent chart for making rapid time-of-flight calculations for the ballistic missile. In fact, since five variables have been plotted on the figure, most ballistic missile problems can be solved completely using just this chart.

The free-flight time is read from the chart as the ratio $t_{ff}/\mathbb{P}_{cs}$, where $\mathbb{P}_{cs}$ is the period of a fictitious circular satellite orbiting at the burnout altitude. Values for $\mathbb{P}_{cs}$ may be calculated from

$$\mathbb{P}_{cs} = 2\pi\sqrt{\frac{r_{bo}^3}{\mu}} \tag{6-24}$$

or they may read directly from Figure 6-8.

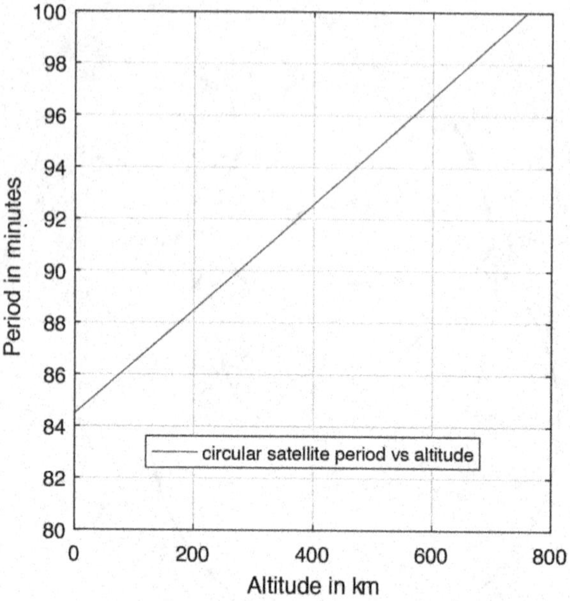

Fig. 6-8 Circular Satellite Period vs. Altitude.

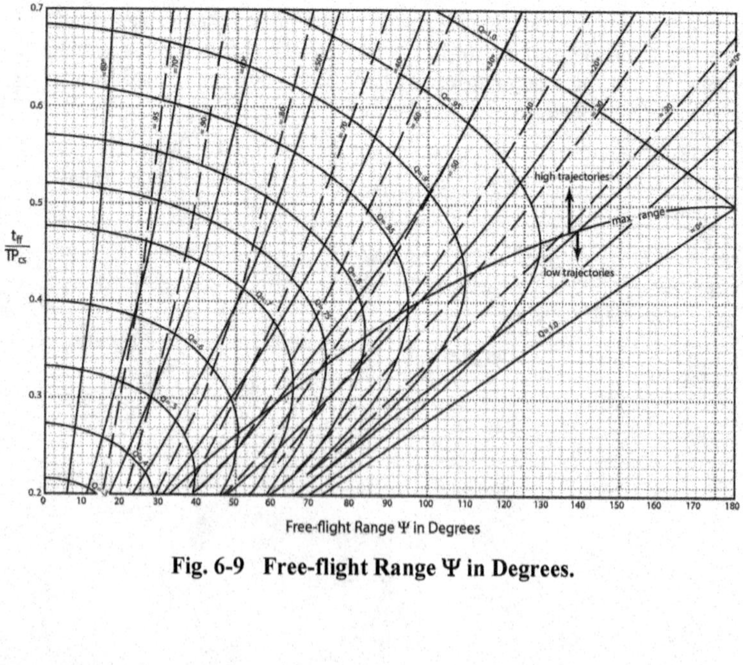

Fig. 6-9 Free-flight Range Ψ in Degrees.

Example Problem. A ballistic missile was observed to have a burnout speed and altitude of 7.4066 km/s and 477.82 km, respectively. What must be the maximum free-flight range capability of this missile?

Calculate Q_{bo}:

$$Q_{bo} = \frac{v_{bo}^2 r_{bo}}{\mu} = 0.95$$

From Figure 6-9, it is rapidly found that

$$\Psi_{max} = 129°$$

and $R_{ff} = 129 \text{ deg} \times 111.12 \text{ km/deg} = \underline{\underline{14{,}334 \text{ km}}}$

Example Problem. It is desired to maximize the payload of a new ballistic missile for a free-flight range of 14,816 km. The design burnout altitude has been fixed at 637.09 km. What should be the design burnout speed?

For a given amount of propellant, range may be sacrificed to increase payload and vice versa. For a fixed burnout altitude, the payload may be maximized by minimizing the burnout speed (minimum Q_{bo}):

$$\Psi_{max} = \frac{14{,}816 \text{ km}}{6{,}378 \text{ km}} = 2.32 \text{ radians} = 133.3°$$

From Equation (6-21),

$$Q_{bo\,min} = \frac{2 \sin 66.7°}{1 + \sin 66.7°} = 0.957$$

From Equation (6-1),

$$v_{bo} = \sqrt{\frac{Q_{bo}\mu}{r_{bo}}} = \underline{\underline{7.376 \text{ km/s}}}$$

6.3 EFFECT OF LAUNCHING ERRORS ON RANGE

Variations in the speed, position and launch direction of the missile at thrust cutoff will produce errors at the impact point. These errors are of two types—errors in the intended plane, which cause either a long or a short hit, and out-of-plane errors, which cause the missile to hit to the right or left of the target. For brevity, we will refer to errors in the intended plane as "down range" errors, and out-of-plane errors as "cross-range" errors.

These are two possible sources of cross-range error and these will be treated first.

6.3.1 Effect of a Lateral Displacement of the Burnout Point

If the thrust cutoff point is displaced by an amount,ΔX perpendicular to the intended plane of the trajectory and all other conditions are nominal, the cross-range error, ΔC, at impact can be determined from spherical trigonometry. In Figure 6-10 we show the ground traces of the intended and actual trajectories. For purposes of this example, suppose the intended burnout point is on the equator and the launch azimuth is due north along a meridian toward the intended target A. The actual burnout point occurs at a point on the equator a distance ΔX to the east but with the correct launch azimuth of due north. As a result, the missile flies up the wrong meridian, impacting at B.

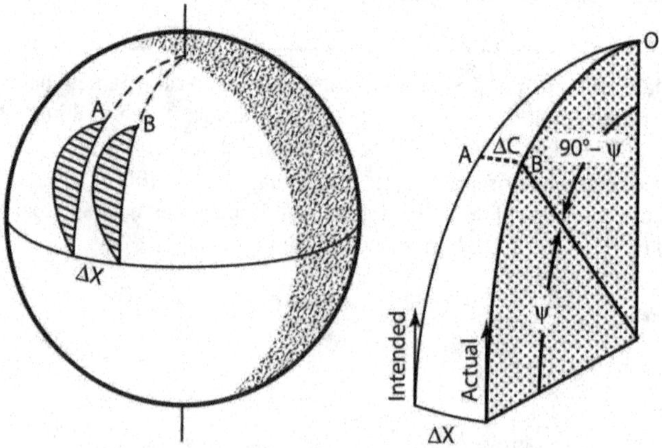

Fig. 6-10 Lateral Displacement of Burnout Point.

The arc length ΔC represents the cross-range error. It is customary in spherical trigonometry to measure arc length in terms of the angle subtended at the center of the sphere so that both ΔX and ΔC may be thought of as angles.

Applying the law of cosines for spherical trigonometry to triangle OAB in Figure 6-10 and noting that the small angle at O is the same as ΔX, we get

$$\cos\Delta C = \sin^2 \Psi + \cos^2 \Psi \cos\Delta X \tag{6-25}$$

Since both ΔX and ΔC will be very small angles, we can use the small-angle approximation

$$\cos x \approx 1 - \frac{x^2}{2} \tag{6-26}$$

to simplify Equation 6-25 to

$$\boxed{\Delta C \approx \Delta X \cos \Psi} \tag{6-27}$$

Both ΔX and ΔC are assumed to be expressed as angles in this equation. If they are in radians you can convert them to arc length by multiplying by the radius of Earth; if they are in degrees, you can use the fact that a 111.12 km (60 nautical miles) arc on the surface of Earth subtends in angle of 1° at the center.

Equation (6-27) tells us that the cross-range error is zero for a free-flight range of 90° (10,000 km or 5,400 nautical miles) regardless of how far the burnout point is displaced out of the intended plane. In Figure 6-10, for example, if the intended target had been the north pole, the actual burnout point could occur anywhere on the equator and we would hit the target as long as launch azimuth and free-flight range, Ψ, are as planned.

6.3.2 Cross-Range Error Resulting from Incorrect Launch Azimuth

If the actual launch azimuth differs from the intended launch azimuth by an amount of $\Delta\beta$, a cross-range error, ΔC, will result. Figure 6-11 illustrates the geometry of an azimuth error. The ground trace of the actual and intended trajectory are shown. Since all launch conditions are assumed to be nominal except launch azimuth, the free-flight range of both the actual and the intended trajectories is Ψ. The third side of the spherical triangle shown in Figure 6-11 is ΔC, the cross-range error. As before, we will consider ΔC to be expressed in terms of the angle it subtends at the center of Earth.

From the law of cosines for spherical triangles we get

$$\cos \Delta C = \cos^2 \Psi + \sin^2 \Psi \cos \Delta\beta \tag{6-28}$$

If we assume that both $\Delta\beta$ and ΔC will be very small angles we can use the approximation $\cos x \approx 1 - x^2/2$ to simplify Equation (6-28) to

$$\boxed{\Delta C \approx \Delta\beta \sin \Psi} \tag{6-29}$$

This time we see that the cross-range error is a maximum for a free-flight range of 90° (or 270°) and goes to zero if Ψ is 180°. In other words, if you have a missile that will travel exactly halfway around Earth, it really doesn't matter in what direction you launch it; it will hit the target anyway.

6.3.3 Effect of Downrange Displacement of the Burnout Point

An error in downrange position at thrust cutoff produces an equal error at impact. The effect may be visualized by rotating the trajectory in Figure 6-1 about the center of Earth. If the actual burnout point is 1 km farther downrange than was intended, the missile will overshoot the target by exactly 1 km.

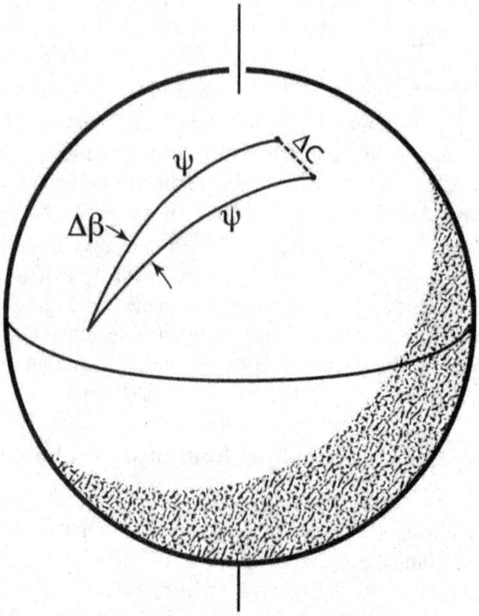

Fig. 6-11 Azimuth Error.

6.3.4 Errors in Burnout Flight-Path Angle, ϕ_{bo}.

In Figure 6-12 we show a typical plot of free-flight range versus flight-path angle for a fixed value of r_{bo} and v_{bo}. The intended flight path and intended range, Ψ, are shown by a solid line in the figure. If the actual ϕ_{bo} differs from the intended value by an amount $\Delta\phi_{bo}$, the actual range will be different by an amount ΔΨ. This ΔΨ will represent a downrange error, causing the missile to undershoot or overshoot the target.

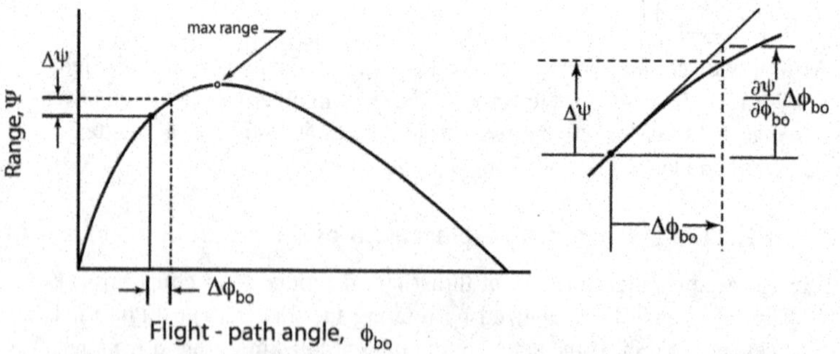

Fig. 6-12 Effect of Flight-path Angle Errors on Range.

We could get an approximate value for $\Delta\Psi$ if we knew the slope of the curve, which is $(\partial\Psi)/(\partial\phi_{bo})$ at the point corresponding to the intended trajectory. The diagram on the right of Figure 6-12 illustrates the fact that

$$\Delta\Psi \approx \frac{\partial\Psi}{\partial\phi_{bo}}\Delta\phi_{bo} \tag{6-30}$$

which is a good approximation for very small values of $\Delta\phi_{bo}$.

The expression for $\partial\Psi/\partial\phi_{bo}$ may be obtained by implicit partial differentiation of the free-flight range equation. But first we will derive an alternate form of the free-flight range equation.

The free-flight range equation was derived as Equation (6-12) of this chapter. If we call the numerator of this expression α and the denominator β, then

$$\cos\frac{\Psi}{2} = \frac{\alpha}{\beta} \tag{6-31}$$

and

$$\cot\frac{\Psi}{2} = \frac{\alpha}{\sqrt{\beta^2 - \alpha^2}} \tag{6-32}$$

Substituting for α and β, we get

$$\cot\frac{\Psi}{2} = \frac{1 - Q_{bo}\cos^2\phi_{bo}}{Q_{bo}\cos\phi_{bo}\sqrt{1-\cos^2\phi}} \tag{6-33}$$

Since $\sqrt{1-\cos^2\phi_{bo}} = \sin\phi_{bo}$,

$$\cot\frac{\Psi}{2} = \frac{1 - Q_{bo}\cos^2\phi_{bo}}{Q_{bo}\cos\phi_{bo}\sin\phi_{bo}} \tag{6-34}$$

But since cos x sin x = (1/2) sin 2x, we can simplify further to obtain

$$\boxed{\cot\frac{\Psi}{2} = \frac{2}{Q_{bo}}\csc 2\phi_{bo} - \cot\phi_{bo}} \tag{6-35}$$

We now have another form of the free-flight range equation, which is much simpler to differentiate. Let us first express Equation (6-35) in terms of r_{bo}, v_{bo} as constants:

$$\boxed{\cot\frac{\Psi}{2} = \frac{2\mu}{v_{bo}^2 r_{bo}}\csc 2\phi_{bo} - \cot\phi_{bo}} \tag{6-36}$$

We proceed to differentiate Equation (6-36) implicitly with respect to ϕ_{bo}, considering r_{bo} and v_{bo} as constants, obtaining

$$-\frac{1}{2}\csc^2\frac{\Psi}{2}\frac{\partial\Psi}{\partial\phi_{bo}} = \frac{2\mu}{v_{bo}^2 r_{bo}}\left(-2\cot 2\phi_{bo}\csc 2\phi_{bo}\right) + \csc^2\phi_{bo} \qquad (6\text{-}37)$$

Substituting from Equation (6-36), we get

$$-\frac{1}{2}\csc^2\frac{\Psi}{2}\frac{\partial\Psi}{\partial\phi_{bo}} = -2\cot 2\phi_{bo}\left(\cot\frac{\Psi}{2} + \cot\phi_{bo}\right) + \csc^2\phi_{bo}$$

$$= 2\left(1 - \cot 2\phi_{bo}\cot\frac{\Psi}{2}\right)$$

Solving for $\partial\Psi/\partial\phi_{bo}$

$$\frac{\partial\Psi}{\partial\phi_{bo}} = 4\left(\cot 2\phi_{bo}\sin\frac{\Psi}{2}\cos\frac{\Psi}{2} - \sin^2\frac{\Psi}{2}\right)$$

$$= 4\left(\frac{1}{2}\cot 2\phi_{bo}\sin\Psi - \sin^2\frac{\Psi}{2}\right)$$

$$= 2(\sin\Psi\cot 2\phi_{bo} + \cos\Psi - 1)$$

$$= 2\frac{\sin\Psi\cot 2\phi_{bo} + \cos\Psi\sin 2\phi_{bo}}{\sin 2\phi_{bo}} - 2$$

which finally reduces to

$$\frac{\partial\Psi}{\partial\phi_{bo}} = \frac{2\sin\left(\Psi + 2\phi_{bo}\right)}{\sin 2\phi_{bo}} - 2 \qquad (6\text{-}38)$$

This partial derivative, when used in the manner described above, is called an influence coefficient since it influences the size of the range error resulting from a particular burnout error.

While we need Equation (6-38) to evaluate the magnitude of the flight-path angle influence coefficient, the general effect of errors in flight-path angle at burnout are apparent from Figure 6-12.

The maximum range condition separates the low trajectories from the high trajectories. For all low trajectories the slope of the curve $\partial\Psi/\partial\phi_{bo}$ is positive, which means that too high a flight-path angle (a positive $\Delta\phi_{bo}$) will cause a positive $\Delta\Psi$ (overshoot); a flight-path angle that is lower than intended (a negative $\Delta\phi_{bo}$) will cause a negative $\Delta\Psi$ (undershoot).

Just the opposite effect occurs for all high trajectories where $\partial\Psi/\partial\phi_{bo}$ is always negative. Too high a ϕ_{bo} on the high trajectory will cause the missile to fall short and too low a ϕ_{bo} will cause an overshoot. If this seems strange, remember that water from a garden hose behaves in exactly the same way.

Figure 6-12 also reveals that the high trajectory is less sensitive than the low trajectory to flight-path angle errors. For a typical ICBM fired over a 9,260 km

(5,000 nautical miles) range, an error of 1 minute (1/60 of a degree) causes a miss of about 1.85 km (1 nautical miles) on the high trajectory and about 5.6 km (3 nautical miles) on the low trajectory.

The maximum range trajectory is the least sensitive to flight-path angle errors. Since $\partial\Psi/\partial\phi_{bo} = 0$ for the maximum range case, Equation (6-30) tells us that $\Delta\Psi$ will be approximately zero for small values of $\Delta\phi_{bo}$. In fact, the actual range error on a 6,667 km (3,600 nautical miles) ICBM flight resulting from a 1 minute error in ϕ_{bo} is only 1.2 m (4 ft)!

6.3.5 Downrange Errors Caused by Incorrect Burnout Height

We can use exactly the same approach to errors in burnout height as we used in the previous section. A plot of range versus burnout radius, however, is not particularly interesting since it reveals just what we might suspect—if burnout occurs higher than intended, the missile overshoots; if burnout occurs too low, the missile falls short of the target in every case.

Following the arguments in the last section we can say that

$$\boxed{\Delta\Psi \approx \frac{\partial\Psi}{\partial r_{bo}}\Delta r_{bo}} \tag{6-39}$$

for small values of Δr_{bo}.

The partial derivative with respect to r_{bo} is much simpler. Again, differentiating the range Equation (6-36) implicitly gives

$$-\frac{1}{2}\csc^2\frac{\Psi}{2}\frac{\partial\Psi}{\partial r_{bo}} = \frac{-2\mu}{v_{bo}^2 r_{bo}^2}\csc 2\phi_{bo} \tag{6-40}$$

Solving for $\partial\Psi/\partial r_{bo}$, we get

$$\frac{\partial\Psi}{\partial r_{bo}} = \frac{4\mu}{v_{bo}^2 r_{bo}^2}\frac{\sin^2\frac{\Psi}{2}}{\sin 2\phi_{bo}} \tag{6-41}$$

A burnout error of 1.85 km (1 nautical miles) in height on a 9,260 km (5,000 nautical miles) range trajectory will cause a miss of about 3.7 km (2 nautical miles) on the high trajectory and about 9.26 km (5 nautical miles) on the low trajectory.

6.3.6 Downrange Errors Caused by Incorrect Speed at Burnout

Speed at burnout affects range in just the way we would expect—too fast and the missile overshoots; too slow and the missile falls short. The magnitude of the error is

$$\boxed{\Delta\Psi \approx \frac{\partial\Psi}{\partial v_{bo}}\Delta v_{bo}} \tag{6-42}$$

where $\partial\Psi/\partial v_{bo}$ is given by implicit differentiation of Equation (6-36) as

$$\boxed{\frac{\partial\Psi}{\partial v_{bo}} = \frac{8\mu}{v_{bo}^3 r_{bo}} \frac{\sin^2\frac{\Psi}{2}}{\sin 2\phi_{bo}}} \tag{6-43}$$

A rough rule to follow is that an error of 1 m/s will cause a miss of about 5 km over a typical ICBM range.

An analysis of Equation (6-43) would reveal that, like the other two influence coefficients, $\partial\Psi/\partial v_{bo}$ is larger on the low trajectory than on the high. Most ICBM are programmed for the high trajectory for the simple reason that is revealed here—the guidance requirements are less stringent and the accuracy is better.

Example Problem. A ballistic missile has the following nominal burnout conditions:

$$v_{bo} = 7.1543\,\text{km/s}, r_{bo} = 7,016\,\text{km}, \ \phi_{bo} = 30°$$

The following errors exist at burnout:

$$\Delta v_{bo} = -3.9527 \times 10^{-4}\ \text{km/s}, \Delta r_{bo} = 3.1891\,\text{km}$$

$$\Delta\phi_{bo} = -10^{-4}\ \text{radians}$$

By how far will the missile miss the target? What will be the direction of the miss relative to he trajectory plane?

We will find total downrange error. There is no cross-range error. We will need Q_{bo} (0.901 for this case) and Ψ ($\Psi \approx$ 100° from Figure 6-9). That total downrange error is

$$\Delta\Psi_{TOT} = \frac{\partial\Psi}{\partial r_{bo}}\Delta r_{bo} + \frac{\partial\Psi}{\partial v_{bo}}\Delta v_{bo} + \frac{\partial\Psi}{\partial\phi_{bo}}\Delta\phi_{bo}$$

$$\frac{\partial\Psi}{\partial r_{bo}} = \frac{4\mu}{v_{bo}^2 r_{bo}^2}\frac{\sin^2\frac{\Psi}{2}}{\sin 2\phi_{bo}} = 4.2881 \times 10^{-4}\ \text{radians/km}$$

$$\frac{\partial \Psi}{\partial v_{bo}} = \frac{2\, r_{bo}}{v_{bo}} \left[\frac{\partial \Psi}{\partial r_{bo}} \right] = 0.8410 \text{ radians/km/s}$$

$$\frac{\partial \Psi}{\partial \phi_{bo}} = \frac{2 \sin 160°}{\sin 60°} - 2 = -1.21$$

$$\Delta\Psi_{TOT} = \left(4.2881 \times 10^{-4}\right)(3.1891) + 0.841\left(-3.9527 \times 10^{-5}\right) - 1.21\left(-10^{-4}\right)$$

$$= 12.132 \times 10^{-4} \text{ radians}$$

$$\Delta R_{ff} = \left(12.132 \times 10^{-4}\right)(6{,}378.137) \approx \underline{\underline{7.74 \text{ km overshoot}}}$$

6.4 THE EFFECT OF EARTH ROTATION

Up to this point, Earth has been considered as non-rotating. In the two sections that follow we will see what effect Earth rotation has on the problem of sending a ballistic missile from one fixed point on Earth to another.

Earth rotates once on its axis in 23 hr 56 min producing a surface velocity at the equator of 465.1 m/s. The rotation is from west to east.

The free-flight portion of a ballistic missile trajectory is inertial in character. That is, it remains fixed in the **XYZ** inertial frame while Earth runs under it. Relative to this inertial **XYZ** frame, both the launch point and the target are in motion.

We will compensate for motion of the launch site by recognizing that the "true velocity" of the missile at burnout is the velocity relative to the launch site (which could be measured by radar) plus the initial eastward velocity of the launch site as a result of Earth rotation.

We will compensate for motion of the target by "leading it" slightly. That is, we will send our missile on a trajectory that passes through the point in space where the target will be when our missile arrives. If we know the time of flight of the missile, we can compute how much Earth (and the target) will turn in that time and can aim for a point the proper distance to the east of the target.

6.4.1 Compensating for the Initial Velocity of the Missile Resulting from Earth Rotation

We describe the speed and direction of a missile at burnout in terms of its speed, v, its flight-path angle, ϕ, and its azimuth angle, β. If measurements of these three quantities are made from the surface of the rotating Earth (by radar, for example), then all three measurements are erroneous in that they do not indicate the "true" speed and direction of the missile. A simple example should clarify this point.

Suppose we set up a cannon on a train that is moving along a straight section of track in a northerly direction. If we point the cannon due east (perpendicular to the track) and upward at a 45° angle and then fire it, an observer on the moving train would say that the projectile had a flight-path angle ϕ, of 45°, an azimuth, β, of 90° (east), and a speed of, say, 330 m/s.

An observer at rest would not agree. The observer would correctly see that the "true" velocity of the projectile includes the velocity of the train relative to the "fixed frame." The vector diagram at the right of Figure 6-13 illustrates the true situation and shows that the speed is actually somewhat greater than 330 m/s, the flight-path angle slightly less than 45° and the azimuth considerably less than 90°.

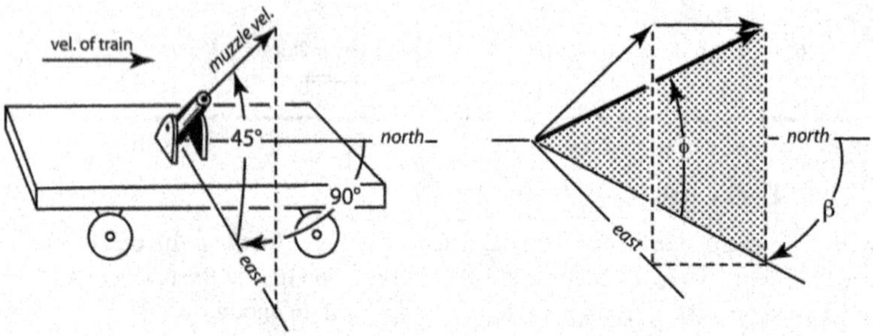

Fig. 6-13 "True" Velocity and Direction.

Like the observer on the moving train we make our measurements in a moving reference frame. This frame is called the topocentric-horizon system. Hereafter, we will refer to measurements of burnout direction in this frame as ϕ_e and β_e.

Since a point on the equator has a speed of 465.1 m/s in the eastward direction, we can express the speed of any launch point on the surface of Earth as

$$\boxed{v_0 = 465.1 \cos L_o \text{ m/s}} \qquad (6\text{-}44)$$

where L_o is the latitude of the launch site. The subscript "o" is used to indicate that this is the initial eastward velocity of the missile even while it is on the launch pad.

We can obtain the south, east and up components of the true velocity, $\mathbf{v}$, by breaking up $\mathbf{v_e}$ into its components and adding $\mathbf{v_o}$ to the eastward (**E**) component (see Figure 6-14). Thus,

$$\boxed{\begin{aligned} v_S &= -v_e \cos\phi_e \cos\beta_e \\ v_E &= v_e \cos\phi_e \sin\beta_e + v_o \\ v_Z &= v_e \sin\phi_e \end{aligned}} \qquad (6\text{-}45)$$

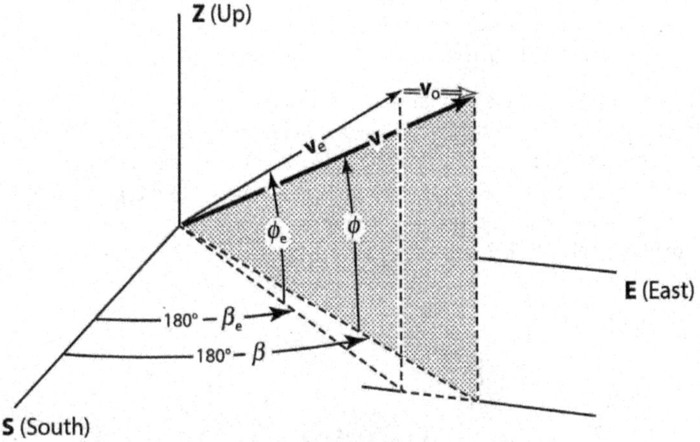

Fig. 6-14 "True" Speed and Direction at Burnout.

The true speed, flight-path angle and azimuth can then be found from

$$v = \sqrt{v_S^2 + v_E^2 + v_Z^2} \tag{6-46}$$

$$\sin\phi = \frac{v_Z}{v} \tag{6-47}$$

$$\tan\beta = \frac{-v_E}{v_S} \tag{6-48}$$

The inverse problem of determining v_e, ϕ_e and β_e if you are given **v**, ϕ and β can be handled in a similar manner; first, break up **v** into its components, then subtract v_o from the eastward component to obtain the components of v_e. Once you have the components of v_e, finding v_e, ϕ_e and β_e is easy.

One word of caution is in order: you will have to determine in which quadrant the azimuth β lies. If you can draw even a crude sketch and visualize the geometry of the problem this will not be difficult.

It is, of course, **v** at burnout that determines the missile's trajectory. The rocket booster only has to add $\mathbf{v_e}$ to the initial velocity $\mathbf{v_0}$. If the desired launch velocity is eastward the rocket will not have to provide as much speed as it would for a westward launch.

6.4.2 Compensating for Movement of the Target Resulting from Earth Rotation

In Figure 6-15 we show Earth at the instant a missile is launched. The trajectory goes from the launch point at A to an "aiming point" at B. This aiming point does not coincide with the target at the time the missile is launched. Rather, it is a point at the same latitude as the target but east of it by an amount equal to the number of degrees Earth will turn during the total time the missile is in flight. Hopefully, when the missile arrives at point B, the target will be there also.

The latitude and longitude coordinates of the launch point are L_0 and N_0, respectively, so the arc length OA in Figure 6-15 is just $90° - L_0$. If the coordinates of the target are (L_t, N_t) then the latitude and longitude of the aiming point should be $(L_t, N_t + \omega_{\oplus} t_{\Lambda})$. The term $\omega_{\oplus} t_{\Lambda}$ represents the number of degrees Earth turns during the time t_{Λ}. The angular rate $\omega_{\oplus}$ at which Earth turns is approximately 15°/hr.

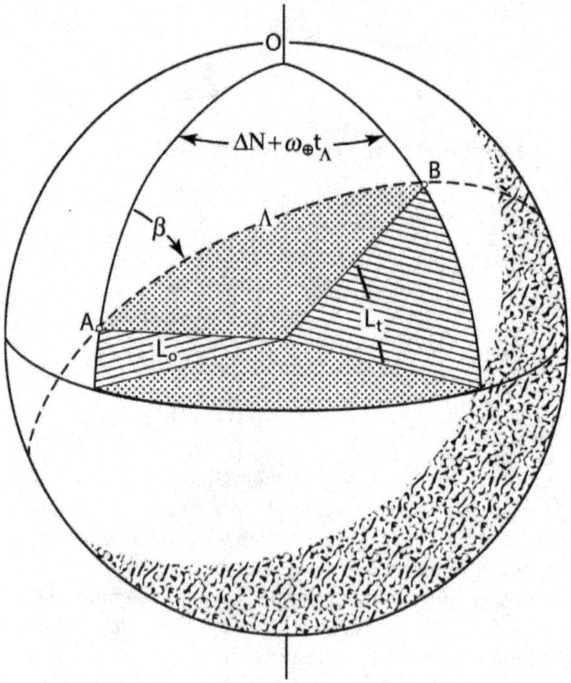

Fig. 6-15 Launch Site and "Aiming Point" at the Instant of Launch.

Arc length OB is simply $90° - L_t$, and the third side of the spherical triangle (the dashed line) is the ground trace of the missile trajectory that subtends the angle L. The included angle at A is the launch azimuth, β. The angle formed at O is just the difference in longitude between the launch point and the aiming point, $\Delta N + \omega_{\oplus} t_{\Lambda}$, where ΔN is the difference in longitude between launch point and target.

If we assume that we know the coordinates of the launch point and target and the total time of flight, t_Λ, we can use the law of cosines for spherical triangles to obtain

$$\boxed{\cos\Lambda = \sin L_0 \sin L_t + \cos L_0 \cos L_t \cos(\Delta N + \omega_\oplus t_\Lambda)} \tag{6-49}$$

In applying this equation we must observe certain precautions. The longitude difference, ΔN, should be measured from the launch point *eastward* toward the target. The equation yields two solutions for Λ—one angle between 0° and 180°, and another between 180° and 360°. These two solutions represent the two great-circle paths between the launch point and aiming point. Whether you select one or the other depends on whether you want to go the short way or the long way around to the target.

Once we know the total range angle, Λ, we can solve for the required launch azimuth, β, by applying the law of cosines to the triangle in Figure 6-15 again—this time considering β as the included angle:

$$\sin L_t = \sin L_o \cos\Lambda + \cos L_o \sin\Lambda \cos\beta \tag{6-50}$$

Solving for $\cos\beta$, we get

$$\boxed{\cos\beta = \frac{\sin L_t - \sin L_o \cos\Lambda}{\cos L_o \sin\Lambda}} \tag{6-51}$$

Again, this equation yields two solutions for β—one between 0° and 180° and the other between 180° and 360°. A simple rule exists for determining which value of β is correct: if you are going the "short way" to the target, if $\Delta N + \omega_\oplus t_\Lambda$ lies between 0° and 180° then so does β. If you are firing the missile the "long way" around, then a value of $\Delta N + \omega_\oplus t_\Lambda$ between 0° and 180° requires that β lie between 180° and 360°.

It is worth returning for a moment to look at the equations for total range, Λ, and launch azimuth, β. In order to compute β we must know the coordinates of the launch point and target as well as the range, Λ. In order to compute Λ we must know the time of flight, t_Λ. But, how can we know the time of flight if we do not know the range being flown?

The situation is not entirely desperate. In actual practice you would begin by "guessing" a reasonable time for t_Λ. This value would be used to compute an initial estimate for Λ, which in turn would allow you to get a first estimate of t_{ff}. By adding the times of powered flight and reentry (which also depend somewhat on Λ) to t_{ff} you get a value of t_Λ that you can use as your second "guess." The whole process is then repeated until the computed value of t_Λ agrees with the estimated value. Needless to say, the digital computer is more suited for this type of trial-and-error computation than the average student whose patience is limited.

Example Problem. A ballistic missile launched from 29°N, 79.3°W burns out at 30°N, 80°W after developing an increase in velocity of 7.1148 km/s. Its elevation and azimuth relative to Earth are

$$\phi_e = 30°, \beta_e = 330°$$

Assuming a rotating Earth and a r_{bo} of 7,016 km, what are the coordinates of the reentry point? Assume a symmetrical trajectory.

The inertial speed, flight-path angle and azimuth may be found using Equations (6-45) through (6-48):

$$v_S = -v_{bo} \cos30° \cos330° = -5.3361 \text{ km/s}$$

$$v_E = v_{bo} \cos30° \sin330° + 0.46452 \cos29° = -2.6745 \text{ km/s}$$

$$v_Z = v_{bo} \sin30° = 3.5574 \text{ km/s}$$

$$v = \sqrt{v_s^2 + v_E^2 + v_Z^2} = 6.9486 \text{ km/s}$$

$$\sin\phi_{bo} = \frac{v_Z}{v} = \frac{0.45}{0.879} = 0.51195$$

$$\phi_{bo} = 30.8°$$

$$\tan\beta = \frac{-v_E}{v_S} = -\frac{(2.6745)}{-5.3361} = -0.5022$$

$$\beta = 333.33°$$

From Figure 6-9, using $Q_{bo} = 0.85$ and $\phi_{bo} = 31°$ we get

$$\Psi = 90°$$

Using the law of cosines from spherical trigonometry give

$$\sin L_{re} = \sin L_{bo} \cos\Psi + \cos L_{bo} \sin\Psi \cos(360° - \beta)$$

$$= \sin30° \cos90° + \cos30° \sin90° \cos(26.67°)$$

$$= 0.7739$$

$$L_{re} = \underline{\underline{50°42'N}} \text{ (reentry latitude)}$$

Using the law of cosines again, we have

$$\cos\Psi = \sin L_{re} \sin L_{bo} + \cos L_{re} \cos L_{bo} \cos(\Delta N + \omega_\oplus t_{ff})$$

$$\cos90° = \sin50°\,42' \sin30°$$

$$+ \cos50°42' \cos30° \cos(\Delta N + w_\oplus t_{ff})$$

$$\cos(\Delta N + \omega_\oplus t_{ff}) = -0.7055$$

$$\Delta N + \omega_{\oplus} t_{ff} = 135° 8'$$

From Figure 6-9, using $Q_{bo} = 0.85$ and $\phi_{bo} = 30.78°$ we get

$$\frac{t_{tt}}{\mathbb{P}_{cs}} \approx 0.46$$

$$\text{so } \mathbb{P}_{cs} = 2\pi\sqrt{\frac{r_{bo}^3}{\mu}} = 5{,}848.4 \text{ s}$$

Therefore, $t_{ff} = 2{,}690.3$ s

$$\Delta N = 135° - \omega_{\oplus} t_{ff} = 123.77°$$

$$N_{bo} + \Delta N = -80° - 123.77° = -203.77°$$

$$N_{bo} = N_{re} = \underline{\underline{156.23°\text{E}}} \text{ (reentry longitude)}$$

Exercises

6.1 The following measurements are obtained during the testing of an ICBM:

$v_{bo} = 7.3204$ km/s, $r_{bo} = 6{,}697.0$ km

$\phi_{bo} = 10°$, $R_p = 120.0$ km

$R_{re} = 560.0$ km

What is R_t?

6.2 A ballistic missile is launched from a submarine in the Atlantic (30° N, 75° W) on an azimuth of 135°. The burnout speed relative to the submarine is 5 km/s and at an angle of 30° to the local horizontal. Assume the submarine lies motionless in the water during the firing. What is the true speed of the missile relative to the center of the rotating Earth?

(Answer: v = 4.764 km/s)

6.3 For a ballistic missile having

$r_{bo} = 7{,}000$ km

$v_{bo} = 4.0$ km/s

what will be the maximum rangeand ϕ_{bo}?

6.4 What value of Q_{bo} may be used in Equation (6-20)? Why?

6.5 A ballistic missile's burnout point is at the end of the semi-minor axis of an ellipse. Assuming burnout altitude equals reentry altitude, and a spherical Earth, what will the value of Q be at reentry?

6.6 What is the minimum velocity required for a ballistic missile to travel a distance measured on the surface of Earth of 9,400 km? Neglect atmosphere and assume $r_{bo} = 6{,}400$ km.

6.7 In general, how many possible trajectories are there for a given range and ($Q_{bo} < 1$)? What is the exception to this rule?

6.8 An enterprising younger engineer was able to increase a certain rocket's Q_{bo} from 0.98 to 1.02. Using the equations

$$\sin\frac{\Psi}{2} = \frac{Q_{bo}}{2 - Q_{bo}}$$

$$\phi_{bo} = \frac{1}{4}(180° - \Psi)$$

the engineer was unable to obtain a new maximum range ϕ_{bo} for Q_{bo} = 1.02. Why?

6.9 A ballistic missile is capable of achieving a burnout velocity of 6.5615 km/s at a radius of 6,760 km. What is the maximum free-flight range of this missile in kilometers? Assume a symmetrical trajectory. Do not use charts.

(Answer: $R_{ff} = 7{,}814.50$ km)

6.10 During a test flight, an ICBM is observed to have the following position and velocity at burnout:

$\mathbf{r}_{bo} = 6{,}400\mathbf{I} - 4{,}800\mathbf{J}$ km
$\mathbf{v}_{bo} = 1.600\mathbf{J} + 6.200\mathbf{K}$ km/s

What is the maximum range capability of this missile in km?

(Answer: $R_{ff} = 20{,}531.77$ km)

6.11 A rocket testing facility located at 30° N, 100° W launches a missile to impact at a latitude of 70° S. A lateral displacement, ΔX, in the launch causes the rocket to burn out east of the intended burnout point. In what direction will the error at impact be?

6.12 Assuming that the maximum allowable cross-range error at the impact point of a ballistic missile is 2.0 km where the free-flight range of the ballistic missile is 10,000 km, how large can Δx and $\Delta\beta$ be?

6.13 A malfunction causes the flight-path angle of a ballistic missile to be greater than nominal. How will this affect the missile's free-flight range? Consider three separate cases:

a. ϕ_{bo} was less than ϕ_{bo} for maximum range.

b. ϕ_{bo} was greater than ϕ_{bo} for maximum range.

c. ϕ_{bo} was equal to ϕ_{bo} for maximum range.

6.14 The ballistic missile has been launched with the following burnout errors:

$\Delta r = 1.2757$ km

$\Delta v = -7.9053$ m/s

$\Delta \phi = 0.25°$

where the influence coefficients have been calculated to be:

$$\frac{\partial \Psi}{\partial \phi} = 1.5$$

$$\frac{\partial \Psi}{\partial r} = 4.7036 \times 10^{-4} \text{ km}^{-1}$$

$$\frac{\partial \Psi}{\partial v} = 0.63248 \text{ km/s}$$

a. Determine the error at the impact point in km.

(Answer: 13.682 km long)

b. Is this a high, low or maximum range trajectory? Why?

6.15 In general, will a given $\Delta\phi_{bo}$ cause a larger error in a high or low trajectory? Why?

6.16 Assuming r_{bo} = 6,400 km for a ballistic missile, what is the minimum burnout velocity required to achieve a free-flight range of 3,400 km?

6.17 A ballistic missile whose maximum free-flight range is 6,700 km is to be launched from the equator on the Greenwich meridian toward a target located at 45° N, 30° E, using a minimum time-of-flight trajectory. What should the flight-path angle be at burnout? Neglect the atmosphere and assume $Q_{bo} = Q_{bo}$ maximum.

6.18 Show that for maximum range $Q_{bo} = 1 - e^2$, where e is the eccentricity.

6.19 An ICBM is to be flight-tested over a total range R_t of 8,700 km using a high trajectory. Burnout will occur 80 km downrange at a Q of 0.8 and an altitude of 280 km. The reentry range is calculated at 290 km. What will be the time of free flight?

(Answer: t_{ff} = 39.973 min)

6.20 A ballistic missile that burned out at 45°N, 150°E, at an altitude of 637.82 km will re enter at 45° N, 120° W, at the same altitude, using a "backdoor" trajectory ($\Psi > 180°$). If the velocity at burnout was 8.6959 km/s, what was the flight-path angle at burnout?

6.21 A ballistic missile's trajectory is a portion of an ellipse whose apogee radius is 9,567.2 km and whose perigee radius is 3,189.1 km. Assuming burnout occurred at sea level on a spherical Earth, what is the free-flight range expressed in km? Assume a symmetrical trajectory. Do not use charts.

6.22 The range error equation could be written as:

$$\Delta\Psi = \frac{\partial\Psi}{\partial Q_{bo}}\Delta Q_{bo} + \frac{\partial\Psi}{\partial\phi_{bo}}\Delta\phi_{bo}$$

Derive an expression for $\partial\Psi/\partial Q_{bo}$ in terms of Q_{bo}, Ψ and ϕ_{bo} and analyze the result, i.e., determine whether the influence coefficient is always positive or negative and if so what it means.

6.23 A ballistic missile has the following nominal burnout conditions:

$v_{bo} = 7.1626$ km/s

$r_{bo} = 7{,}007.4$ km

$\phi_{bo} = 30°$

The following errors exist at burnout:

$\Delta v_{bo} = -0.39624$ m/s

$\Delta r_{bo} = 3.1854$ km

$\Delta\phi_{bo} = -0.0057° = -10^{-4}$ radians

a. By how far will the missile miss the target?

(Answer: 5.113 km)

b. Is the shot long or short?

c. Is this a high, low or maximum range trajectory?

6.24 A rocket booster is programmed for a true velocity relative to the center of Earth at burnout of

$v_{bo} = -0.3188\mathbf{S} - 3.2335\mathbf{E} + 6.3351\mathbf{Z}$ km/s

What must the speed, elevation and azimuth relative to the launch site be at burnout? Launch site coordinates are 28° N, 120° W. Do not assume a non-rotating Earth.

(Partial answer: $\phi_e = 60.02°$)

6.25 *An ICBM is to be flight tested. It is desired that the missile display the following nominal parameters at burnout:

h = 63.782 km

$v_e = 6.2452$ km/s

$\beta_e = 315.2°$

$\phi_e = 44.5°$

All angles and velocities are measured relative to the launch site located at 30° N, 120° W.

During the actual firing, the following errors were measured: $\Delta\phi_e = 30'$, $\Delta\beta_e = -12'$, and down-range displacement of burnout point = 1.3520 km.

What are the coordinates of the missile's reentry point? Assume a symmetrical trajectory and do not assume a non-rotating Earth.

(Answer: $L_{re} = 54.37°$ N, $N_{re} = 161.53°$ W)

6.26 *An ICBM located at 60° N, 160° E is programmed against a target located on the equator at 114°W using a minimum velocity trajectory. What should the time of flight be? Assume a spherical, rotating, atmosphereless Earth, and $r_{bo} \approx r_{\oplus}$.

(Answer: $t_{ff} = 32.57$ min)

6.27 *A requirement exists for a ballistic missile with a total range of 13,700 km where

$R_p = 260$ km

$R_{RE} = 120$ km

$r_{bo} = 6{,}600$ km

a. Assuming a symmetric orbit, what is the minimum burnout velocity required to reach a target at this range?

b. What is the required ϕ_{bo}?
(Answer: $\phi_{bo} = 15.09°$)

c. What will be the time of free flight (t_{ff})?

d. In order to overshoot the impact point would you increase or decrease the elevation angle? Explain.

6.28 *A ballistic missile burns out at an altitude of 320 km with a Q = 1. The maximum altitude achieved during the ensuing flight is 3,000 km. What was the free-flight range, in km? Assume a symmetrical trajectory.

6.29 *Derive the flight-path angle Equation (6-16) from the free-flight range Equation (6-12). (Hint: See Section 6.3.4.)

6.30 *A ballistic missile is targeted with the following parameters: $Q_{bo} = 4/3$, $R_{ff} = 20{,}000$ km, and altitude at bo = 127.5 km. What will be the time of free flight? Do not use charts.

6.31 *The approximation:

$$\Delta\Psi \approx \frac{\partial\Psi}{\partial\phi_{bo}}\Delta\phi_{bo} + \frac{\partial\Psi}{\partial r_{bo}}\Delta r_{bo} + \frac{\partial\Psi}{\partial v_{bo}}\Delta v_{bo}$$

is only useful for small values of the in-plane errors $\Delta\phi_{bo}$, Δr_{bo} and Δv_{bo}. How could this equation be modified to accommodate larger errors? (Hint: Use a Taylor series expansion, and truncate all terms of third degree and higher.)

List of References

1. Burgess, Eric. *Long-Range Ballistic Missiles*. New York, NY, Macmillan, 1962.
2. Dornberger, Walter. *V-2*. New York, NY, Viking Press, 1955.
3. Schwiebert, Ernest G. *A History of the U.S. Air Force Ballistic Missiles*. New York, NY, Frederick A. Praeger, 1965.
4. Gantz, Lt. Col. Kenneth, ed. *The United States Air Force Report on the Ballistic Missile*. Garden City, NY, Doubleday & Company, Inc., 1958.
5. Wheelon, Albert D. "Free Flight," Chapter 1 of *An Introduction to Ballistic Missiles*. Vol. 2, Los Angeles, CA, Space Technology Laboratories, Inc., 1960.

Chapter 7

Lunar Trajectories

> Which is more useful, the Sun or the Moon?...The Moon is the more useful, since it gives us light during the night, when it is dark, whereas the Sun shines only in the daytime, when it is light anyway.
>
> -Fictitious philosopher created by George Gamow[1]

7.1 HISTORICAL BACKGROUND

According to the British astronomer Sir Richard A. Proctor, "Altogether the most important circumstance in what may be called the history of the Moon is the part which she has played in assisting the progress of modern exact astronomy. It is not saying too much to assert that if Earth had no satellite the law of gravitation would never have been discovered."[2]

The first complete explanation of the irregularities in the motion of the Moon was given by Newton in Book I of the *Principia* where he states:

> For the Moon, though principally attracted by Earth, and moving round it, does, together with Earth, move around the Sun once a year, and is, according as she is nearer or farther from the Sun, drawn by him more or less than the center of Earth, about which she moves; whence arise several irregularities in her motion, of all which, the Author in this book, with no less subtility than industry, has given a full account.[3]

Newton nevertheless regarded the lunar theory as very difficult and confided to Halley in despair that it "made his head ache and kept him awake so often that he would think of it no more."

In the 18th century lunar theory was developed analytically by Euler, Clairaut, D'Alembert, Lagrange and Laplace. Much of the work was motivated by the offer of substantial cash prizes by the English Government and numerous scientific societies to anyone who could produce accurate lunar tables for the use of navigators in determining their position at sea.

A more exact theory based on new concepts and developed by new mathematical methods was published by G. W. Hill in 1878 and was finally brought to perfection by the research of E. W. Brown. Jean Meeus, a well-known Belgian astronomer specializing in celestial mechanics, developed widely used tables accurately describing the position of the Moon. In the first part of this chapter we will describe the Earth–Moon system together with some of the irregularities of the Moon's motion which have occupied astronomers since Newton's time. In the second part we will look at the problem of launching a vehicle from Earth to the Moon.

The motion of near-Earth satellites or ballistic missiles may be described by two-body orbital mechanics where Earth is the single point of attraction. Even interplanetary trajectories, which are the subject of the next chapter, may be characterized by motion that is predominately shaped by the presence of a single center of attraction, in this case the Sun. What distinguishes these situations from the problem of lunar trajectories is that the vast majority of the flight time is spent in the gravitational environment of a single body.

A significant feature of lunar trajectories is not merely the presence of two centers of attraction but the relative sizes of Earth and the Moon. Although the mass of the Moon is only about 1/80th the mass of Earth, this ratio is far larger than any other binary system in our solar system. Thus the Earth–Moon system is a rather singular event, not merely because we find our abode on Earth, but because it comes close to being a double planet.

7.2 THE EARTH–MOON SYSTEM

The notion that the Moon revolves about Earth is somewhat erroneous; it is more precise to say that both Earth and the Moon revolve about their common center of mass. The mean distance between the center of Earth and the center of the Moon is 384,400 km[4] and the mass of the Moon is 1/81.30[4] of the mass of Earth. This puts the center of the system 4,671 km from the center of Earth or about 3/4 of the way from the center to the surface. Describing the motion of the Earth–Moon system is a fairly complex business and begins by noting that the center of mass revolves around the Sun once per year (by definition). Earth and the Moon both revolve about their common center of mass once in 27.3 days. As a result, the longitude of an object such as the Sun or a nearby planet exhibits fluctuations with a period of 27.3 days arising from the fact that we observe it from Earth and not from the center of mass of the Earth–Moon system. These periodic fluctuations in longitude were, in fact, the most reliable source for determining the Moon's mass until Ranger 5 flew within 725 km of the Moon in October 1962.

The orbital period of the Moon is not constant but is slowly increasing at the same time the distance between Earth and the Moon is increasing. According to one theory advanced by G. H. Darwin, son of the great biologist Charles Darwin, the Moon was at one time much closer to Earth than at present. The slow recession of the Moon can be explained by the fact that the tidal bulge in Earth's oceans raised by the Moon is carried eastward by Earth's rotation. This shifts the center of gravity of Earth to the east of the line joining the centers of mass of Earth and the Moon and gives the Moon a small acceleration in the direction of its orbital motion, causing it to gainenergy and slowly spiral outward.

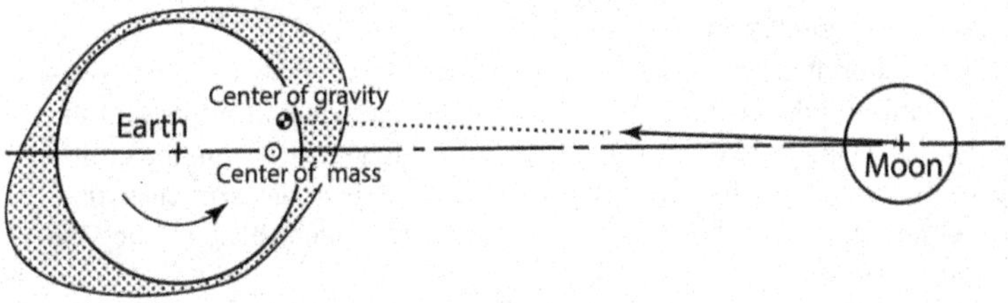

Fig. 7-1 Acceleration of the Moon caused by Earth's Tidal Bulge.

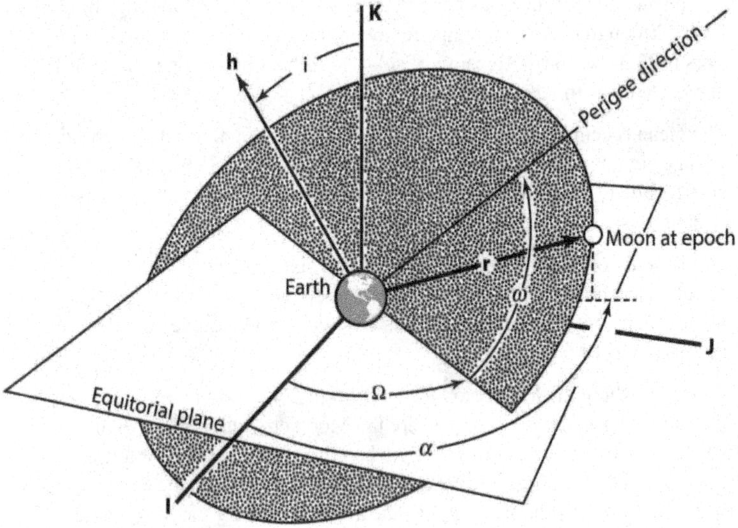

Fig. 7-2 Lunar Orbital Elements.

7.2.1 Orbital Elements of the Moon

When viewed from the center of Earth the Moon's orbit can be described by six classical orbital elements:

a–semi-major axis

e–eccentricity

i–inclination

Ω–longitude of the ascending node

ω–argument of perigee

α–right ascension at epoch

The first five of these elements are defined and discussed in Chapter 2 and should be familiar to you. The right ascension at epoch is the angle measured eastward from the vernal equinox to the projection of the Moon's position vector on the equatorial plane.

In Chapter 2 orbital elements are studied in the context of the "restricted two-body problem" and the elements were found to be constants. In the case of the Moon's orbit, owing primarily to the perturbative effect of the Sun, the orbital elements are constantly changing with time; their value at any particular time can be obtained from a lunar ephemeris such as is published in the *The Astronomical Almanac.*[7]

We will mention some of the principal perturbations of the Moon's motion to illustrate its complexity:

a. The mean value of the semi-major axis is 384,400 km. The average time for the Moon to make one complete revolution around Earth relative to the stars is 27.31661 days. Because of solar perturbations the sidereal period may vary by as much as 7 hours.

b. The mean eccentricity of the Moon's orbit is 0.054900489. Small periodic changes in the orbital eccentricity occur at intervals of 31.8 days. This effect, called "evection," was discovered more than 2,000 years ago by Hipparchus.

c. The Moon's orbit is inclined to the ecliptic (plane of Earth's orbit) by about 5°8′. The line of nodes, which is the intersection of the Moon's orbital plane with the ecliptic, rotates westward, making one complete revolution in 18.6 years.

   The node where the Moon crosses the ecliptic from south to north is called the ascending node; the other, where the Moon crosses from north to south, is the descending node. Only when the Moon is at one of these nodal crossing points can eclipses occur, for only then can the Sun, Earth and the Moon be suitably aligned. The average time for the Moon to go around its orbit from node to (the same) node is 27.21222 days and is called the "draconitic period," in reference to the superstition that a dragon was supposed to swallow the Sun at a total eclipse.

d. The inclination of the Moon's orbit to the ecliptic actually varies between 4°59′ and 5°18′; its mean value is 5°8′. Earth's equator is inclined to the ecliptic by 23°27′, and, except for the slow precession of Earth's axis of rotation with a period of 26,000 years, the equatorial plane is relatively stationary.

   From Figure 7-3 we can see that the angle between the equator and the Moon's orbital plane varies because of the rotation of the Moon's line of nodes. When the Moon's ascending node coincides with the vernal equinox direction, the inclination of the Moon's orbit to the equator is a maximum, being the sum of 5°8′ and 23°27′ or 28°35′. When the descending node is at the vernal equinox, the inclination of the Moon's orbit to the equator is the difference, 23°27′ – 5°8′ = 18°19′. Thus, the inclination relative to the equator varies between 18°19′ and 28°35′ with a period of 18.6 years.

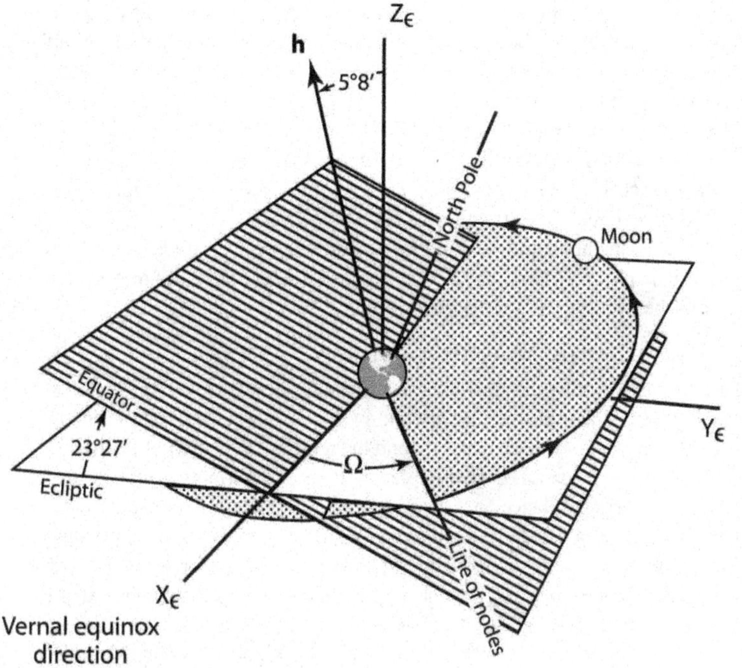

Fig. 7-3 Rotation of the Moon's Line of Nodes.

Both the slight variation in inclination relative to the ecliptic and the regression of the line of nodes were first observed by Flamsteed about 1670.

e. The line of apsides (line joining perigee and apogee) rotates in the direction of the Moon's orbital motion, causing ω to change 360° in about 8.9 years. Newton tried to explain this effect in the *Principia* but his predictions accounted for only about half the observed apsidal rotation. In 1749 the French mathematician Clairaut was able to derive the correct result from theory, but more than a century later, in 1872, the correct calculations were also discovered among Newton's unpublished papers: he had detected his own error but had never bothered to correct it in print!

7.2.2 Lunar Librations

The Moon's period of revolution around Earth is exactly equal to its period of rotation on its axis, so it always keeps the same face turned toward Earth. If the Moon's orbit were circular and if its axis of rotation were perpendicular to its orbit, we would see exactly half of its surface.

Actually we see, at one time or another, about 59% of the lunar surface because of a phenomenon known as "lunar libration." The libration or "rocking motion" of the Moon has two causes. The geometrical libration in latitude occurs because the Moon's equator is inclined 6.5° to the plane of its orbit. At one time during the month the Moon's north pole is tipped toward Earth and half a month later the south pole is tipped toward us, allowing us to see slightly beyond each pole in turn.

The geometrical libration in longitude is due to eccentricity of the orbit. The rotation of the Moon on its axis is uniform but its angular velocity around its orbit is not since it moves faster near perigee and slower near apogee. This allows us to see about 7.75° around each limb of the Moon.

In addition to the apparent rocking motion described above, there is an actual rocking called "physical libration" caused by the attraction of Earth on the long diameter of the Moon's triaxial ellipsoid figure.

7.3 SIMPLE EARTH–MOON TRAJECTORIES

The computation of a precision lunar trajectory can only be done by numerical integration of the equations of motion, taking into account the oblate shape of Earth, solar perturbations, solar pressure and the terminal attraction of the Moon, among other things. Because of the complex motions of the Moon, actual mission planning places heavy reliance on a lunar ephemeris, which is a tabular listing of the Moon's position at regular intervals of chronological time. As a result, lunar missions are planned on an hour-by-hour, day-by-day and month-by-month basis.

The general procedure is to assume the initial conditions, $\mathbf{r}_0$ and $\mathbf{v}_0$, at the injection point and then use a Runge–Kutta or similar numerical method to determine the subsequent trajectory. Depending on how well we select the values of $\mathbf{r}_0$ and $\mathbf{v}_0$, the trajectory may hit the Moon or miss it entirely The idea is to adjust the injection conditions by trial and error until a suitable lunar arrival occurs.

While a computer can compute an individual trajectory quickly, it is helpful to have a good analytical understanding of the problem in order to intelligently select regions of attractive launch and mission options.

7.3.1 Some Simplifying Assumptions

To study the basic dynamics of lunar trajectories, we will assume that the Moon's orbit is circular with a radius of 384,400 km. Since the mean eccentricity of the actual orbit is only about 0.0549, this will not introduce significant errors. We will also assume that we can neglect the terminal attraction of the Moon and simply look at some trajectories that intersect the Moon's orbit.

In the analysis that follows we will also assume that the lunar trajectory is coplanar with the Moon's orbit. In a precision trajectory calculation the launch time is selected so that this is approximately true in order to minimize the Δv required for the mission since plane changes are expensive in terms of velocity.

7.3.2 Time of Flight Versus Injection Speed

With the assumptions stated above we can proceed to investigate the effect of injection speed on the time of flight of a lunar probe. We can compute the energy and angular momentum of the trajectory from

$$E = \frac{v_0^2}{2} - \frac{\mu}{r_0}$$

$$h = r_0 v_0 \cos\phi_0$$

The parameter, semi-major axis, and eccentricity are then obtained from

$$p = \frac{h^2}{\mu}$$

$$a = \frac{-\mu}{2E}$$

$$e = \sqrt{1 - p/a}$$

Solving the polar equation of a conic for the true anomaly, we get

$$\cos\nu = \frac{p - r}{er}$$

If we let $r = r_0$ in this expression, we can solve for ν_0; if we let r equal the radius of the Moon's orbit, we can find the true anomaly upon arrival at the Moon's orbit.

We now have enough information to determine the time of flight from Earth to the Moon for any set of injection conditions using the equations presented in Chapter 4.

In Figure 7-4 we have plotted time of flight versus injection speed for an injection altitude of 320 km and a flight-path angle of 0°. Actually, the curve is nearly independent of flight-path angle at injection.

We see from this curve that a significant reduction in time of flight is possible with only modest increases in injection speed. For missions with human crews life-support requirements increase with mission duration, so the slightly higher injection speed required to achieve a shorter flight time pays for itself up to a point. It is interesting to note that the flight time chosen for the Apollo lunar landing missions was about 72 hours.

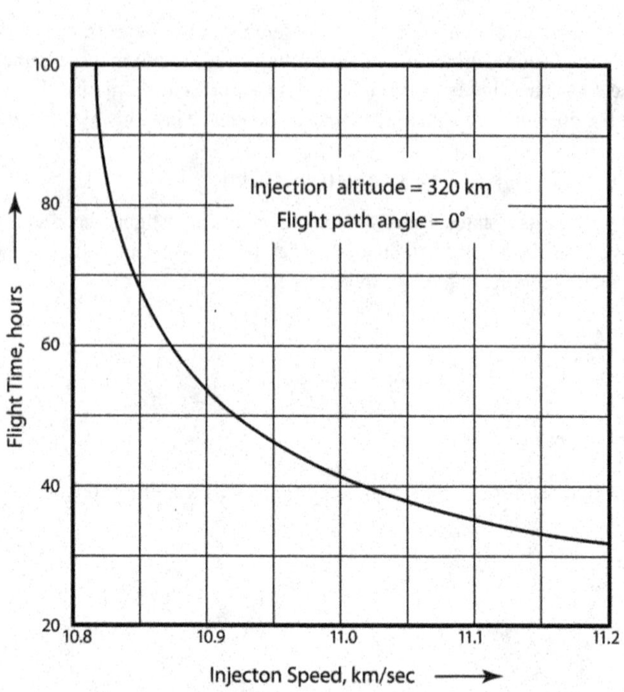

Fig. 7-4 Lunar Flight Time vs. Injection Speed.

7.3.3 The Minimum Energy Trajectory

If we assume that injection into the lunar trajectory occurs at perigee where $\phi_0 = 0°$, then it is easy to see what effect injection speed has on the orbit. In Figure 7-5 we have shown a family of orbits corresponding to different injection speeds.

For the limiting case where the injection speed is infinite, the path is a straight line with a time of flight of zero. As we lower the injection speed the orbit goes from hyperbolic, to parabolic to elliptical in shape and the time of flight increases. Eventually, if we keep reducing the injection speed, we will arrive at the dashed orbit in Figure 7-5, whose apogee just barely reaches to the distance of the Moon. If we assume an injection altitude of 0.05 Earth radii (320 km), this *minimum injection* speed is 10.82 km/s. If we give our lunar probe less than this speed, it will fail to reach the Moon's orbit like the dotted path in Figure 7-5.

The time of flight for this minimum energy lunar trajectory is 7,172 minutes or about 120 hours. Because all other trajectories that reach to the Moon's orbit have shorter flight times, this represents an approximate *maximum time of flight* for a lunar mission; if we try to take longer by going slower, we will never reach the Moon at all.

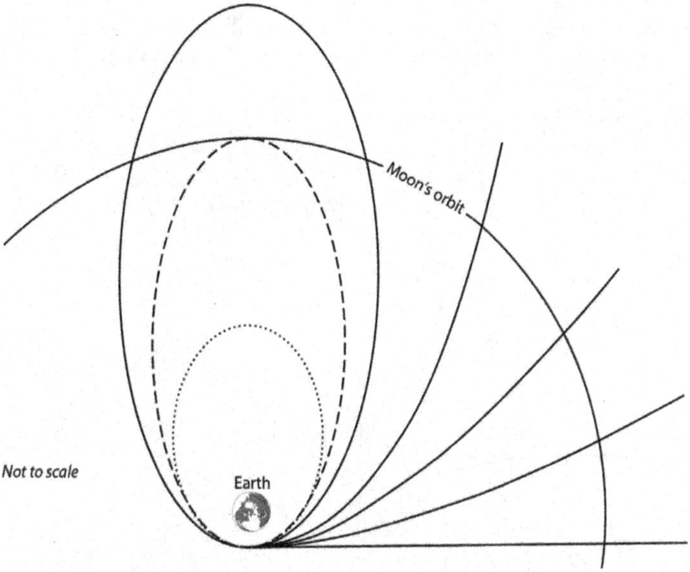

Fig. 7-5 Effect of Injection Speed on Trajectory Shape.

The eccentricity of the minimum energy trajectory is 0.966 and represents the *minimum eccentricity* for an elliptical orbit that reaches as far as the Moon from our assumed injection point.

In the absence of lunar gravity, the speed upon arrival at the Moon's orbit for the minimum energy trajectory is 0.188 km/s. This represents the slowest approach speed possible for our example. Since the Moon's orbital speed is about 1 km/s, the Moon would literally run into our probe from behind, resulting in an impact on the "leading edge" or eastern limb of the Moon. Probes that have a higher arrival speed would tend to impact somewhere on the side of the Moon facing us.

From Figure 7-5 we see that, as the injection speed is decreased, the geocentric angle swept out by the lunar probe from injection to lunar intercept increases from 0° to 180° for the minimum energy case. In general, the sweep angle, which we can call ψ, is a function of injection speed for a fixed injection altitude and flight-path angle; an increase in sweep angle (up to 180°) corresponds to a decrease in injection speed. If we are trying to minimize injection speed, we should try to select an orbit that has a sweep angle close to 180°. We will make use of this general principle later in this chapter.

7.3.4 Miss Distance at the Moon Caused by Injection Errors

In trying for a direct hit on the Moon we would time our launch so that the probe crosses the Moon's orbit just at the instant the Moon is at the intercept point. Using our simplified model of the Earth–Moon system and neglecting lunar gravity, we can get some idea of how much we will miss the center of the Moon if, because of errors in guidance or other factors, the injection conditions are not exactly as specified.

In such a case, both the sweep angle and the time of flight will differ from their nominal values and the trajectory of the probe will cross the Moon's orbit at a different point and time than predicted. In the case of a direct (eastward) launch, the effects of such injection errors tend to cancel. This may be seen from Figure 7-6.

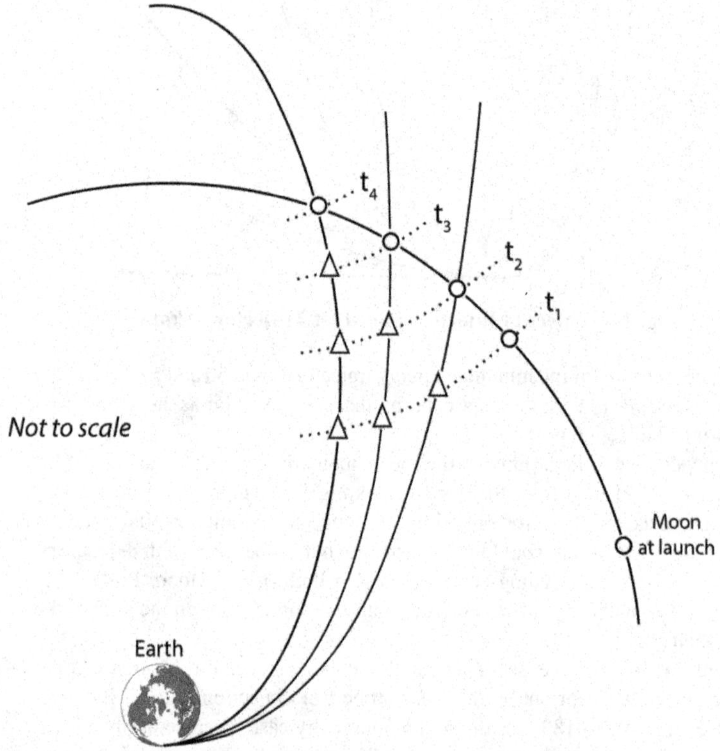

Fig. 7-6 Effects of Launch Error may Tend to Cancel.

If, for example, the initial velocity is too high, the geocentric sweep angle will be smaller than predicted; that is, the probe will cross the Moon's orbit west of the predicted point by some amount $\Delta\psi$. But, the time of flight will be shorter by an amount Δt, so the Moon will be west of the predicted intercept point by an amount $\omega_m \Delta t$, where ω_m is the angular velocity of the Moon in its orbit. If we neglect lunar gravity, the angular miss distance along the Moon's orbit is the difference between $\Delta\psi$ and $\omega_m \Delta t$.

It is possible to show that for an injection speed of about 11.0 km/s, which results in a sweep angle of about 160°, the effects of errors in injection altitude and speed exactly cancel and, for all practical purposes, the miss distance at the Moon is a function only of errors in the flight-path angle. For this condition an error of 1° in flight-path angle produces a miss of about 1,300 km.[5] The effect of lunar gravity is to reduce this miss distance.

7.4 THE PATCHED-CONIC APPROXIMATION

We can take lunar gravity into account and still see two-body orbital mechanics by the simple expedient of considering the probe to be under the gravitational influence of Earth alone until it enters the gravitational "sphere of influence" of the Moon and then assuming that it moves only under the gravitational influence of the Moon. In effect, we pick a point in the vicinity of the Moon where we "turn off Earth and turn on the Moon." Before we show how this is done it should be clear to you that what we are about to do is an approximation. The transition from geocentric motion to selenocentric motion (i.e., motion based on the Moon as the center) is a gradual process that takes place over a finite arc of the trajectory where both Earth and the Moon affect the path equally. There is, however, evidence to show that this simple strategy of patching two conics together at the edge of the Moon's sphere of influence is a sufficiently good approximation for preliminary mission analysis.[5] *However, the patched-conic analysis is not valid for the calculation of the return trajectory to Earth because of errors in the Moon's sphere of influence. It will also be in error for the same reason in calculating the perilune altitude and lunar trajectory orientation. It is good primarily for outbound Δ-v calculations.*

For the analysis that follows we will adopt the definition of sphere of influence suggested by Laplace. This is a sphere centered at the Moon and having a radius R_S given by the expression

$$R_s = D\left(\frac{m_{Moon}}{m_\oplus}\right)^{2/5} \tag{7-1}$$

where D is the distance from Earth to the Moon. A full derivation of Equation (7-1) can be found in Battin.[6] To give at least an elementary idea of its origin, consider the equation of motion viewed from an Earth frame,

$$\mathbf{a} = \begin{pmatrix}\text{central}\\ \text{acceleration}\\ \text{due to Earth}\end{pmatrix} + \begin{pmatrix}\text{perturbing}\\ \text{acceleration}\\ \text{due to Moon}\end{pmatrix}$$

$$= \mathbf{a}_e + \mathbf{a}_{pm}$$

Then, the equation of motion viewed from the Moon is

$$\mathbf{A} = \begin{pmatrix} \text{central} \\ \text{acceleration} \\ \text{due to Moon} \end{pmatrix} + \begin{pmatrix} \text{perturbing} \\ \text{acceleration} \\ \text{due to Earth} \end{pmatrix}$$

$$= \mathbf{A}_m + \mathbf{A}_{pe}$$

The sphere of influence is the approximation resulting from equating the ratios

$$\frac{a_{pm}}{a_e} = \frac{A_{pe}}{A_m}$$

Equation (7-1) yields the value

$$R_S = 66{,}300 \text{ km}$$

or about 1/6 the distance from the Moon to Earth.

7.4.1 The Geocentric Departure Orbit

Figure 7-7 shows the geometry of the geocentric departure orbit. The four quantities that completely specify the geocentric phase are

$$\mathbf{r}_0, \mathbf{v}_0, \phi_0 \text{ and } \gamma_0$$

where γ_0 is called the "phase angle at departure."

The difficulty with selecting these four quantities as the independent variables is that the determination of the point at which the geocentric trajectory crosses the lunar sphere of influence involves an iterative procedure in which time of flight must be computed during each iteration. This difficulty may be bypassed by selecting three initial conditions and one arrival condition as the independent variables. A particularly convenient set is

$$\mathbf{r}_0, \mathbf{v}_0, \phi_0 \text{ and } \lambda_1$$

where the angle λ_1 specifies the point at which the geocentric trajectory crosses the lunar sphere of influence.

Given these four quantities, we can compute the remaining arrival conditions, $\mathbf{r}_1$, $\mathbf{v}_1$, ϕ_1 and γ_1. We will assume that the geocentric trajectory is direct and that lunar arrival occurs prior to apogee of the geocentric orbit. The energy and angular momentum of the orbit can be determined from

$$E = \frac{v_0^2}{2} - \frac{\mu}{r_0} \tag{7-2}$$

$$h = r_0 v_0 \cos\phi_0 \tag{7-3}$$

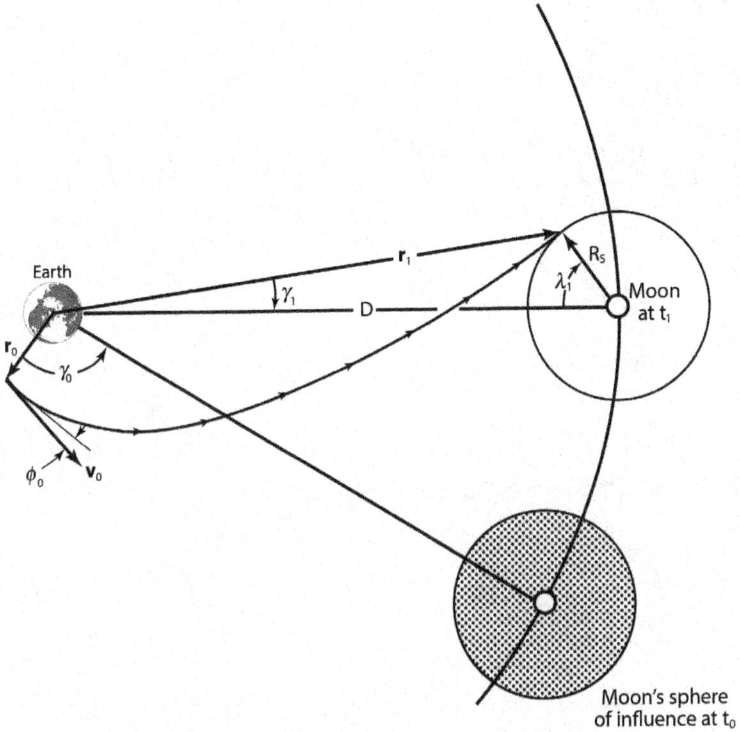

Fig. 7-7 Geocentric Transfer to the Lunar Sphere of Influence.

From the law of cosines, the radius, r_1, at lunar arrival is

$$r_1 = \sqrt{D^2 + R_S^2 - 2DR_S \cos \lambda_1} \tag{7-4}$$

The speed and flight path angle at arrival follow from conservation of energy and momentum:

$$v_1 = \sqrt{2(E + \mu / r_1)} \tag{7-5}$$

$$\cos \phi_1 = \frac{h}{r_1 v_1} \tag{7-6}$$

where ϕ_1 is known to lie between 0° and 90° since arrival occurs prior to apogee.

Finally, from geometry

$$\sin \gamma_1 = \frac{R_s}{r_1} \sin \lambda_1 \tag{7-7}$$

The time of flight, $t_1 - t_0$, from injection to arrival at the lunar sphere of influence can be computed once ν_0 and ν_1 are determined. Before the true anomalies can be found we must determine p, a and e of the geocentric trajectory from

$$p = \frac{h^2}{\mu} \tag{7-8}$$

$$a = \frac{-\mu}{2E} \tag{7-9}$$

$$e = \sqrt{1 - p/a} \tag{7-10}$$

Then ν_0 and ν_1 follow from the polar equation of a conic:

$$\cos \nu_0 = \frac{p - r_0}{r_0 e} \tag{7-11}$$

$$\cos \nu_1 = \frac{p - r_1}{r_1 e} \tag{7-12}$$

Next, we can determine the eccentric anomalies, E_0 and E_1 from

$$\cos E_0 = \frac{e + \cos \nu_0}{1 + e \cos \nu_0} \tag{7-13}$$

$$\cos E_1 = \frac{e + \cos \nu_1}{1 + e \cos \nu_1} \tag{7-14}$$

Finally, time of flight is obtained from

$$t_1 - t_0 = \sqrt{\frac{a^3}{\mu}} \left[\left(E_1 - e \sin E_1 \right) - \left(E_0 - e \sin E_0 \right) \right] \tag{7-15}$$

The Moon moves through an angle $\omega_m (t_1 - t_0)$ between injection and arrival at the lunar sphere of influence, where ω_m is the angular velocity of the Moon in its orbit. Based on our simplified model of the Earth–Moon system,

$$\omega_m = 2.649 \times 10^{-6} \text{ radians/s}$$

The phase angle at departure, γ_0, is then determined from

$$\gamma_0 = \nu_1 - \nu_0 - \gamma_1 - \omega_m (t_1 - t_0) \tag{7-16}$$

Actually, the time of flight and phase angle at departure need not be computed until we have verified that the values $\mathbf{r_0}$, $\mathbf{v_0}$, ϕ_0 and λ_1, which we chose arbitrarily, result in a satisfactory lunar approach trajectory or impact. We will do this in the sections that follow. If the lunar trajectory is not satisfactory, then we will adjust the values of $\mathbf{r_0}$, $\mathbf{v_0}$, ϕ_0 and λ_1 until it is. Only after this trial-and-error procedure is complete should we perform the computations embodied in Equations (7-8) through (7-16) above.

Before proceeding to a discussion of the patch condition, a few remarks are in order concerning Equation (7-5). The energy in the geocentric departure orbit is completely determined by r_0 and v_0. The geocentric radius, r_1, at arrival is completely determined by λ_1. It may happen that the trajectory is not sufficiently energetic to reach the specified point on the lunar sphere of influence as determined by λ_1. If so, the quantity under the radical sign in Equation (7-5) will be negative and the whole computational process fails.

7.4.2 Conditions at the Patch Point

We are now ready to determine the trajectory inside the Moon's sphere of influence where only lunar gravity is assumed to act on the spacecraft. Since we must now consider the Moon as the central body, it is necessary to find the speed and direction of the spacecraft *relative to the center of the Moon.* In Figure 7-8 the geometry of the situation at arrival is shown in detail.

If we let the subscript 2 indicate the initial conditions relative to the Moon's center, then the selenocentric radius, r_2, is

$$r_2 = R_S \tag{7-17}$$

The velocity of the spacecraft relative to the center of the Moon is

$$\mathbf{v_2} = \mathbf{v_1} - \mathbf{v_m}$$

where $\mathbf{v_m}$ is the velocity of the Moon relative to the center of Earth. The orbital speed of the Moon for our simplified Earth–Moon model is

$$v_m = 1.018 \text{ km/s} \tag{7-18}$$

The selenocentric arrival speed, v_2, may be obtained by applying the law of cosines to the vector triangle in Figure 7-8:

$$v_2 = \sqrt{v_1^2 + v_m^2 - 2 v_1 v_m \cos(\phi_1 - \gamma_1)} \tag{7-19}$$

The angle ε_2 defines the direction of the initial selenocentric velocity relative to the Moon's center. Equating the components of $\mathbf{v_2}$ perpendicular to $\mathbf{r_2}$ yields

$$v_2 \sin \varepsilon_2 = v_m \cos \lambda_1 - v_1 \cos(\lambda_1 + \gamma_1 - \phi_1)$$

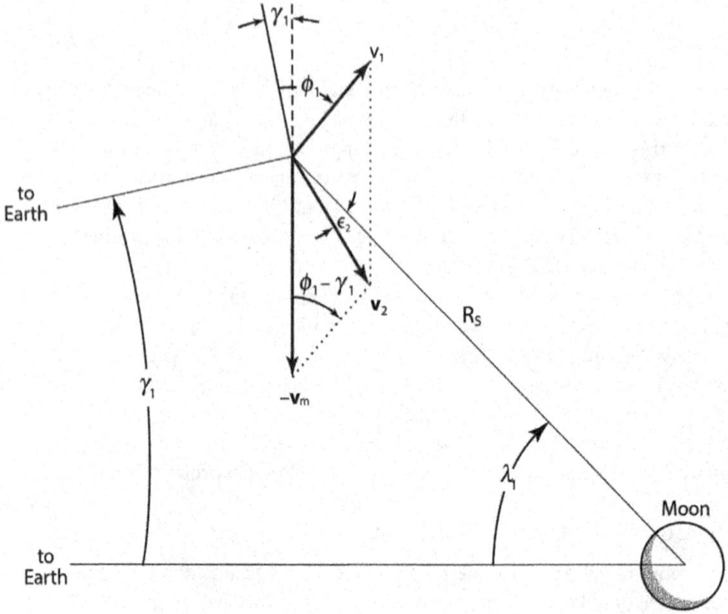

Fig. 7-8 The Patch Condition.

from which we obtain

$$\varepsilon_2 = \sin^{-1}\left[\frac{V_m}{V_2}\cos\lambda_1 - \frac{V_1}{V_2}\cos(\lambda_1 + \gamma_1 - \phi_1)\right] \tag{7-20}$$

It is obvious that for a dead-center hit on the Moon ε_2 must be exactly zero.

7.4.3 The Selenocentric Arrival Orbit

The selenocentric initial conditions $\mathbf{r}_2$, $\mathbf{v}_2$, and ε_2 are now known and so we can compute terminal conditions at other points on the trajectory. There are a number of terminal conditions of interest depending on the nature of the mission:

1. Lunar impact. In this case we wish to determine whether the periselenium radius is less than the radius of the Moon. In case $r_p < r_m$ we may wish to compute the speed at impact. The radius of the Moon, r_m, may be taken as 1,738 km.

2. Lunar orbit. In this case we may wish to compute the speed increment necessary to produce a circular lunar satellite at the periselenium altitude.

3. Circumlunar flight. In this case we probably would want to compute both the periselenium conditions and the conditions upon exit from the lunar sphere of influence.

In any case, the conditions at periselenium will certainly be of interest and are probably the best single measure of the trajectory.

The energy and momentum relative to the center of the Moon are given by

$$E = \frac{v_2^2}{2} - \frac{\mu_m}{r_2} \tag{7-21}$$

$$h = r_2 v_2 \sin \varepsilon_2 \tag{7-22}$$

where μ_m is the gravitational parameter of the Moon. Since the mass of the Moon is 1/81.3 of Earth's mass, μ_m may be determined from

$$\mu_m = \frac{\mu_\oplus}{81.3}$$

$$\mu_m = 4.9028 \times 10^3 \text{km}^3/\text{s}^2$$

The parameter and eccentricity of the selenocentric orbit can be computed from

$$p = \frac{h^2}{\mu_m} \tag{7-23}$$

$$e = \sqrt{1 + 2Eh^2/\mu_m^2} \tag{7-24}$$

The conditions at periselenium are then obtained from

$$r_p = \frac{p}{1+e} \tag{7-25}$$

$$v_p = \sqrt{2\left(E + \mu_m/r_p\right)} \tag{7-26}$$

If the periselenium conditions are not satisfactory, either the injection conditions, $\mathbf{r_0}$, $\mathbf{v_0}$ and ϕ_0 or the angle λ_1 should be adjusted by trial and error until the trajectory is acceptable.

Example Problem. A lunar probe is sent to the Moon on a trajectory with the following injection conditions:

$r_0 = 6{,}697.0$ km

$v_0 = 10.8462$ km/s

$\phi_0 = 0°$

Upon arrival at the Moon's sphere of influence, $\lambda_1 = 30°$.

Calculate the initial phase angle γ_0 and the altitude at closest approach to the Moon for the probe.

From Equations (7-2) and (7-3),

$$E = -0.6995 \text{ km}^2/\text{s}^2$$

$$h = 72{,}637 \text{ km}^2/\text{s}$$

We know that

$$D = 384{,}400 \text{ km}$$

$$R_S = 66{,}300 \text{ km}$$

Using Equation (7-4) gives

$$r_1 = \sqrt{D^2 + R_S^2 - 2DR_S \cos\lambda_1} = 328{,}660 \text{ km}$$

and from Equation (7-5) we get

$$v_1 = \sqrt{2\left(E + \frac{\mu}{r_1}\right)} = 1.1033 \text{ km/s}$$

From Equation (7-6),

$$\phi_1 = \cos^{-1}\left(\frac{h}{r_1 v_1}\right) = 77.401°$$

($\phi_1 < 90°$ since arrival at the Moon's sphere of influence occurs prior to apogee.) From Equation (7-7), we get the phase angle of the Moon at arrival,

$$\gamma_1 = \sin^{-1}\left(\frac{R_s}{r_1} \sin\lambda_1\right) = 5.789^\circ = 0.101 \text{ radians}$$

To calculate the line of flight to the Moon's sphere of influence, we need the parameters p, a, e, E_0 and E_1 for the geocentric trajectory. Using Equations (7-8) through (7-14) we obtain:

$$p = 13{,}237 \text{ km},\ a = 284{,}930 \text{ km},\ e = 0.9765$$

$\nu_0 = 0$ since $\phi_0 = 0°$ (the probe burns out at perigee)

$$\nu_1 = 169.37° = 2.956 \text{ radians}$$

$E_0 = 0°$ since $\nu_0 = 0°$

$$E_1 = 97.043° = 1.7286 \text{ radians}$$

Using Equation (7-15) we get the time of flight:

$$t_1 - t_0 = \sqrt{\frac{a^3}{\mu}}\left[\left(E_1 - e\sin E_1\right) - \left(E_0 - e\sin E_0\right)\right]$$

$$= 1.8411\times 10^5\ \text{s} = 51.1416\ \text{hr}$$

From Equation (7-16) we get the phase angle at departure,

$$\gamma_0 = \nu_1 - \nu_0 - \gamma_1 - \omega_m(t_1 - t_0) = \underline{\underline{2.3673\ \text{radians}}} = \underline{\underline{135.64°}}$$

where $\omega_m = 2.6847 \times 10^{-6}$ radians/s.

At the Moon's sphere of influence it is necessary to convert v_1 and R_S to units based on the Moon as the gravitational center of attraction. So

$$\mu_m = \frac{\mu_\oplus}{81.3} = 4.903\left(10^3\right)\ \text{km}^3/\text{s}^2$$

$$v_m = 1.018\ \text{km/s}$$

Using Equation (7-19), we have

$$v_2 = 1.198\ \text{km/s}$$

From Equation (7-20),

$$\varepsilon_2 = 8.9387°$$

Using Equations (7-20) through (7-25), we can determine the minimum distance of approach to the Moon as follows:

$$h = 13{,}036\ \text{km}^2/\text{s}$$

$$E = \frac{v_2^2}{2} - \frac{\mu_m}{R_S} = 0.7267\ \text{km}^2/\text{s}^2\quad \text{(Why is E> 0?)}$$

$$p = \frac{h^2}{\mu_m} = 34{,}662\ \text{km}$$

$$e = \sqrt{1 + \frac{2Eh^2}{m_\mu^2}} = 3.3580$$

$$r_p = 7{,}953.8\ \text{km}$$

$$h_p = r_p - r_m = \underline{\underline{6{,}215.8\ \text{km}}}$$

This is the minimum distance of approach.

7.5 NONCOPLANAR LUNAR TRAJECTORIES

The preceding analysis has been based on the assumption that the lunar trajectory lies in the plane of the Moon's orbit. The inclination of the Moon's orbit varies between about 18.2° and 28.5° over a period of 18.6 years. Since it is impossible to launch an Earth satellite into an orbit whose inclination is less than the latitude of the launch site, a coplanar trajectory originating from Cape Canaveral, whose latitude is 28.5°, is possible only when the inclination of the Moon's orbit is at its maximum value. This occurred in the early part of 1988 and again in 2007.

Launches that occur at times other than these must necessarily result in noncoplanar trajectories. In the sections that follow we will examine noncoplanar trajectories and develop a method for selecting acceptable launch dates and injection conditions.

7.5.1 Some Typical Constraints on Lunar Trajectories

If there are no restrictions on the launch conditions of a spacecraft or on the conditions at lunar approach, then there are no limitations on the time of the lunar month at which the spacecraft can approach the Moon. Practical considerations, such as launch site location, missile-range safety, accuracy tolerances, and the limited range of attainable injection, can all be accommodated. It is interesting to examine the limitations on the possible launch times for lunar missions that are imposed by some of these restrictions.

A typical design restriction for lunar missions is the specification of the lighting conditions on the surface of the Moon as determined by the phase of the Moon. For a particular year, the declination of the Moon at a given phase varies between maximum and minimum values that correspond approximately to the mean inclination of the Moon's orbit for that year.

Another typical restriction concerns the permissible directions of launch from a particular site.

In the analysis that follows, the launch site is assumed to be Cape Canaveral, which has a latitude of 28.5°. The launch azimuth, β_0, must be between 40° and 115° as specified by Eastern Test Range safety requirements.

7.5.2 Determining the Geocentric Sweep Angle

An important parameter in determining acceptable launch times is the total geocentric angle swept out by the spacecraft from launch to lunar intercept. The total sweep angle, ψ_t, consists of the free-flight sweep angle, ψ_{ff}, from injection to intercept plus the geocentric angular travel from launch to injection, ψ_c. Depending on the launch technique used, ψ_c may be simply the burning arc plus the angular distance traveled during a coasting period prior to injections.

While the angle ψ_c may be selected arbitrarily, the free-flight sweep angle, ψ_{ff}, is determined by the injection conditions r_0, v_0 and ϕ_0. The angle ψ_{ff} is just the difference in the true anomaly between injection and lunar intercept and may be computed from the equations in Section 7.3.2. In Figure 7-10 we have plotted the free-flight sweep angle for several values of v_0 and ϕ_0 and a fixed injection altitude of 320 km or about 200 miles. Lunar intercepts are assumed to occur at a distance of 384,400 km.

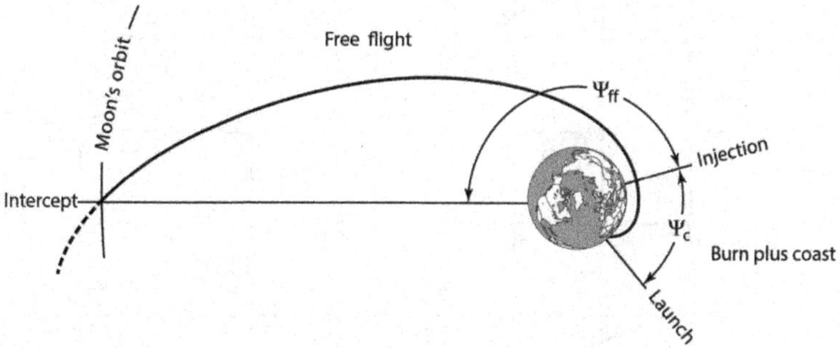

Fig. 7-9 Geocentric Sweep Angle.

By selecting the injection conditions r_0, v_0 and ϕ_0 we can determine ψ_{ff}. If we arbitrarily select ψ_c, we can obtain the total sweep angle from

$$\Psi_t = \Psi_{ff} + \Psi_c \tag{7-27}$$

Since the latitude or declination of the launch site is known, we may determine the declination of the spacecraft after it has traversed an arc ψ_t if we know the launch azimuth, β_0. This is essentially a problem in spherical trigonometry and is illustrated in Figure 7-11.

From the law of cosines for spherical triangles, we obtain

$$\sin\delta_1 = \sin\delta_0 \cos\Psi_t + \cos\delta_0 \sin\Psi_t \cos\beta_0 \tag{7-28}$$

We have plotted values of Δ_1 obtained from Equation (7-28) versus total sweep angle for launch azimuths between 40° and 115° in Figure 7-12. Since the Moon's declination at intercept is limited between +28.5° and –28.5° (during 1969) and because of the range safety restrictions on launch azimuth, those portions of the graph that are shaded represent impossible launch conditions.

There are several interesting features of Figure 7-12 worth noticing. Perhaps the most interesting is the fact that a sweep angle of 180° is possible only if we intercept the Moon at its maximum southern declination of –28.5°. For this condition any launch azimuth is correct.

From Figure 7-10 we see that free-flight sweep angles of less than about 120° are impossible to achieve without going to very high injection speeds or large flight path angles, both of which are undesirable. As a result, if we are interested in minimizing injection speed, we must intercept the Moon when it is near its

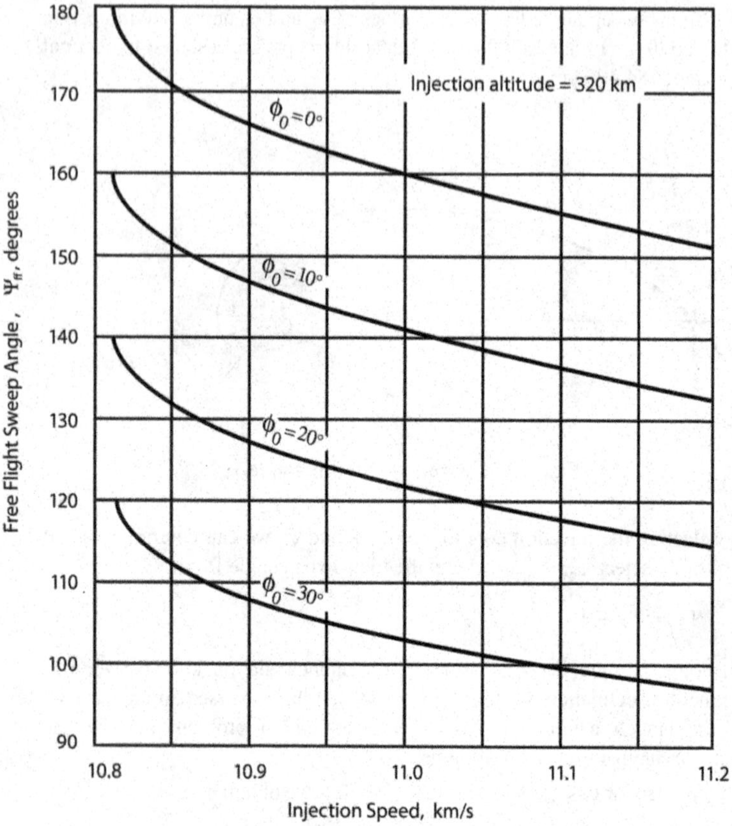

Fig. 7-10 Free-Flight Sweep Angle vs. Injection Speed.

maximum southern declination for direct-ascent launches or launches where the coasting period is small enough to keep ψ_t less than 180°.

However, if we add a coasting arc, ψ_t, large enough to make ψ_t greater than 180°, we can intercept the Moon at any point along its orbit.

It should be obvious from Equation (7-28) and Figure 7-12 that launch azimuth and lunar declination at intercept are not independent; once ψ_t is known, selecting either launch azimuth or declination at intercept determines the other uniquely. If lunar declination at intercept is specified, then launch azimuth may be read directly from Figure 7-12 or computed more accurately from

$$\cos\beta_0 = \frac{\sin\delta_1 - \sin\delta_0\cos\Psi_t}{\cos\delta_0\sin\Psi_t} \tag{7-29}$$

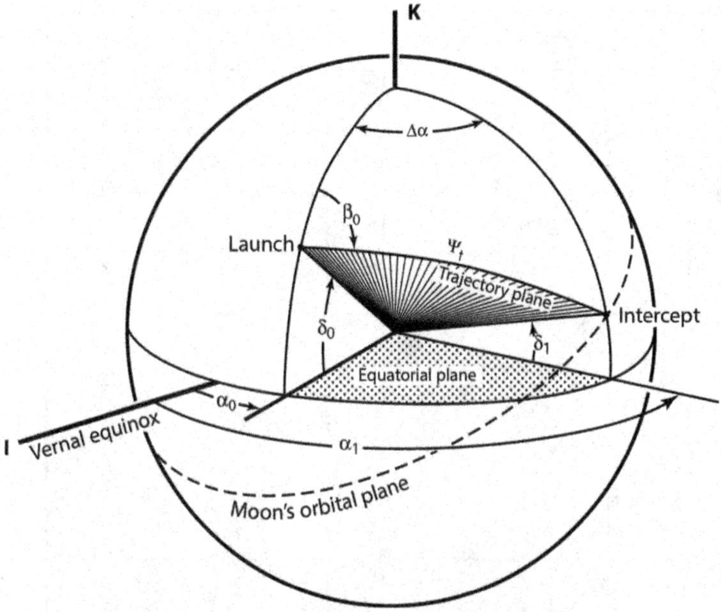

Fig. 7-11 Lunar Interception Angular Relationships.

7.5.3 Selecting an Acceptable Launch Date

Once the lunar declination at intercept is determined the next step is to search through a lunar ephemeris to find a time when the Moon is at the correct declination. If lighting conditions are important, we must find a time when both declination and phase are simultaneously correct. NASA's Jet Propulsion Laboratory (JPL) publishes updated planetary and lunar ephemeris on a continuing basis and is the recommended source of data for lunar mission planning.

Suppose we now select a time, t_1, for lunar intercept that meets the declination and lighting constraints. The right ascension of the Moon at t_1 can be obtained from the ephemeris tables. This is the angle α_1 in Figure 7-11.

The difference in right ascension, $\Delta\alpha$, between launch and intercept is fixed by the geometry of Figure 7-11. Applying the law of cosines to the spherical triangle in Figure 7-11, we get

$$\cos \Delta\alpha = \frac{\cos \psi_t - \sin \delta_0 \ \sin \delta_1}{\cos \delta_0 \ \cos \delta_1} \tag{7-30}$$

The next step is to establish the exact time of launch, t_0, and the right ascension of the launch site, α_0, at this launch time. To establish t_0 we need to compute the total time from launch to intercept. This consists of the free-flight

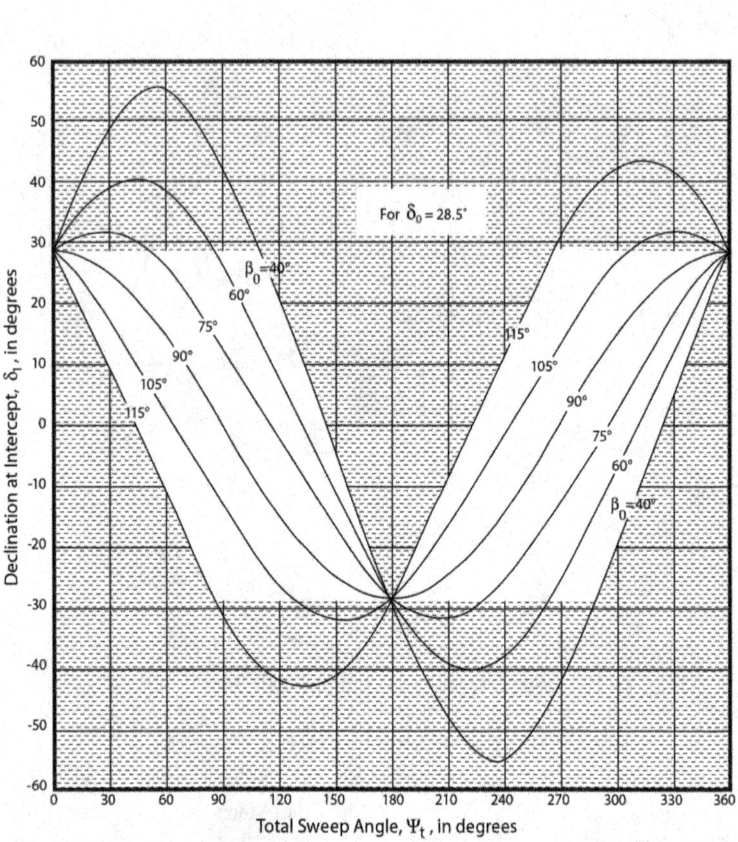

Fig. 7-12 Intercept Declination vs. Sweep Angle for Launch Azimuths between 40° and 115°.

time from injection to intercept, which may be computed from the injection conditions, plus the time to traverse the burning and coasting arc, ψ_c. Thus, the total time, t_t, is

$$t_t = t_{ff} + t_c \tag{7-31}$$

where t_{ff} is the free-flight time and t_c is the burn-plus-coast time.

The launch time, t_0, can now be obtained from

$$t_0 = t_1 - t_t \tag{7-32}$$

The right ascension of the launch point, α_0, is the same as the "local sidereal time," θ, at the launch point (see Figure 2-21) and may be obtained from Equation (2-70):

$$\alpha_0 = \theta = \theta_g + \lambda_E \tag{7-33}$$

where λ_E is the east longitude of the launch site and θ_g is the Greenwich sidereal time at t_0. Values of θ_g are tabulated in the ephemeris tables for 0^h UT on every day of the year and may be obtained by interpolation for any hour or by the method outlined in Section 2.9.

It would be sheer coincidence if the difference $\alpha_1 - \alpha_0$ were the same as the $\Delta\alpha$ calculated from Equation (7-30). Since the right ascension of the launch point changes by justover 360° in a day, it is possible to adjust t_1 slightly and recompute t_0 and α_0 until $\alpha_1 - \alpha_0$ and the required $\Delta\alpha$ from Equation (7-30) agree. The lunar declination at intercept will change very slightly as t_1 is adjusted and it may be necessary to return to Equation (7-29) and redetermine the launch azimuth β_0.

We now know the injection conditions, r_0, v_0, ϕ_0 and β_0, and the exact day and hour of launch and lunar intercept that will satisfy all of the constraints set forth earlier. These launch conditions should provide us with sufficiently accurate initial conditions to begin the computation of a precision lunar trajectory using numerical methods.

Exercises

7.1 Calculate the burnout velocity required to transfer a probe between the vicinity of Earth (assume r_{bo} = 6,700 km) and the Moon's orbit using Hohmann transfer. What additional Δv would be required to place the probe in the same orbit as that of the Moon? Neglect the Moon's gravity in both parts.

7.2 For a lunar vehicle that is injected at perigee near the surface of Earth, determine the eccentricity of the trajectory that just reaches the sphere of influence of the Moon.

7.3 For a value of $\varepsilon = +20°$ determine the maximum value of v_2 that will allow lunar impact.

7.4 The following quantities are with respect to the Moon:

$\lambda_1 = 30°$, $h_m = 33{,}084$ km²/s

$\varepsilon_2 = -30°$, $p_m = 224{,}000$ km

$v_1 = 1$ km/s, $e_m = 6.3$

$E_m = 0.426$ km²/s²

Determine λ of perilune.

7.5 It is desired that a lunar vehicle have its perilune 262 km above the lunar surface with direct motion about the Moon.

a. At the sphere of influence $v_2 = 500$ m/s. Calculate the initial value of ε_2 that must exist to satisfy the conditions.

b. Determine the $\Delta \mathbf{v}$ necessary to place the probe in a circular orbit in which perilune is an altitude of 262 km.

7.6 A space probe was sent to investigate the planet Mars. On the way it crosses the Moon's orbit. The burnout conditions are

$$r_{bo} = 7972 \text{ km}, \qquad \phi_{bo} = 0°$$

$$v_{bo} = 1600 \text{ m/s}$$

For $v_{Moon} = 1024$ m/s and D = 384,400 km:

a. What is the speed of the probe as it is at the distance D from Earth?

b. What is the elevation angle of the probe at D?

c. What is the angle through which the Moon will have moved 30 hours after launch of the probe?

7.7 A lunar vehicle arrives at the sphere of influence of the Moon with $\lambda_1 = 0°$. The speed of the vehicle relative to Earth is 200 m/s and the flight-path angle relative to Earth is 80°. The vehicle is in direct motion relative to Earth. Find v_2 relative to the Moon and ε_2. Is the vehicle in retrograde or direct motion relative to the Moon?

7.8 Given a lunar declination at intercept of 15° and a total geocentric sweep angle of 150° and find the launch azimuth, is this an acceptable launch azimuth?

7.9 A sounding rocket is fired vertically from the surface of Earth so that its maximum altitude occurs at the distance of the Moon's orbit. Determine the velocity of the sounding rocket at apogee relative to the Moon (neglecting Moon gravity) if the Moon were nearby.

7.10 Discuss the change in energy with respect to Earth as the result of a satellite passing by the Moon in front of or behind the Moon's orbit (retrograde or direct orbit with respect to the Moon). Which would be best for a lunar landing?

7.11 Design a computer program that will solve the basic patched-conic lunar transfer problem. Specifically, given that you wish to arrive at perilune at a specified altitude, select $\mathbf{r}_0$, $\mathbf{v}_0$ and ϕ_0 to meet this requirement. One approach is to select a reasonable $\mathbf{r}_0$, $\mathbf{v}_0$ and ϕ_0 and iterate on λ_1 until the lunar orbit conditions are met. A suggested set of data is h_{bo} = 200 km, ϕ_0 = 0°; varying v_0 to show the effect of insufficient energy to reach the Moon. Choose a reasonable value of maximum available velocity (of which v_0 is a part) and consider having enough velocity remaining at perilune to inject the probe into a circular lunar orbit at perilune altitude.

7.12 *A "free-return" lunar trajectory is one that passes around the Moon in such a manner that it will return to Earth with no additional power to change its orbit. Determine such an orbit using the patched-conic method. Find the v_{bo}, ϕ_{bo} and γ_0 such that the return trajectory will have $h_{perigee}$ = 100 km and thus ensure re-entry. Use h_{bo} = 200 km. Attempt to have a perilune altitude of about 180 km. (Hint: Use a computer and don't expect an exact solution. There are several orbits that may meet the criteria. You will have to extend the iterative technique of this chapter.)

List of References

1. Gamow, George. *The Moon.* London and New York, NY, Abelard-Schuman, 1959.
2. Fisher, Clyde. *The Story of the Moon.* Garden City, NY, Doubleday, Doran and Company, Inc., 1943.
3. Newton, Sir Isaac. *Principia.* Motte's translation revised by Cajori. Berkeley and Los Angeles, University of California Press, 1962.
4. Baker, Robert M. L. and Maud W. Makemson. *An Introduction to Astrodynamics.* Second Edition. New York, NY, Academic Press, 1967.
5. *Inertial Guidance for Lunar Probes.* ARMA Document DR 220-16, November 1959.
6. Battin, Richard H. *Astronautical Guidance.* New York, NY, McGraw-Hill Book Company, 1964.
7. *The Astronomical Almanac.* http://asa.usno.navy.mil/

Chapter 8

Interplanetary Trajectories

There are seven windows in the head, two nostrils, two eyes, two ears, and a mouth; so in the heavens there are two favorable stars, two unpropitious, two luminaries, and Mercury alone undecided and indifferent. From which and many other similar phenomena of nature, such as the seven metals, etc., which it were tedious to enumerate, we gather that the number of planets is necessarily seven.

-Francesco Sizzi
(Argument against Galileo's
discovery of the satellites of Jupiter)[1]

8.1 Historical Background

The word "planet" means "wanderer." That the naked-eye planets wander among the stars was one of the earliest astronomical observations. At first it was not understood. The ancient Greeks gradually saw its significance; indeed, Aristarchus of Samos realized that the planets must revolve around the Sun as the central body of the solar system. But the tide of opinion ebbed, and an Earth-centered system held the field until Copernicus rediscovered the heliocentric system in the 16th century.

Copernicus, who was born in Polish Prussia in 1473, compiled tables of the planetary motions that remained useful until superseded by the more accurate measurements of Tycho Brahe. By 1507 Copernicus was convinced that the planets revolved around the Sun and in 1530 he wrote a treatise setting forth his revolutionary concept. It is not well known that this work was dedicated to Pope Paul III and that a cardinal paid for the printing; indeed, during the life time of Copernicus his work received the approval of the Church. Not until 1616 was it declared "false, and altogether opposed to Holy Scripture," in spite of the fact that Kepler published his first two laws of planetary motion in 1609.

Such was the atmosphere of the Dark Ages that even the telescopic observations in Galileo of 1610 failed to change the Church's position.

Galileo's data seemed to point decisively to the heliocentric hypothesis: the moons of Jupiter were a solar system in miniature. Galileo's books, which set forth cogent and unanswerable astronomical arguments in favor of the Copernican theory, were suppressed and Galileo himself at the age of 70, was forced by the Inquisition to renounce what he knew to be true. After swearing that Earth was "fixed" at the center of the solar system he is said to have murmured under his breath "it does move, nevertheless."

The publication of Newton's *Principia* in 1687 laid to rest forever the Earth-centered concept of the solar system. His formulas greatly improved understanding of the motions within the solar system. Over the ensuing centuries, however, as those planetary orbits were measured and calculated with ever greater accuracy, there remained a tiny deviation of about 43 arcsec per century between the measured precession of perihelion of Mercury's orbit and the value as predicted by Newtonian mechanics. Much effort went into looking for a hypothetical planet Vulcan, thought to be a leading cause for this unexplained perturbation. It took until the early 20th century and Einstein's insight into relativity to remove this discrepancy. The calculated relativistic effect of the Sun's mass in close proximity to Mercury exactly matches the additional perturbation.

All the planets are affected by this relativistic effect, but for Earth it is only about four arcsec per century, and it is even less for the outer planets. These effects are small enough to safely ignore for the calculation of trajectories in this book.

8.2 THE SOLAR SYSTEM

The Sun is attended by an enormous number of lesser bodies, the members of the solar system. Most conspicuous are the eight planets—Mercury, Venus, Earth, Mars, Jupiter, Saturn, Uranus, and Neptune. When the first edition of Fundamentals of Astrodynamics was written, Pluto was considered a planet since its discovery on February 18, 1930, by Astronomer Clyde W. Tombaugh at the Lowell Observatory in Arizona. On August 24, 2006, the International Astronomical Union reclassified Pluto as a dwarf planet shortly before the New Horizons mission was preparing to blast off in January 2007 for a flyby of Pluto and its several moons on July 14, 2015—fifty years to the day after humans first explored Mars with NASA's Mariner 4 on July 14, 1965.

Between Mars and Jupiter circulate many minor planets, or asteroids, that vary in size from a few hundred miles to a few feet in diameter. In addition, comets, some of which pass near the Sun, are spread much more widely throughout the system. There is now a new category, dwarf planets, that are similar to regular planets. They have enough mass and gravity to be nearly round—unlike odd-shaped asteroids, and they travel through space in a path around the Sun. However, unlike planets, a dwarf planet's path around the Sun is full of other objects such as asteroids and comets while a regular planet has a clear path around the Sun. Most of the major impacts with other objects in its orbit happened billions of years ago. There may be dozens of dwarf planets in our solar system and only a few have been classified.

Closest to home, Ceres is the largest resident of the main asteroid belt between Jupiter and Mars. It is the first dwarf planet to be visited by a spacecraft. NASA's Dawn mission entered orbit around Ceres in 2015 after mapping the giant asteroid Vesta. Astronomers discovered Eris, a Pluto-sized world, in 2003. It takes icy Eris 557 Earth years to complete a single orbit around our Sun. Two more dwarf planets have confirmed orbits around the Sun in the icy zone beyond Neptune—Haumea and Makemake.

8.2.1 Bode's Law

The mean distances of the principal planets from the Sun show a simple relationship, which is known as Bode's law after the man who formulated it in 1772. If we write down the series 0, 3, 6, 12..., add 4 to each number and divide by 10, the numbers thus obtained represent the mean distances of the planets from the Sun in Astronomical Units (AUs). An AU is the mean distance from the Sun to Earth. The "law," as may be seen in Table 8-1, predicts fairly well the distances of all the planets except Neptune.

Table 8-1: Bode's Law

Planet	Bode's Law Distance	Actual Distance
Mercury	0.4	0.39
Venus	0.7	0.72
Earth	1.0	1.0
Mars	1.6	1.52
Asteroids (ave)	2.8	2.65
Jupiter	5.2	5.20
Saturn	10.0	9.52
Uranus	19.6	19.28
Neptune	38.8	30.17

Whether Bode's law is an empirical accident or is somehow related to the origin and evolution of the solar system by physical laws is a question that remains unanswered. However, new discoveries of the evolution of the solar system indicate that the orbits of the planets have changed dramatically over billions of years and Bode's law is probably just a numerical curiosity.

8.2.2 Orbital elements and Physical Constants

Except for Mercury, the orbits of the planets are nearly circular and lie nearly in the plane of the ecliptic. Dwarf planet Pluto's orbit is so eccentric that the perihelion point lies inside the orbit of Neptune; this suggests that Pluto may be an escaped satellite of Neptune. In 2013 the International Astronomical Union (IAU) reclassified Pluto as one of perhaps hundreds to thousands of objects now categorized as dwarf planets. The Kuiper Belt includes a region of heliocentric space past Neptune's orbit and has at least 100 objects approximately the size of Pluto. These objects are also referred to as Kuiper Belt Objects (KBOs).

The size, shape and orientation of the planetary orbits is described by five classical orbital elements that remain relatively fixed except for slight perturbations caused by the mutual attraction of the planets. The sixth orbital element, which defines the position of the planet in its orbit, changes with time and may be obtained for any date from *The Astronomical Almanac.*[4] A complete set of orbital elements for the epoch J2000 is presented in Table 8-2.

Table 8-3 summarizes some of the important physical characteristics of the planets.

8.3 THE PATCHED-CONIC APPROXIMATION

An interplanetary spacecraft spends most of its flight time moving under the gravitational influence of a single body—the Sun. Only for brief periods, compared with the total mission duration, is its path shaped by the gravitational field of the departure or arrival planet. The perturbations caused by the other planets while the spacecraft is pursuing its heliocentric course are negligible.

Just as in lunar trajectories, the computation of a precision orbit is a trial-and-error procedure involving numerical integration of the complete equations of motion where all perturbation effects are considered. For preliminary mission analysis and feasibility studies it is sufficient to have an approximate analytical method for determining the total Δv required to accomplish an interplanetary mission. The best method available for such an analysis is called the patched-conic approximation and was introduced in Chapter 7.

The patched-conic method enables us to ignore the gravitational influence of the Sun until the spacecraft is a great distance from Earth (perhaps a million kilometers). At this point its speed relative to Earth is very nearly the "hyperbolic excess speed" referred to in Chapter 1. If we now switch to a heliocentric frame of reference, we can determine both the velocity of the spacecraft relative to the Sun and the subsequent heliocentric orbit. The same procedure is followed in reverse upon arrival at the target planet's sphere of influence.

The first step in designing a successful interplanetary trajectory is to select the heliocentric transfer orbit that takes the spacecraft from the sphere of influence of the departure planet to the sphere of influence of the arrival planet.

Table 8-2: Orbital Elements of the Planets for the Mean Ecliptic and Equinox of J2000.[10]

Planet	Semi-Major Axis a [AU]	Orbital Eccentricity e	Inclination to Ecliptic i	Longitude of Ascending Node Ω	Longitude of Perihelion Π	True Longitude at Epoch ℓ_o
Mercury	0.3871	0.2056	7.005°	48.332°	77.450°	252.251°
Venus	0.7233	0.0068	3.394°	76.681°	131.533°	181.980°
Earth	1.0000	0.0167	0.000°	−11.261°	102.948°	100.464°
Mars	1.524	0.0934	1.851°	49.579°	335.041°	355.453°
Jupiter	5.203	0.0484	1.305°	100.556°	14.754°	34.404°
Saturn	9.537	0.0542	2.484°	113.715°	92.432°	49.944°
Uranus	19.191	0.0472	0.770°	74.230°	170.964°	313.232°
Neptune	30.069	0.00859	1.769°	131.722°	44.971°	304.880°

Table 8-3: Physical Characteristics of the Sun and Planets.[6]

Planet	Orbital Period (Years)	Semi-Major Axis (10^6 km)	Orbital Speed (km/s)	Relative Mass (Earth = 1)	μ (km^3/s^2)	Equatorial Radius (km)	Inclination of Equator to Orbit
Sun	--	--	--	333,432	1.327×10^{11}	696,000	7.5°
Mercury	0.241	57.91	47.87	0.055	2.203×10^4	2439	7.005°
Venus	0.615	108.2	35.02	0.815	3.2486×10^5	6052	177.36°
Earth	1.000	149.6	29.79	1	3.986×10^5	6378	23.45°
Mars	1.881	227.9	24.13	0.107	4.305×10^4	3397	25.19°
Jupiter	11.86	778.3	13.07	318.0	1.268×10^8	71,492	3.12°
Saturn	29.42	1429	9.67	95.16	3.794×10^7	60,268	26.73°
Uranus	83.75	2875	6.83	14.50	5.794×10^6	25,559	97.86°
Neptune	163.7	4504	5.48	17.2	6.809×10^6	24,764	29.56°

8.3.1 The Heliocentric Transfer Orbit

For transfers to most of the planets, we may consider that the planetary orbits are both circular and coplanar. In Chapter 3 we discussed the problem of transferring between coplanar circular orbits and found that the most economical method, from the standpoint of Δv required, was the Hohmann transfer. A Hohmann transfer between Earth and Mars is pictured in Figure 8-1.

While it is always desirable that the transfer orbit be tangential to Earth's orbit at departure, it may be preferable to intercept Mars' orbit prior to apogee, especially if the spacecraft is to return to Earth. The Hohmann transfer, if continued past the destination planet, would not provide a suitable return trajectory. For a one-way trip this is irrelevant; however, for a probe that is to be recovered or for a manned mission this consideration is important. The outbound trip to Mars on the Hohmann trajectory takes between 8 and 9 months. If the spacecraft continued its flight it would return to the original point of departure only to find Earth nearly on the opposite side of its orbit. Therefore, either the spacecraft must loiter in the vicinity of Mars for nearly 6 months or the original trajectory must be modified so that the spacecraft will encounter Earth at the point where it recrosses Earth's orbit.

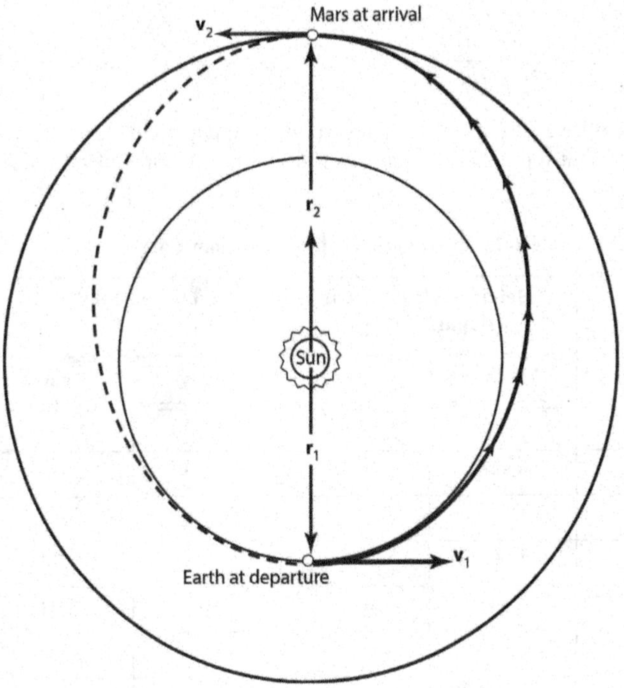

Fig. 8-1 Hohmann Transfer to Mars.

Nevertheless, the Hohmann transfer provides us with a convenient yardstick for determining the minimum Δv required for an interplanetary mission. The energy of the Hohmann transfer orbit is given by

$$E_t = -\mu_\odot / (r_1 + r_2) \tag{8-1}$$

where $\mu_\odot$ is the gravitational parameter of the Sun, r_1 is the radius of the departure planet's orbit, and r_2 is the radius of the arrival planet's orbit.

The heliocentric speed, v_1, required at the departure point is obtained from

$$v_1 = \sqrt{2\left(\frac{\mu_\odot}{r_1} + E_t\right)} \tag{8-2}$$

The time of flight for the Hohmann transfer is just half the period of the transfer orbit since departure occurs at perihelion and arrival occurs at aphelion. If t_1 is the time when the spacecraft departs and t_2 is the arrival time, then, for the Hohmann transfer,

$$t_2 - t_1 = \pi \sqrt{\frac{a_t^3}{\mu_\odot}}$$

$$= \pi \sqrt{\frac{(r_1 + r_2)^3}{8\mu_\odot}} \tag{8-3}$$

In Table 8-4 we have listed the heliocentric speed required at departure and the time of flight for Hohmann transfers to all of the principal planets of the solar system.

Table 8-4: Hohmann Trajectories from Earth.

	Heliocentric Speed at Departure, v_1		Time of Flight $t_2 - t_1$	
Planet	**AU/TU$_\odot$**	**km/s**	**Days**	**Years**
Mercury	0.748	22.28	105.5	--
Venus	0.916	27.28	146.1	--
Mars	1.099	32.73	258.9	--
Jupiter	1.295	38.57	--	2.74
Saturn	1.345	40.05	--	6.04
Uranus	1.379	41.07	--	16.16
Neptune	1.391	41.42	--	30.78

The orbital speed of Earth is 1 AU/TU$_\odot$ or 29.78 km/s. From Table 8-4 it is obvious that transfers to the inner planets require that the spacecraft be launched in the direction opposite Earth's orbital motion so as to cancel some of Earth's orbital velocity; transfers to the outer planets require that the spacecraft depart Earth in the same direction as Earth's orbital velocity.

In any case, the difference between the heliocentric speed, v_1, and Earth's orbital speed represents the speed of the spacecraft *relative to Earth* at departure, which we will call Δv_1.

The method for computing Δv_1 for other than Hohmann transfers is presented in Section 3.4.3 and should be reviewed.

The important thing to remember is that Δv_1 represents the speed of the spacecraft relative to Earth *upon exit from Earth's sphere of influence.* In other words, it is the hyperbolic excess speed: the speed left over after the spacecraft has escaped Earth.

Example Problem. Calculate the heliocentric departure speed and time of flight for a Hohmann transfer from Earth to Mars. Assume both planets are in circular coplanar orbits. Neglect the phasing requirements resulting from their relative positions at the time of transfer.

From Table 8-2 the radius to Mars from the Sun, $r_♂$, and the radius to Earth from the Sun, $r_\oplus$, are, respectively,

$$r_♂ = 1.524 \text{ AU}$$

$$r_\oplus = 1.0 \text{ AU}$$

From Equation (8-1) the energy of the Hohmann transfer trajectory is

$$E_t = -\mu_\odot/(r_1 + r_2) = -\mu_\odot/(r_\oplus + r_♂) = -0.396 \text{ AU}^2/\text{TU}_\odot^2$$

From Equation (8-2) the heliocentric speed, v_1, required at the departure point is:

$$v_1 = \sqrt{2\left(\frac{\mu_\odot}{r_\oplus} + E_t\right)} = 1.099 \text{AU/TU}_\odot$$

$$= \underline{\underline{32.73 \text{ km/s}}}$$

From Equation (8-3) the time of flight from Earth to Mars for the Hohmann transfer is

$$t_2 - t_1 = \pi\sqrt{\frac{a_t^3}{\mu_\odot}} = \pi\sqrt{\frac{(r_\oplus + r_♂)^3}{8\mu_\odot}}$$

$$= \underline{\underline{4.4538 \text{ TU}_\odot = 0.7088 \text{ years} = 258.9 \text{ days}}}$$

8.3.2 Phase Angle at Departure

If the spacecraft is to encounter the target planet at the time it crosses the planet's orbit then obviously Earth and the target planet must have the correct angular relationship at departure. The angle between the radius vectors to the departure and arrival planets is called γ_1, the phase angle at departure, and is illustrated in Figure 8-2 for a Mars trajectory.

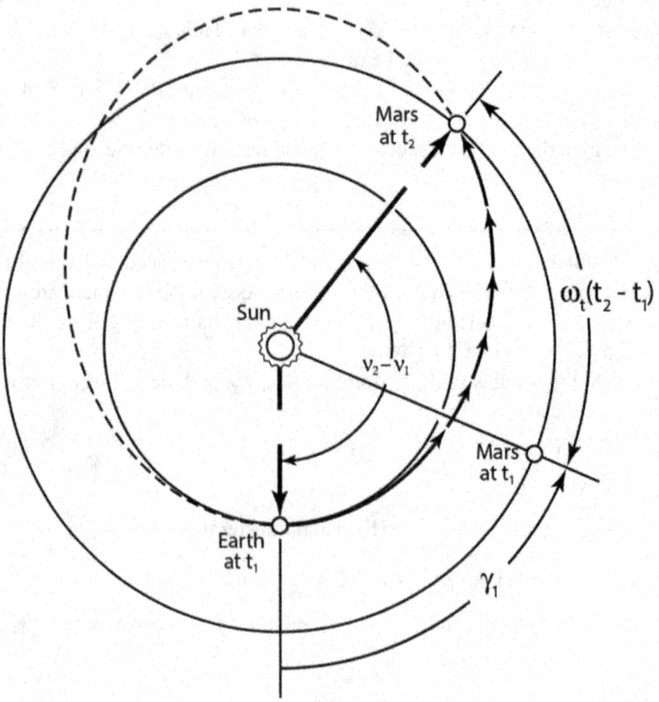

Fig. 8-2 Phase Angle at Departure, γ_1.

The total sweep angle from departure to arrival is just the difference in the true anomaly at the two points, $\nu_2 - \nu_1$, which may be determined from the polar equation of a conic once p and e of the heliocentric transfer orbit have been selected:

$$\cos \nu_2 = \frac{p - r_2}{e r_2} \tag{8-4}$$

$$\cos \nu_1 = \frac{p - r_1}{e r_1} \tag{8-5}$$

The time of flight may be determined from the Kepler equation, which was derived in Chapter 4. The target planet will move through an angle $\omega_t\,(t_2 - t_1)$ while the spacecraft is in flight, where ω_t is the angular velocity of the target planet. Thus, the correct phase angle at departure is

$$\gamma_1 = (\nu_2 - \nu_1) - \omega_t\,(t_2 - t_1) \tag{8–6}$$

The requirement that the phase angle at departure be correct severely limits the times when a launch may take place. The heliocentric longitudes of the planets are tabulated in the *The Astronomical Almanac* and these may be used to determine when the phase angle will be correct. The heliocentric longitude of Earth may be obtained by adding to the value for the geocentric longitude of the Sun from the *Astronomical Almanac*. This publication is now available online at www.asa.usno.navy.mil and is the preferred reference for the most up-to-date information.

If we miss a particular launch opportunity, how long will we have to wait until the correct phase angle repeats itself? The answer to this question depends on the "synodic period" between Earth and the particular target planet. The synodic period is defined as the time required for any phase angle to repeat itself.

In a time τ_S, Earth moves through an angle $\omega_\oplus \tau_S$ and the target planet advances by $\omega_t \tau_S$. If τ_S is the synodic period, then the angular advance of one will exceed that of the other by 2π radians and the original phase angle will be repeated, so

$$\omega_\oplus \tau_S - \omega_t \tau_S = \pm 2\pi$$

$$\tau_S = \frac{2\pi}{|\omega_\oplus - \omega_t|} \tag{8-7}$$

The synodic periods of all the planets relative to Earth are given in Table 8-5.

Table 8-5: Synodic Periods of the Planets with Respect to Earth.

Planet	ω_t (radians/year)	Synodic Period (years)
Mercury	26.071	0.32
Venus	10.217	1.60
Mars	3.340	2.13
Jupiter	0.530	1.09
Saturn	0.213	1.04
Uranus	0.075	1.01
Neptune	0.038	1.01

It is clear that for the two planets nearest Earth, Mars and Venus, the time between the reoccurrence of a particular phase angle is quite long. Thus, if a Mars or Venus launch is postponed, we must either compute a new trajectory or wait a long time for the same launch conditions to occur again.

8.3.3 Escape from Earth's Sphere of Influence

Once the heliocentric transfer orbit has been selected and Δv_1 determined, we can proceed to establish the injection or launch conditions near the surface of Earth that will result in the required hyperbolic excess speed. Since Earth's sphere of influence has a radius of about 10^6 km, we assume that

$$v_\infty \approx \Delta v_1 \tag{8-8}$$

Since energy is constant along the geocentric escape hyperbola, we may equate E at injection and E at the edge of the sphere of influence where $r = r_\infty$:

$$E = \frac{v_0^2}{2} - \frac{\mu}{r_0} = \frac{v_\infty^2}{2} - \frac{\mu}{r_\infty} \tag{8-9}$$

Solving for v_0, we get

$$v_0 = \sqrt{v_\infty^2 + 2\mu/r_0} \tag{8-10}$$

The hyperbolic excess speed is extremely sensitive to small errors in the injection speed. This may be seen by solving Equation (8-9) for v_∞^2 and taking the differential of both sides of the resulting equation assuming that r_0 is constant:

$$v_\infty^2 = v_0^2 - \frac{2\mu}{r_0}$$

$$2v_\infty \, dv_\infty = 2\, v_0 \, dv_0$$

The relative error in v_∞ may be expressed as

$$\frac{dv_\infty}{v_\infty} = \left(\frac{v_0}{v_\infty}\right)^2 \frac{dv_0}{v_0} \tag{8-11}$$

For a Hohmann transfer to Mars, $v_\infty = 2.98$ km/s and $v_0 = 11.6$ km/s, so Equation (8-11) becomes

$$\frac{dv_\infty}{v_\infty} = 15.2 \frac{dv_0}{v_0}$$

which says that a 1% error in injection speed results in a 15.2% error in hyperbolic excess speed.

For transfer to one of the outer planets, the hyperbolic excess velocity should be parallel and in the same direction as Earth's orbital velocity, as shown in Figure 8-3. If we assume that injection occurs at perigee of the departure hyperbola, the angle η, between Earth's orbital velocity vector and the injection radius vector may be determined from the geometry of the hyperbola.

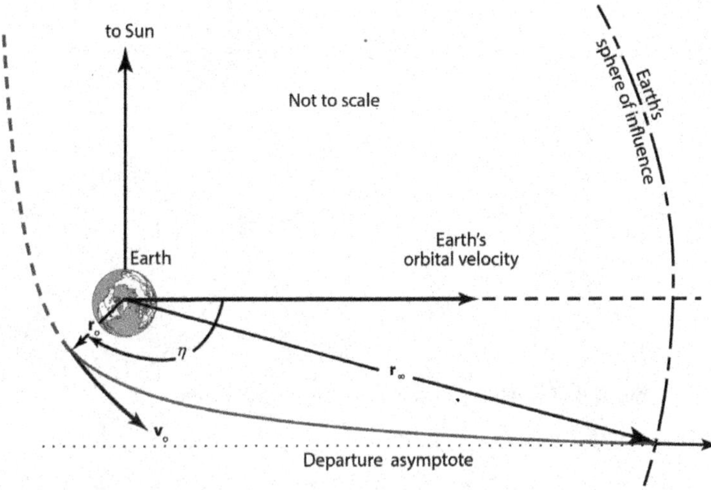

Fig. 8-3 Escape Hyperbola.

From Figure 8-4 we see that the angle, η, is given by

$$\cos \eta = -\frac{a}{c} \tag{8-12}$$

but since e = c/a for any conic, we may write

$$\cos \eta = -\frac{1}{e} \tag{8-13}$$

where the eccentricity, e, is obtained directly from the injection conditions:

$$E = \frac{v_0^2}{2} - \frac{\mu}{r_0} \tag{8-14}$$

$$h = r_0 v_0 \quad \text{(for injection at perigee)} \tag{8-15}$$

$$e = \sqrt{1 + 2Eh^2/\mu^2} \tag{8-16}$$

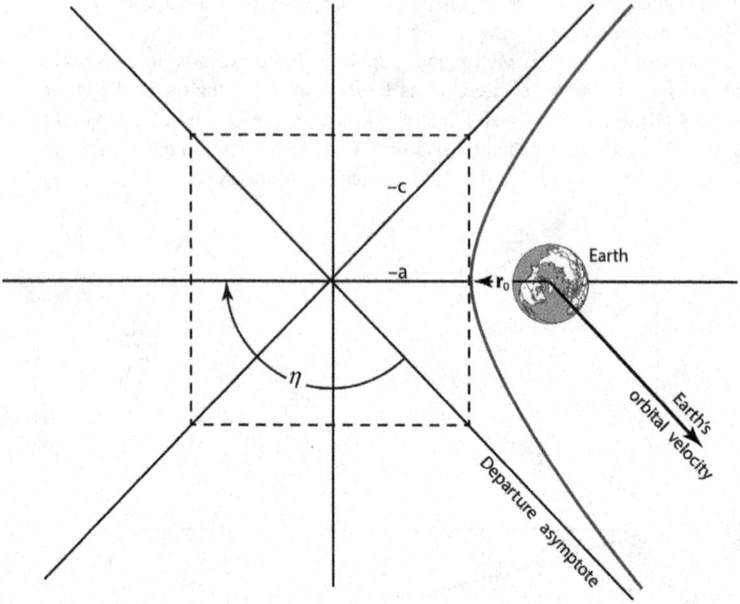

Fig. 8-4 Geometry of the Departure Hyperbola.

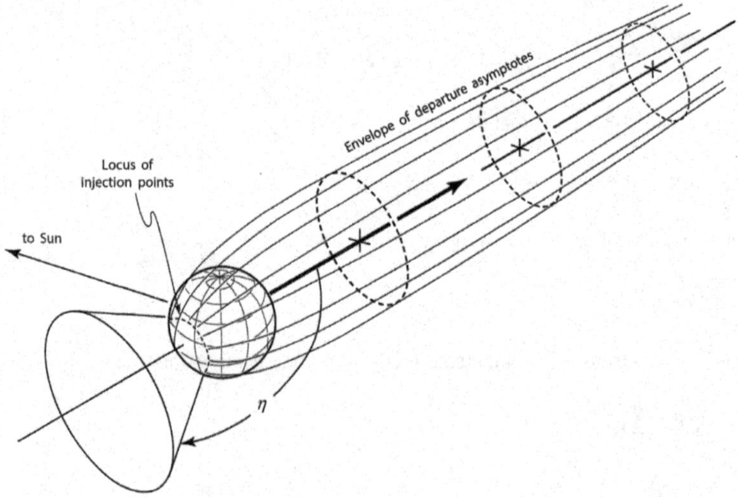

Fig. 8-5 Interplanetary Launch.

It should be noted at this point that it is not necessary that the geocentric departure orbit lie in the plane of the ecliptic but only that the hyperbolic departure asymptote be parallel to Earth's orbital velocity vector.

The locus of possible injection points forms a circle on the surface of Earth. Probes are typically launched into a parking orbit, and when its path crosses the locus, the injection firing can occur.

8.3.4 Arrival at the Target Planet

Generally, the heliocentric transfer orbit will be tangent to Earth's orbit at departure in order to take full advantage of Earth's orbital velocity. At arrival, however, the heliocentric transfer orbit usually crosses the target planet's orbit at some angle, ϕ_2, as shown in Figure 8-6.

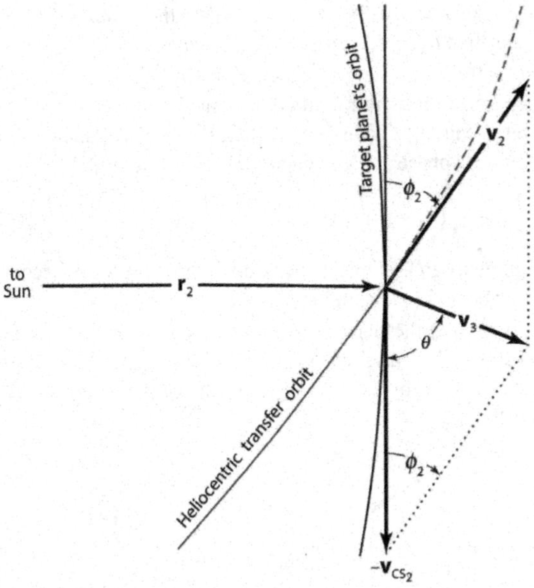

Fig. 8-6 Relative Velocity at Arrival.

If E_t and h_t are the energy and angular momentum of the heliocentric transfer orbit, respectively, then v_2 and ϕ_2 may be determined from

$$v_2 = \sqrt{2\left(\mu_\odot / r_2 + E_t\right)} \tag{8-17}$$

$$\cos \phi_2 = h_t / r_2 v_2 \tag{8-18}$$

If we call the velocity of the spacecraft *relative to the target planet* $\mathbf{v_3}$, then, from the law of cosines,

$$v_3^2 = v_2^2 + v_{cs_2}^2 - 2v_2 v_{cs_2} \cos \phi_2 \tag{8-19}$$

where v_{cs_2} is the orbital speed of the target planet.

The angle, θ, in Figure 8-6 may be determined from the law of sines as

$$\sin \theta = \frac{v_2}{v_3} \sin \phi_2 \tag{8-20}$$

If a dead-center hit on the target planet is planned then the phase angle at departure, γ_1, should be selected from Equation (8-6). This ensures that the target planet will be at the intercept point at the same time the spacecraft is there. It also means that the relative velocity vector, $\mathbf{v_3}$, upon arrival at the target planet's sphere of influence will be directed toward the center of the planet, resulting in a straight-line hyperbolic approach trajectory.

If it is desired to fly by the target planet instead of impacting it, then the phase angle at departure must be modified so that the spacecraft crosses the target planet's orbit ahead of or behind the planet. If the miss distance along the orbit is called x, then the phase angle at departure should be

$$\gamma_1 = \nu_2 - \nu_1 - \omega_t (t_2 - t_1) \pm x/r_2 \tag{8-21}$$

where the plus or minus sign is shown depending on whether the spacecraft is to cross ahead of (–) or behind (+) the target planet.

If the spacecraft crosses the planet's orbit a distance x ahead of the planet, then the vector $\mathbf{v_3}$, which represents the hyperbolic excess velocity on the approach hyperbola, is offset by a distance y from the center of the target planet as shown in Figure 8-7.

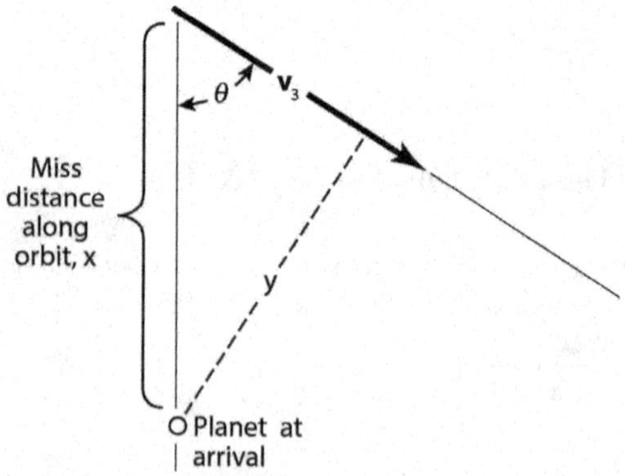

Fig. 8-7 Offset Miss Distance at Arrival.

From Figure 8-7 we see that

$$y = x \sin\theta \tag{8-22}$$

Once the offset distance is known, the distance of closest approach or periapsis radius may be computed.

8.3.5 Effective Collision Cross Section

Figure 8-8 shows the hyperbolic approach trajectory, where $\mathbf{v_3}$ is the hyperbolic excess velocity upon entrance to the target planet's sphere of influence and **y** is the offset distance.

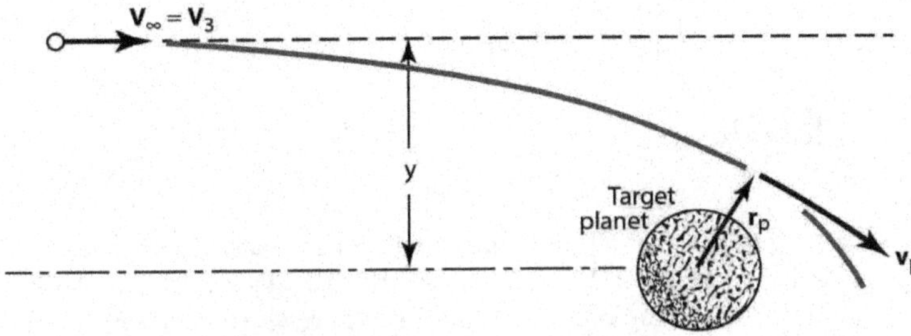

Fig. 8-8 Hyperbolic Approach Orbit.

From the energy equation we know that the energy in the hyperbolic approach trajectory is

$$E = \frac{v_3^2}{2} - \cancelto{0}{\frac{\mu_t}{r_\infty}} \tag{8-23}$$

The angular momentum is obtained from

$$h = y v_3 \tag{8-24}$$

The parameter and eccentricity of the approach trajectory follow from

$$p = h^2 / \mu_t \tag{8-25}$$

$$e = \sqrt{1 + 2Eh^2 / \mu_t^2} \tag{8-26}$$

where μ_t is the gravitational parameter of the target planet. We may now compute the periapsis radius from

$$r_p = \frac{p}{1 + e} \tag{8-27}$$

Because angular momentum is conserved, the speed at periapsis is simply

$$v_p = \frac{y v_3}{r_p} \tag{8-28}$$

The usual procedure is to specify the desired periapsis radius and then compute the required offset distance, y. Solving Equation (8-28) for y yields

$$y = \frac{r_p v_p}{v_3} \tag{8-29}$$

Equating the energy at periapsis with the energy upon entrance to the sphere of influence, we obtain

$$E = \frac{v_3^2}{2} = \frac{v_p^2}{2} - \frac{\mu_t}{r_p}$$

and solving for v_p and substituting it into Equation (8-29) we get

$$y = \frac{r_p}{v_3}\sqrt{v_3^2 + \frac{2\mu_t}{r_p}} \tag{8-30}$$

It is interesting to see what offset distance results in a periapsis radius equal to the radius of the target planet, r_t. This particular offset distance is called b, the "impact parameter," since any offset less than this will result in a collision. In this connection it is convenient to assign to the target planet an impact size that is larger than its physical size. The concept is also employed by nuclear physicists and is called "effective collision cross section." The radius of the effective collision cross section is just the impact parameter, b.

We can determine the impact parameter by setting $r_p = r_t$ in Equation (8-30):

$$b = \frac{r_t}{v_3}\sqrt{v_3^2 + \frac{2\mu_t}{r_t}} \tag{8-31}$$

The effective cross section of the planet represents a rather large target, as shown in Figure 8-9. If we wish to take advantage of atmospheric braking, then a much smaller target must be considered; only a thin annulus of radius b and width db. The cross section for hitting the atmosphere of the target planet is called the "reentry corridor" and may be very small indeed.

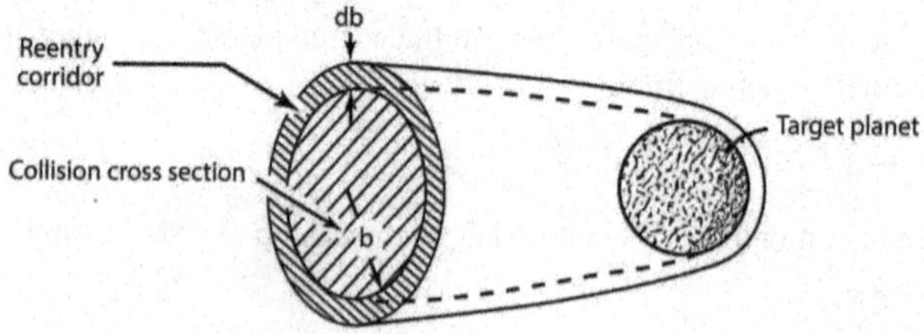

Fig. 8-9 Effective Cross Section.

Example Problem. It is desired to send an interplanetary probe to Mars on a Hohmann ellipse around the Sun. The launch vehicle burns out at an altitude of 0.05 DU. Determine the burnout velocity required to accomplish this mission.

The given information is

$$h_{bo} = 0.05\ DU_{\oplus}$$

Therefore, $r_{bo} = r_{\oplus} + h_{bo} = 1.05$ DU.

From Table 8-4, we have

$v_1 = 1.099$ AU/TU on the Hohmann ellipse at the region of Earth

$$\text{Therefore, } \Delta v_1 = 1.099 - v_{\oplus} = 1.099 - 1 = 0.099 \text{ AU/TU}$$

$$= 0.373 \text{ DU/TU}$$

At Earth's sphere of influence, $v_{\infty} = \Delta v_1 = 0.373$ DU/TU and we can write the energy equation as

$$\frac{v_{\infty}^2}{2} - \cancelto{0}{\frac{\mu}{r_{\infty}}} = \frac{v_{bo}^2}{2} - \frac{\mu}{r_{bo}}$$

Hence

$$v_{bo}^2 = v_{\infty}^2 + \frac{2\mu}{r_{bo}} = (0.373)^2 + \frac{2(1)}{1.05}$$

$$= 0.139 + 1.905 = 2.044$$

$$\text{so } v_{bo} = 1.43 \text{ DU/TU} = \underline{\underline{11.307 \text{ km/s}}}$$

8.4 NONCOPLANAR INTERPLANETARY TRAJECTORIES

Up to now we have assumed that the planetary orbits all lie in the plane of the ecliptic. From Table 8-2 it may be seen that some of the planetary orbits are inclined several degrees to the ecliptic. Fimple[4] has shown that a good procedure to use when the target planet lies above or below the ecliptic plane at intercept is to launch the spacecraft into a transfer orbit that lies in the ecliptic plane and then make a simple plane change during mid course when the true anomaly change remaining to intercept is 90°. This minimizes the magnitude of the plane change required and is illustrated in Figure 8-10.

Since the plane change is made 90° short of intercept, the required inclination change is just equal to the ecliptic latitude, β_2, of the target planet at the time of intercept, t_2.

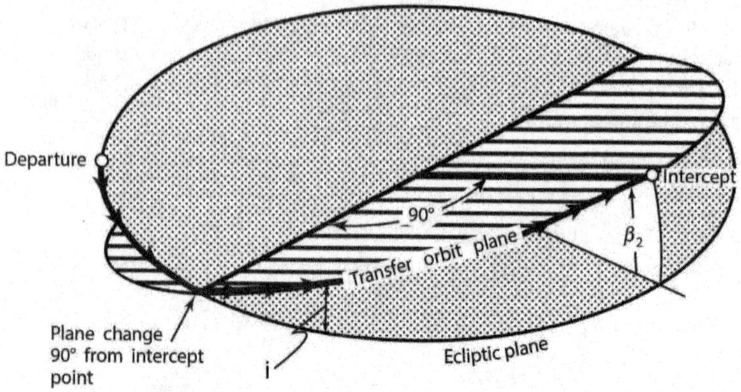

Fig. 8-10 Optimum Plane Change Point.

The Δ**v** required to produce this midcourse plane change depends on the speed of the spacecraft at the time of the plane change and is

$$\Delta v = 2v \sin \frac{i}{2} \tag{8-32}$$

as derived in Equation (3-22).

8.5 PLANETARY FLYBYS

The previous sections have all ignored the presence of other planets in determining interplanetary trajectories and mission design. However, it has been proven to be feasible and very beneficial to use other planets to gravity-assist spacecraft on numerous interplanetary missions. Battin,[7] Schaub,[8] and Prussing[9] have written extensively about the theory and techniques used for "planetary flybys" and the approach is summarized here as an extension of interplanetary trajectories.

Section 8.3.4 derived the geometry for the arrival at a planet when the spacecraft is in an interplanetary trajectory. Figure 8-6 shows the geometry of the arrival in the vicinity of the target planet. In this section we assume that the target planet is to be used for a gravity-assisted maneuver to change the orbit (heliocentric trajectory) of the spacecraft.

Figures 8-6 and 8-8 are used to introduce variable definitions for the planetary flyby example. It is important to remember that during the transfer orbit the spacecraft is modeled using the heliocentric frame of reference. Once the spacecraft is within the planet's sphere of influence, then the velocity of interest is the spacecraft velocity in the planet's frame of reference. That velocity is the incoming hyperbolic excess velocity, $\mathbf{v}_{\infty i}$. Similarly the departing hyperbolic excess velocity is $\mathbf{v}_{\infty o}$. In the absence of external forces such as atmospheric effects or a thruster burn by the spacecraft, the incoming and outgoing hyperbolic velocities have the same magnitude in the planet's frame of reference.

However, we are concerned about the change in the heliocentric velocity during the planetary flyby. The heliocentric velocities of the spacecraft in the heliocentric frame of reference are $\mathbf{v}_{sci}$ and $\mathbf{v}_{sco}$. If the planet's velocity in the heliocentric frame of reference is $\mathbf{v}_p$ then:

$$\mathbf{v}_{\infty i} + \mathbf{v}_p = \mathbf{v}_{sci} \qquad \mathbf{v}_{\infty o} + \mathbf{v}_p = \mathbf{v}_{sco} \tag{8-33}$$

The change in velocity from the planetary flyby, $\Delta\mathbf{v}_{FB}$, is

$$\begin{aligned} \Delta\mathbf{v}_{FB} &= \mathbf{v}_{sco} - \mathbf{v}_{sci} \\ &= \mathbf{v}_{\infty o} - \mathbf{v}_{\infty i} \end{aligned} \tag{8-34}$$

Figure 8-11 shows the incoming spacecraft on the side of the planet opposite the planet velocity vector, so the spacecraft is "behind" the planet. Therefore, in the heliocentric frame of reference, the planetary flyby has changed the spacecraft velocity vector and increased the magnitude of the spacecraft's velocity. The additional energy in the satellite's heliocentric orbit comes from, of course, the orbital energy of the planet. Passing in "front" of the planet would decrease the velocity of the spacecraft in the heliocentric frame of reference.

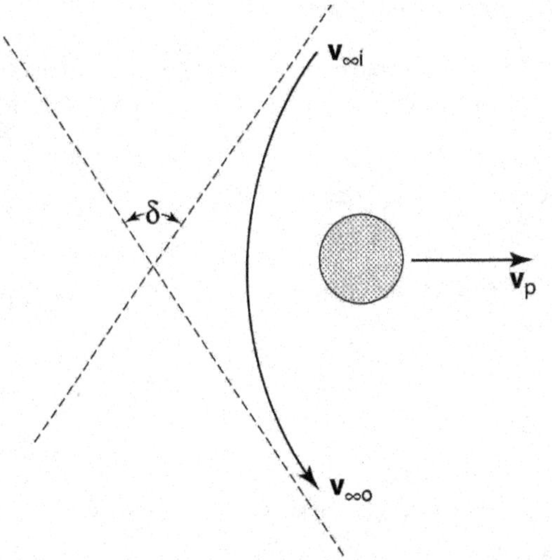

Fig. 8-11 Flyby Trajectory of Interplanetary Spacecraft.

Figure 8-12 shows the vector geometry during the planetary flyby. By using Equations (8-33) and (8-34) and the patched-conic techniques of the previous sections to calculate the hyperbolic excess velocity in the planetary frame, the exit velocity of the spacecraft in the heliocentric frame can be calculated. The magnitude of that change is

$$\Delta v_{FB} = 2v_{\infty} \sin\frac{\delta}{2} \tag{8-35}$$

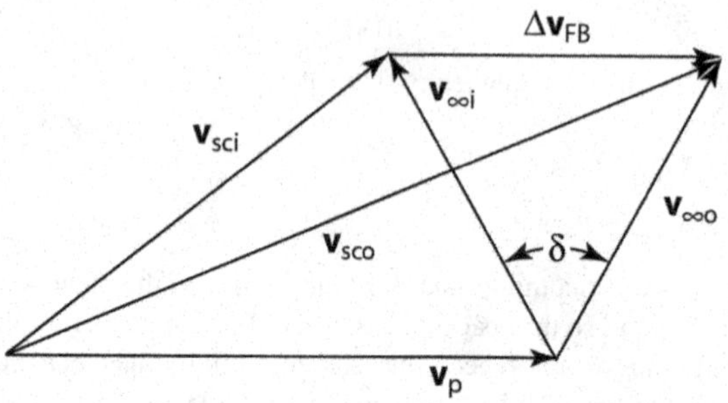

Fig. 8-12 Geometry of Planetary Flyby.

Exercises

8.1 Verify the synodic period of Venus in Table 8-5.

8.2 Calculate the velocity requirement to fly a solar probe into the surface of the Sun. Use a trajectory that causes the probe to fall directly into the Sun (a degenerate ellipse).

8.3 Calculate the radius of Earth's sphere of influence with respect to the Sun (see Equation (7-1)).

8.4 Discuss the advantages and disadvantages of the use of a Hohmann transfer for interplanetary travel.

8.5 Repeat the example problem at the end of Section 8.3.5 for the planet Jupiter.

8.6 Calculate the escape speed from the top of the visible atmosphere of Jupiter in km/s and in Earth canonical units.

(Answer: 7.62 $DU_{\oplus}/TU_{\oplus}$)

8.7 Find the distance from the Sun at which a space station must be placed in order that a particular phase angle between the station and Earth will repeat itself every 4 years.

8.8 The following figure illustrates the general coplanar interplanetary transfer:

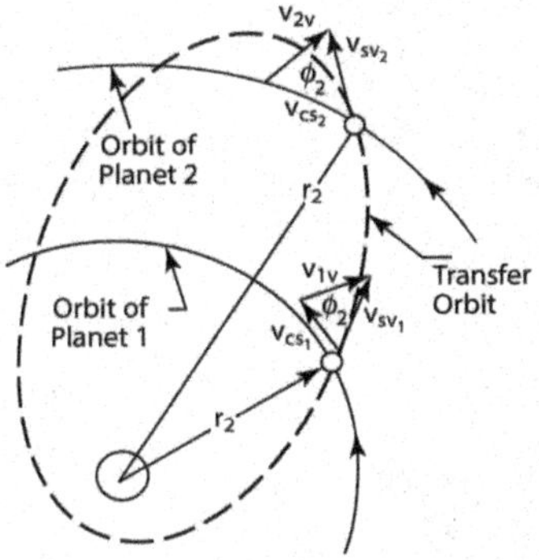

Show that

$$v_{iv} = \left\{ \mu_\odot \left[\frac{3}{r_i} - \frac{(1 - e_\odot^2)}{p_\odot} - 2\sqrt{\frac{p_\odot}{r_i^3}} \right] \right\}^{\frac{1}{2}} \text{ AU/TU}$$

or

$$v_{iv} = \left\{ \mu_\odot \left[\frac{3}{r_i} + \frac{2E_\odot}{\mu_\odot} - \frac{2h_\odot}{\mathbb{P}_i \mu_\odot} \right] \right\}^{\frac{1}{2}} \text{ AU/TU}$$

where v_{iv} is the speed of the space vehicle relative to planet i,

r_i is the heliocentric orbit radius of planet i,

$e_\odot$ is the eccentricity of the heliocentric transfer orbit,

$p_\odot$ is the parameter of the heliocentric transfer orbit,

$E_\odot$ is the specific mechanical energy of the heliocentric transfer orbit,

$h_\odot$ is the specific angular momentum of the heliocentric transfer orbit, and

$\mathbb{P}_i$ is the period of planet i in *sidereal years*.

Note that $v_{iv} = v_\infty$ at planet i. Refer to material in Chapters 1 and 7 for treating the orbit inside the sphere of influence.

8.9 In preliminary planing for any space mission, it is necessary to see whether we have the capability of actually producing the velocity required to accomplish this mission.

a. The space vehicle is in a circular orbit around Earth of 7,020 km radius. What is the speed necessary to place the vehicle on a parabolic escape path? What is the Δv required?

(Answer: $v_{esc} = 10.656$ km/s, $\Delta v = 3.121$ km/s)

b. Actually, we want a hyperbolic excess speed of 1.525 km/s. What must our speed be as we leave the circular orbit? What is the Δv?

(Answer: $v = 10.765$ km/s, $\Delta v = 3.230$ km/s)

c. It can be shown that $\Delta v = c \ln M$ where c is the effective exhaust velocity of the engine and $M = \dfrac{m_o}{m_{bo}}$. If $c = 2.7432$ km/s what is the ratio of the initial mass to burnout mass?

(Answer: $M = 3.246$)

8.10 A Venus probe departs from a $2DU_\oplus$ radius circular parking orbit with a burnout speed of 1.1 $DU_\oplus/TU_\oplus$. Find the hyperbolic excess speed in geocentric and heliocentric speed units. What is v_∞ in km/s?

(Answer: $v_\infty = 0.458\ DU_\oplus/TU_\oplus = 0.122\ AU/TU_\odot = 3.6271$ km/s)

8.11 A space probe is to be sent from Earth to Mars on a heliocentric transfer orbit with $p = 2/3\ DU_\odot$ and $e = 2/3$.

a. What will be the speed of the probe relative to Earth after it has escaped Earth's gravity?

(Answer: 0.731 $DU_\odot/TU_\odot$)

b. What burnout speed is required near the surface of Earth to inject the probe into its heliocentric orbit?

(Answer: 3.09 $DU_\oplus/TU_\oplus$)

8.12 A space probe is in an elliptical orbit with an apogee of 3 $DU_\oplus$ and a perigee of 1 $DU_\oplus$.

a. Determine the energy in the orbit before and after an impulsive Δv of 0.1 $DU_\oplus/TU_\oplus$ is applied at apogee. What is the ΔE?

(Answer: $\Delta E = 0.046\ DU_\oplus^2/TU_\oplus^2$)

b. Find the ΔE that results if the same Δv is applied at perigee.

(Answer $\Delta E = 0.128\ DU_\oplus^2/TU_\oplus^2$)

c. What is the Δv required to achieve escape speed from the original elliptical orbit at apogee?

d. What is the Δv required to escape from the above orbit at perigee?

e. Is it more efficient to apply a Δv at apogee or perigee?

8.13 (This problem is fairly long, so work carefully and follow any suggestions.) We wish to travel from Earth to Mars. The mission will begin in a 1.5 $DU_\oplus$ circular orbit. The burnout speed after thrust application in the circular orbit is to be 1.5 $DU_\oplus/TU_\oplus$. The thrust is applied at the perigee of the escape orbit. The transfer orbit has an energy (with respect to the Sun) of –0.28 $AU^2/TU_\odot^2$.

a. Find the hyperbolic excess speed, v_∞.

(Answer: 0.956 $DU_\oplus/TU_\oplus$)

b. Find the hyperbolic excess speed in heliocentric units.

(Answer: 0.254 $AU/TU_\odot$)

c. Find the velocity of the satellite with respect to the Sun at its departure from Earth.

(Answer: 1.2 $AU/TU_\odot$)

d. Find v_{sv_2} at arrival at the Mars orbit.

(Answer: 0.867 $AU/TU_\odot$)

e. Find the hyperbolic excess speed upon arrival at Mars. (Hint: Find ϕ_1 from the law of cosines, then find h, ϕ_2 and v_{mv} in that order.)

(Answer: v_{mv} = 0.373 $AU/TU_\odot$)

f. What will be the reentry speed at the surface of Mars?

(Answer: 3.39 $DU_♂/TU_♂$)

8.14

a. What will be the time of flight to Mars on the heliocentric orbit of Problem 8.11 if the radius of Mars' orbit is 1.5 AU?

(Answer: 0.87 $TU_\odot$)

b. Compute the phase angle at departure, γ_1, for the above transfer.

(Answer: 0°)

8.15 To accomplish certain measurements of phenomena associated with sunspot activity, it is necessary to establish a heliocentric orbit with a perihelion of 0.5 AU. The departure from Earth's orbit will be at aphelion. What must the burnout velocity (in $DU_\oplus/TU_\oplus$ and km/s) be at an altitude of 0.2 $DU_\oplus$ to accomplish this mission?

(Answer 1.464 $DU_\oplus/TU_\oplus$ = 11.598 km/s)

8.16 A space probe can alter its flight path without adding propulsive energy by passing by a planet en route to its destination. Upon passing the planet it will be deflected some amount, called the turning angle δ. For a given planet P, show that

$$\sin\delta = \frac{1}{1 + \dfrac{Dv^2}{\mu_p}}$$

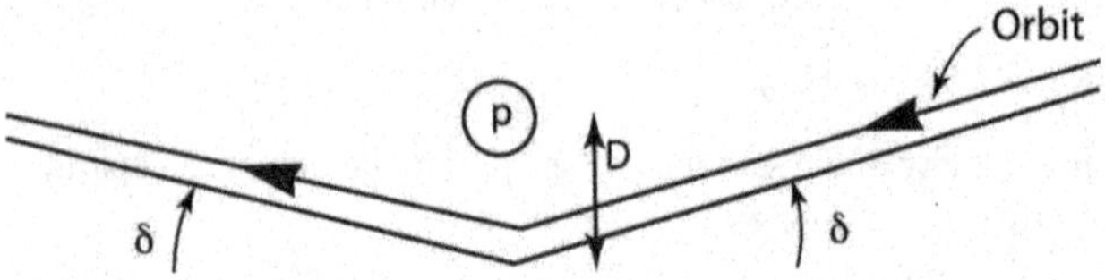

where D and δ (half of the turning angle) are as defined in the drawing and v is the velocity relative to the planet and μ_p is the gravitational parameter of the planet.

8.17 * With the use of Hohmann transfer analysis calculate an estimate of the total Δv required to depart from Earth and soft land a craft on Mars. What would be an estimate of the return Δv? Give the answer in km/s.

8.18 * It is desired to return to an inferior (inner) planet at the earliest possible time from a superior (outer) planet. The return is to be made on an ellipse with the same value of p and e as was used on the outbound journey. Derive an expression for the loiter time at the superior planet in terms of the phase angle, transfer time and the angular rates of the planets. Assume circular coplanar planet orbits.

List of References

1. Sizzi, Francesco. Dianoia astronomica, optica, physica, qua Syderei Nuncij rumor de quatuor planetis à Galilaeo Galilaeo mathematico celeberrimo recens perspicillì cuiusdam ope conspectis, vanus redditur (Understanding of astronomy, optics, and physics, about a rumor in Sidereus Nuncius about the four planets seen by the very celebrated mathematician Galileo Galilei with his telescope, found to be in error), published 1611.

2. *American Ephemeris and Nautical Almanac*, 1969. Washington, DC, U.S. Government Printing Office, 1967.

3. *Explanatory Supplement to the American Ephemeris and Nautical Almanac*, London, Her Majesty's Stationery Office, 1961.

4. Fimple, W. R. "Optimum Midcourse Plane Change for Ballistic Interplanetary Trajectories." *AIAA Journal*, Vol. 1, No. 2, pp 430–434, 1963.

5. http://www.met.rdg.ac.uk/~ross/Astronomy/Planets.html, 4 Jun 2013.

6. Seidelmann, P. Kenneth, ed., *Explanatory Supplement to the Astronomical Almanac*, University Science Books, Mill Valley, CA, 1992.

7. Battin, R. H., *An Introduction to the Mathematics and Methods of Astrodynamics*, AIAA, Inc., New York, NY, 1987.

8. Prussing, J. E. and Conway, C. A. *Orbital Mechanics*. Oxford University Press, New York, NY, 1993.

9. Schaub, H. and Junkins, J. L. *Analytical Mechanics of Space Systems*. AIAA, Reston, VA, 2003.

10. *The Astronomical Almanac*. http://asa.usno.navy.mil/.

Chapter 9

Perturbations

"Life itself is a perturbation of the norm."
"An unperturbed existence leads to dullness of spirit and mind."
"To cause a planet or other celestial body to deviate from a theoretically regular orbital motion."

-Webster

9.1 Historical Background

A perturbation is a deviation from some normal or expected motion. Macroscopically, one tends to view the Universe as a highly regular and predictable scheme of motion. Yet when it is analyzed from accurate observational data, it is found that there seem to be distinct, and at times unexplainable, irregularities of motion superimposed upon the average, or mean, movement of the celestial bodies. After working in great detail in this text with many aspects of the two-body problem, it may come as somewhat of a shock to realize that real-world trajectory problems cannot be solved accurately with the restricted two-body theory presented. The actual path will vary from the theoretical two-body path as the result of perturbations caused by other mass bodies (such as the Moon) and additional forces not considered in Keplerian motion (such as a nonspherical Earth).

We are very familiar with perturbations in almost every area of life. Seldom does anything go exactly as planned, but is perturbed by various unpredictable circumstances. There is always some variation from the norm. An excellent example is an aircraft encountering a wind gust. The power and all directional controls are kept constant, yet the path may change abruptly from the theoretical path. The wind gusts were a perturbation. Fortunately, most of the perturbations we will need to consider in orbital flight are much more predictable and analytically tractable than the above example. Those that are unpredictable (solar flares, meteoroid collisions, etc.) must be treated in a stochastic (probabilistic) manner. Some of the perturbations that could be considered are due to the presence of other attracting bodies, atmospheric drag and lift, the asphericity of

Earth or Moon and solar radiation, magnetic, and relativistic effects. Analytic formulations of some of these will be given in Section 9.7.

It should not be supposed that perturbations are always small, for they can be as large as or larger than the primary attracting force. In fact, many interplanetary missions would miss their target entirely if the perturbing effect of other attracting bodies were not taken into account. Ignoring the effect of the oblateness of Earth on an artificial satellite would cause us to completely fail in the prediction of its position over a long period of time. Without the use of perturbation methods of analysis it would be impossible to explain or predict the orbit of the Moon. The following are some specific examples of the past value of perturbation analysis. Newton had explained most of the variations in the Moon's orbit, except for the motion of its perigee. In 1749 Clairant found that the second-order perturbation terms removed discrepancies between the observed and theoretical values, which had not been treated by Newton. Then, about a century later, the full explanation was found in an unpublished manuscript of Newton's. E.M. Brown's papers of 1897–1908 explain in great detail the perturbative effects of the oblateness of Earth and the Moon on the Moon's orbit and the effects of other planets. The presence of the planet Neptune was deduced analytically by Adams and by Leverriere from analysis of the perturbed motion of Uranus. The first accurate prediction of the return of Halley's Comet in 1759 by Clairant was made from calculation of the perturbations caused by Jupiter and Saturn. He correctly predicted a possible error of one month owing to mass uncertainty and other more distant planets. The shape of Earth was deduced by an analysis of long-period perturbation terms in the eccentricity of Earth's orbit. These are but a few of the examples of application of perturbation analysis. Note that these were studies of bodies over which we have no control. With the capability for orbital flight, knowledge of how to handle perturbations is a necessary skill for applying astrodynamics to real-world problems of getting from one place (or planet) to another. This chapter is a first step to reality from the simplified theoretical foundation laid earlier in this text.

These are two main categories of perturbation techniques. These are referred to as *special perturbations* and *general (or absolute) perturbations*. Special perturbations are techniques that deal with the direct numerical integration of the equations of motion including all necessary perturbing accelerations. This is the main emphasis of this chapter. General perturbation techniques involve an analytic integration of series expansions of the perturbing accelerations. This is a more difficult and lengthy technique than special perturbation techniques but it leads to better understanding of the source of the perturbation. Most of the discoveries of the preceding paragraph resulted from general perturbation studies.

The objective of this chapter is to present some of the more useful and well-known special perturbation techniques in such a way that the reader can determine when and how to apply them to a specific problem. In particular, the Cowell, Encke, and variation of elements techniques are discussed. Since numerical integration is necessary in many uses, a section discussing methods and errors is included. Analytical formulation of some of the more common perturbation accelerations is included to aid application to specific problems. A list of classical and current works on perturbations and integration techniques is included at the end of the chapter for the reader who wishes to study perturbation analysis in greater depth. See Section 1.2 for a table showing relative perturbation accelerations on a typical Earth satellite.

9.2 COWELL'S METHOD

This is the simplest and most straightforward of all the perturbation methods. It was developed by P. H. Cowell in the early 20th century and was used to determine the orbit of the eighth satellite of Jupiter. Cowell and Crommelin also used it to predict the return of Halley's Comet for the period 1759 to 1910. This method has been "rediscovered" many times in many forms. It is especially popular and useful for computer implementation.

The application of Cowell's method is simply to write the equations of motion of the object being studied, including all the perturbations, and then to integrate them step-by-step numerically. For the two-body problem with perturbations, the equation would be

$$\ddot{\mathbf{r}} + \frac{\mu}{r^3}\mathbf{r} = \mathbf{a}_p \tag{9-1}$$

For numerical integration, this would be reduced to first-order differential equations:

$$\begin{aligned} \dot{\mathbf{r}} &= \mathbf{v} \\ \dot{\mathbf{v}} &= \mathbf{a}_p - \frac{\mu}{r^3}\mathbf{r} \end{aligned} \tag{9-2}$$

where **r** and **v** are the radius and velocity of a satellite with respect to the larger central body. For numerical integration, Equation (9-2) would be further broken down into the vector components:

$$\begin{aligned} \dot{r}_x &= v_x, \quad & \dot{v}_x &= a_{px} - \frac{\mu}{r^3}x \\ \dot{r}_y &= v_y, \quad & \dot{v}_y &= a_{py} - \frac{\mu}{r^3}y \\ \dot{r}_z &= v_z, \quad & \dot{v}_z &= a_{pz} - \frac{\mu}{r^3}z \end{aligned} \tag{9-3}$$

where $r = (r_x^2 + r_y^2 + r_z^2)^{1/2}$.

The perturbing acceleration, $\mathbf{a}_p$, could be due to the presence of other gravitational bodies such as the Moon, Sun and planets. If we consider the Moon as the perturbing body, the equations would be

$$\dot{\mathbf{r}} = \mathbf{v}$$

$$\dot{\mathbf{v}} = -\frac{\mu_\oplus}{r^3}\mathbf{r} - \mu_m\left(\frac{\mathbf{r}_{ms}}{r_{ms}^3} - \frac{\mathbf{r}_{m\oplus}}{r_{m\oplus}^3}\right) \quad (9\text{-}4)$$

where

r = radius from Earth to the satellite,

r_{ms} = radius from the Moon to the satellite,

$r_{m\oplus}$ = radius from the Moon to Earth, and

μ_m = gravitational parameter of the Moon.

Once the analytical formulation of the perturbation is established, the state (**r** and **v**) at any time can be found by applying one of the many available numerical integration schemes to Equations (9-3).

The main advantage of Cowell's method is the simplicity of formulation and implementation. Any number of perturbations can be handled at the same time. Intuitively, one would suspect that no method so simple would also be free of shortcomings, and one would be correct. There are some disadvantages of the method. When motion is near a large attracting body, smaller integration steps must be taken, which severely affect time and accumulative error owing to round-off errors. This method is approximately 1/10th the speed of Encke's method. It has been found that Cowell's method is not the best for lunar trajectories.

Classically, Cowell's method has been applied in a Cartesian coordinate system as shown in Equations (9-3). Some improvement can be made in trajectory problems by formulating the problem in polar or spherical coordinates. In this case, **r** will tend to vary slowly and the angle change often will be monotonic. This will often allow for larger integration step sizes for the same truncation error. In spherical coordinates (r, θ, ϕ) the equations of motion are (for an equatorial coordinate system):

$$\ddot{r} - r(\dot{\theta}^2\cos^2\phi + \dot{\phi}^2) = -\frac{\mu}{r^2}$$

$$r\ddot{\theta}\cos\phi + 2\dot{r}\dot{\theta}\cos\phi - 2r\dot{\theta}\dot{\phi}\sin\phi = 0 \quad (9\text{-}5)$$

$$r\ddot{\phi} + 2\dot{r}\dot{\phi} + r\dot{\theta}^2\sin\phi\cos\phi = 0 \quad (9\text{-}6)$$

As a reminder, one should always pick the coordinate system that best accommodates the problem formulation and the solution method.

9.3 ENCKE'S METHOD

Though the method of Encke is more complex than that of Cowell, it appeared over half a century earlier in 1851. Bond had actually suggested it in 1849—two years before Encke's work became known.

In the Cowell method the sum of all accelerations is integrated together. In the Encke method, the difference between the primary acceleration and all perturbing accelerations is integrated. This implies a reference orbit along which the object would move in the absence of all perturbing accelerations. Presumably, this reference orbit would be a conic section in an ideal Newtonian gravitational field. Thus all calculations and states (position and velocity) would be with respect to the reference trajectory. In most works the reference trajectory is called the osculating orbit. "Osculation" is the "scientific" term for kissing. The term connotes the sense of contact; and in our case, it is the contact between the reference (or osculating) orbit and the true perturbed orbit. The osculating orbit is the orbit that would result if all perturbing accelerations could be removed at a particular time (epoch). At that time, or epoch, the osculating and true orbits are in contact or coincide (see Figure 9-1).

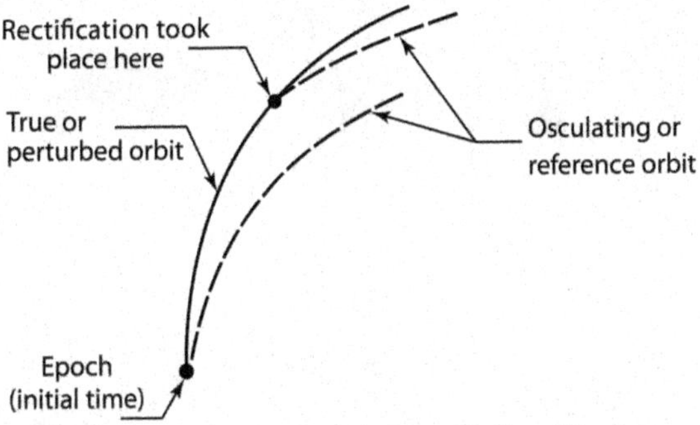

Fig. 9-1 The Osculating Orbit with Rectification.

Any particular osculating orbit is good until the true orbit deviates too far from it. Then a process called *rectification* must occur to continue the integration. This simply means that a new epoch and starting point will be chosen that coincide with the true orbital path. Then a new osculating orbit is calculated from the true radius and velocity vectors, neglecting the perturbations (see Figure 9-1).

Now we will consider the analytic formulation of Encke's method. The basic objective is to find an analytic expression for the difference between the true and reference orbits. Let $\mathbf{r}$ and $\boldsymbol{\rho}$ be the radius vectors to, respectively, the true (perturbed) and osculating (reference) orbits at a particular time $\tau(\tau = t - t_0)$.

Then for the true orbit,

$$\ddot{\mathbf{r}} + \frac{\mu}{r^3}\mathbf{r} = \mathbf{a}_p \tag{9-7}$$

and for the osculating orbit,

$$\ddot{\boldsymbol{\rho}} + \frac{\mu}{\rho^3}\boldsymbol{\rho} = 0 \tag{9-8}$$

Note that, at the epoch $t_0 = 0$,

$$\mathbf{r}(t_0) = \boldsymbol{\rho}(t_0) \text{ and } \mathbf{v}(t_0) = \dot{\boldsymbol{\rho}}(t_0)$$

Let the deviation from the reference orbit, $\delta\mathbf{r}$, be defined as (see Figure 9-2)

$$\delta\mathbf{r} = \mathbf{r} - \boldsymbol{\rho}$$

$$\text{and } \delta\ddot{\mathbf{r}} = \ddot{\mathbf{r}} - \ddot{\boldsymbol{\rho}} \tag{9-9}$$

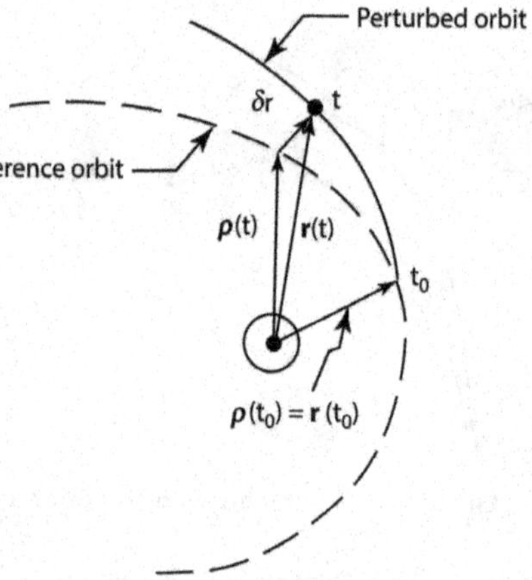

Fig. 9-2 $\delta\mathbf{r}$ Deviation from Reference Orbit.

Substituting Equations (9-7) and (9-8) into (9-9), we obtain

$$\delta\ddot{\mathbf{r}} = \mathbf{a}_p + \left(\frac{\mu}{\rho^3}\boldsymbol{\rho} - \frac{\mu}{r^3}\mathbf{r}\right)$$

$$= \mathbf{a}_p + \left[\frac{\mu}{\rho^3}(\mathbf{r} - \delta\mathbf{r}) - \frac{\mu}{r^3}\mathbf{r}\right]$$

$$\delta\ddot{\mathbf{r}} = \mathbf{a}_p + \frac{\mu}{\rho^3}\left[\left(1 - \frac{\rho^3}{r^3}\right)\mathbf{r} - \delta\mathbf{r}\right] \tag{9-10}$$

This is the desired differential equation for the deviation, $\delta\mathbf{r}$. For a given set of initial conditions, $\delta\mathbf{r}\ (t_0 + \Delta t)$ can be calculated numerically. $\boldsymbol{\rho}$ is a known function of time, so $\mathbf{r}$ can be obtained from $\delta\mathbf{r}$ and $\boldsymbol{\rho}$. So, theoretically, we should be able to calculate the perturbed position and velocity of the object. However, one of the reasons for going to this method from Cowell's was to obtain more accuracy, but the term $\left(1 - \frac{\rho^3}{r^3}\right)$ is the difference of two very nearly equal quantities, which requires many extra digits of computer accuracy on that one operation to maintain reasonable accuracy throughout. A standard method of treating the difference of nearly equal quantities is to define

$$2q = 1 - \frac{r^2}{\rho^2} \tag{9-11}$$

thus

$$\frac{\rho^3}{r^3} = (1-2q)^{-3/2} \tag{9-12}$$

Equation (9-10) then becomes

$$\ddot{\delta\mathbf{r}} = \mathbf{a}_p + \frac{\mu}{\rho^3}\{[1-(1-2q)^{-3/2}]\mathbf{r} - \delta\mathbf{r}\} \tag{9-13}$$

Our problem is not yet solved, however, since q is a small quantity and the accuracy problem seems unsolved. Now some ways of calculating q must be found. In terms of its components,

$$r^2 = x^2 + y^2 + z^2$$
$$= (\rho_x + \delta x)^2 + (\rho_y + \delta y)^2 + (\rho_z + \delta z)^2$$

where δx, δy and δz are the Cartesian components of $\delta\mathbf{r}$.

Expanding, we get

$$r^2 = \rho_x^2 + \rho_y^2 + \rho_z^2 + \delta x^2 + \delta y^2 + \delta z^2 + 2\rho_x\ \delta x + 2\rho_y \delta y + 2\rho_z \delta z$$
$$= \rho^2 + 2\left[\delta x\left(\rho_x + \frac{1}{2}\delta x\right) + \delta y\left(\rho_y + \frac{1}{2}\delta y\right) + \delta z\left(\rho_z + \frac{1}{2}\delta z\right)\right]$$

then

$$\frac{r^2}{\rho^2} = 1 + \frac{2}{\rho^2}\left[\delta x\left(\rho_x + \frac{1}{2}\delta x\right) + \delta y\left(\rho_y + \frac{1}{2}\delta y\right)\right.$$
$$\left. + \delta z\left(\rho_z + \frac{1}{2}\delta z\right)\right]$$

From Equation (9-11) we find that

$$\frac{r^2}{\rho^2} = 1 - 2q$$

So

$$q = -\frac{1}{\rho^2}\left[\delta x\left(\rho_x + \frac{1}{2}\delta x\right) + \delta y\left(\rho_y + \frac{1}{2}\delta y\right) + \delta z\left(\rho_z + \frac{1}{2}\delta z\right)\right] \tag{9-14}$$

Now, at any point where $\boldsymbol{\rho}$ and $\delta\boldsymbol{\rho}$ are known, we can calculate q, but as was noted before the same problem of small differences still exists in Equation (9-13). There are two methods of solution at this point. The first is to expand the term $(1 - 2q)^{-3/2}$ in a binomial series to get

$$1 - (1 - 2q)^{-3/2} = 3q - \frac{3\bullet5}{2!}q^2 + \frac{3\bullet5\bullet7}{3!}q^3 - \ldots \tag{9-15}$$

Before the advent of high-speed digital computers this was a very cumbersome task, so the second method was to define a function f as

$$f = \frac{1}{q}\,[1 - (1 - 2q)^{-3/2}] \tag{9-16}$$

Then Equation (9-13) becomes

$$\delta\ddot{\mathbf{r}} = \mathbf{a}_p + \frac{\mu}{\rho^3}(fq\mathbf{r} - \delta\mathbf{r}) \tag{9-17}$$

Tables of f vs. q have been constructed (see *Planetary Coordinates*, 1960)[1] to avoid hand calculation. It should also be noted that when the deviation from the reference orbit is small (which should be the case most of the time) the δx^2, δy^2 and δz^2 terms can be neglected in Equation (9-14) and

$$q = -\frac{\rho_x\delta x + \rho_y\delta y + \rho_z\delta z}{\rho^2} \tag{9-18}$$

which is easily calculated. Then the tables could be used to find f. The method using Equation (9-15) is recommended for computer use. Although the use of Equation (9-18) may be slightly faster, if Equation (9-18) is used, care must be taken to identify when the approximation is not valid.

The Encke formulation reduces the number of integration steps since the $\delta\mathbf{r}$ presumably changes more slowly than $\mathbf{r}$, so larger step sizes can be taken. The advantages of the method diminish if (1) $\mathbf{a}_p$ becomes much larger than $\frac{\mu}{\rho^3}(fq\mathbf{r} - \delta\mathbf{r})$ or (2) $\frac{\delta\mathbf{r}}{\rho}$ does not remain quite small. In the case of (1) it may be that the reference parameters (or orbit) need to be changed since the perturbations are becoming primary. In the case of (2) a new osculating orbit needs to be chosen using the values of **r** and **v** as described earlier. Rectification should be initiated when $\delta \mathbf{r}/\boldsymbol{\rho}$ is greater than or equal to some small constant (of the reader's selection depending on the accuracy and speed of the computing machinery used, but of the order of magnitude of 0.01).

Encke's method is generally much faster than Cowell's because of the ability to take larger integration step sizes when near a large attracting body. It will be about 10 times faster for interplanetary orbits, but only about 3 times faster for Earth satellites (see Baker, vol. 2, p. 229).[2]

A computational algorithm for Encke's method can be outlined as follows:

a. Given the initial conditions $\mathbf{r}(t_0) = \boldsymbol{\rho}(t_0)$, $\mathbf{v}(t_0) = \dot{\boldsymbol{\rho}}(t_0)$ define the osculating orbit. $\delta \mathbf{r} = 0$ and $\delta \dot{\mathbf{r}} = 0$ at this point.

b. For an integration step, Δt, calculate $\delta \ddot{\mathbf{r}}(t_0 + \Delta t)$, knowing $\boldsymbol{\rho}(t_0)$, $\mathbf{r}(t_0)$; $q(t_0) = 0$.

c. Knowing $\delta \mathbf{r}(t_0 + \Delta t)$ calculate

   1. $\boldsymbol{\rho}(t_0 + \Delta t)$
   2. $q(t_0 + \Delta t)$ from Equation (9-14)
   3. $f(t_0 + \Delta t)$ from Equation (9-16)

d. Integrate another Δt to get $\delta \mathbf{r}(t_0 + k\Delta t)$.

e. If $\delta r/\rho >$ a specified constant, rectify and go to step a. Otherwise, continue.

f. Calculate $\mathbf{r} = \boldsymbol{\rho} + \delta \mathbf{r}$ and $\mathbf{v} = \dot{\boldsymbol{\rho}} + \delta \dot{\mathbf{r}}$.

g. Go to step c with Δt replaced by $k\Delta t$, where k is the step number.

This algorithm requires that the perturbation accelerations be known in analytic form. In a lunar flight another check would be added to determine when the perturbation from the Moon should become the primary acceleration. The Encke method has been found to be quite good for calculating lunar trajectories. See Equation (9-3) for the addition of the Moon's gravitational acceleration. Similarly, from the Encke method, we obtain

$$\delta \ddot{\mathbf{r}} = -\mu_m \left(\frac{\mathbf{r}_{ms}}{r_{ms}^3} - \frac{\mathbf{r}_{m\oplus}}{r_{m\oplus}^3} \right) + \frac{\mu_\oplus}{\rho^3}(fq\mathbf{r} - \delta \mathbf{r}) \tag{9-19}$$

9.4 VARIATION OF PARAMETERS OR ELEMENTS

The method of the variation of parameters was first developed by Euler in 1748 and was the only successful method of perturbations used until the more recent development of the machine-oriented Cowell and Encke methods. The latter are concerned with a calculation of the coordinates whereas the variation of parameters calculates the orbital elements or any other consistent set of parameters that adequately describe the orbit. Though, at the outset, this method may seem more difficult to implement, it has distinct advantages in many problems where the perturbing forces are quite small.

One main difference between the Encke method and the variation of parameters method is that the Encke reference orbit is constant until rectification occurs. In variation of parameters the reference orbit is changing continuously. For instance, if at any two successive points along the perturbed trajectory one were to calculate the eccentricity of a reference trajectory (unperturbed) using the actual **r** and **v**, one would observe a small change in eccentricity because of the perturbing forces.

Then $\dot{e} \approx \Delta e / \Delta t$, which is an approximation for the time rate of change of eccentricity. A similar check could be made on i, a, etc. Thus, the orbital parameters of the reference trajectory are changing with time. In the absence of perturbations, they would remain constant. From previous work we know that **r** and **v** can be calculated from the set of six orbital elements (parameters), so if the perturbed elements could be calculated as a function of time, then the perturbed state (**r** and **v**) would be known.

A two-body orbit can be described by any consistent set of six parameters (or constants) since it is described by three second-order differential equations. The orbital elements are only one of many possible sets (see Baker, vol. 2, p. 246, for a table of parameter sets).[2] The essence of the variation of parameters method is to find how the selected set of parameters vary with time as a result of the perturbations. This is done by finding analytic expressions for the rate of change of the parameters in terms of the perturbations. These expressions are then integrated numerically to find their values at some later time. Integrating the expressions *analytically* is the method of *general perturbations*. It is clear that the elements will vary slowly compared to the position and velocity variations (e.g., the eccentricity may change only slightly in an entire orbit), so larger integration steps may be taken in the cases where the total acceleration is being integrated such as in Cowell's method or where the perturbing acceleration is integrated as in Encke's method. In this section two sets of parameters will be treated. The first will be the classical orbital elements because of their familiarity and universality in the literature. The second will be the f and g expression variations, which are of greater practical value and are easier to implement.

9.4.1 Variation of the Classical Orbital Elements

The standard orbital elements a, e, i, Ω, ω and T (or M) will be used, where

a = semi-major axis,

e = eccentricity,

i = orbit inclination,

Ω = longitude of the ascending node,

ω = argument of periapsis,

T = time of periapsis passage,

(M_0 = mean anomaly at epoch = $M - n(t - t_0)$).

The object is to find analytic expressions for $\frac{da}{dt}, \frac{de}{dt}, \frac{di}{dt}, \frac{d\omega}{dt}, \frac{d\Omega}{dt}$ and $\frac{dM_0}{dt}$. To do this, some reference coordinate system is necessary. Eventually, reference to an "inertial" system may be desired, but any other system can be used in the derivation and the results transferred to the desired coordinate system rather easily. Therefore, it would be best for illustrative purposes to choose a system in which the derivations are the easiest to follow. This derivation partially follows the one described in NASA CR-1005.[3] The coordinate system chosen has its principal axis, **R** (unit vector **R**), along the instantaneous radius vector, **r**. The axis **S** is rotated 90° from **R** in the direction of increasing true anomaly. The third axis, **W**, is perpendicular to both **R** and **S**. Note that this coordinate system is simply rotated ν from the **PQW** perifocal system described in Chapter 2.

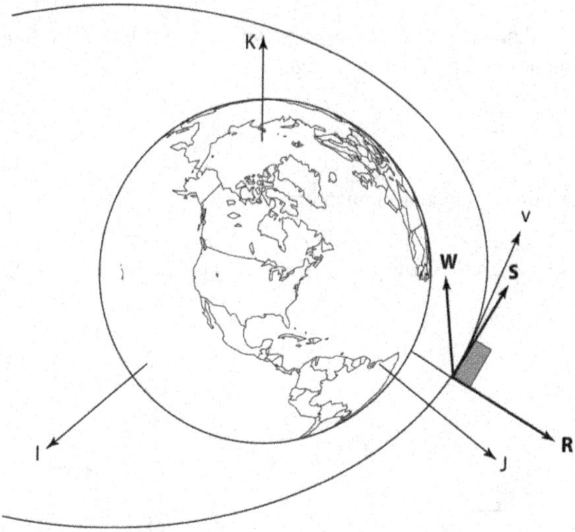

Fig. 9-3 RSW Coordinate System.

In the **RSW** coordinate system the perturbing force is

$$\mathbf{F} = m(F_R\mathbf{R} + F_S\mathbf{S} + F_W\mathbf{W}) \text{ (in terms of specific forces)} \tag{9-20}$$

and

$$\begin{aligned} \mathbf{r} &= r\mathbf{R} \\ \mathbf{v} &= \dot{r}\mathbf{R} + r\dot{\nu}\mathbf{S} \\ &= \dot{\nu}\left(\frac{dr}{d\nu}\mathbf{R} + r\mathbf{S}\right) \end{aligned} \tag{9-21}$$

First, we must consider the derivation of (da/dt).

The time rate of change of energy per unit mass (a constant in the normal two-body problem) is a result of the perturbing force and can be expressed as

$$\frac{dE}{dt} = \frac{\mathbf{F \cdot V}}{m} = \dot{\nu}\left(\frac{dr}{d\nu} F_r + rF_S\right) \tag{9-22}$$

and

$$E = -\frac{\mu}{2a} \quad \text{or} \quad a = -\frac{\mu}{2E} \tag{9-23}$$

Now

$$\frac{da}{dt} = \frac{da}{dE}\frac{dE}{dt} = \frac{\mu}{2E^2}\frac{dE}{dt} \tag{9-24}$$

To express this in known terms, expressions for $\dot{\nu}$ and $(dr)/(d\nu)$ must be found. Differentiating the conic equation gives

$$\frac{dr}{d\nu} = \frac{re \sin \nu}{1 + e \cos \nu} \tag{9-25}$$

From the conservation of angular momentum, we have

$$h = r^2\dot{\nu} = \sqrt{\mu p} = \sqrt{\mu a(1-e^2)} = na^2\sqrt{1-e^2} \tag{9-26}$$

where

$$n = \sqrt{\frac{\mu}{a^3}}$$

Therefore,

$$\dot{\nu} = \frac{na^2}{r^2}\sqrt{1-e^2} \tag{9-27}$$

Substituting in Equation (9-24) from (9-22), (9-23), (9-25) and (9-27), we find

$$\boxed{\frac{da}{dt} = \frac{2e \sin \nu}{n\sqrt{1-e^2}} F_r + \frac{2a\sqrt{1-e^2}}{nr} F_s} \tag{9-28}$$

Consider now the derivation of $\frac{de}{dt}$.

In deriving the remaining variations, the time rate of change of angular momentum is needed. This rate is expressed as the moment of all perturbing forces acting on the system, so

$$\frac{d\mathbf{h}}{dt} = \frac{1}{m}(\mathbf{r} \times \mathbf{F}) = rF_S\mathbf{W} - rF_W\mathbf{S} \tag{9-29}$$

This could also have been developed from

$$\frac{d\mathbf{h}}{dt} = \frac{d}{dt}(\mathbf{r}\times\mathbf{v}) = \left(\frac{d\mathbf{r}}{dt}\times\mathbf{v}\right)^{0} + \mathbf{r}\times\frac{d\mathbf{v}}{dt}$$

$$\text{where } \frac{d\mathbf{v}}{dt} = \frac{\mathbf{F}}{m} = \mathbf{a}_p$$

Since $\mathbf{h} = h\mathbf{W}$, the vector time derivative can be expressed as the change of length in **W** and its transverse component along the plane of rotation of **h**, so

$$\frac{d\mathbf{h}}{dt} = \dot{h}\mathbf{W} + h\frac{d\alpha}{dt}\mathbf{S} \tag{9-30}$$

where α is the angle of rotation. Note that the change of **h** must be in the **S–W** plane since the perturbing force is applied to **r** and the momentum changes the cross product of **r** and (**F**/m). Comparing components of Equations (9-30) and (9-29) we see that

$$\frac{dh}{dt} = rF_S \tag{9-31}$$

From the expression $p = a(1 - e_2)$, we have

$$e = \left(1 - \frac{p}{a}\right)^{1/2} = \left(1 - \frac{h^2}{\mu a}\right)^{1/2} \tag{9-32}$$

Therefore,

$$\begin{aligned}\frac{de}{dt} &= -\frac{h}{2\mu a e}\left(2\frac{dh}{dt} - \frac{h}{a}\frac{da}{dt}\right)\\ &= -\frac{\sqrt{1-e^2}}{2na^2e}\left(2\frac{dh}{dt} - na\sqrt{1-e^2}\frac{da}{dt}\right)\end{aligned} \tag{9-33}$$

Substituting for dh/dt and da/dt then gives

$$\boxed{\frac{de}{dt} = \frac{\sqrt{1-e^2}\sin\nu}{na}F_R + \frac{\sqrt{1-e^2}}{na^2e}\left[\frac{a^2(1-e^2)}{r} - r\right]F_S} \tag{9-34}$$

The derivation of di/dt could be found geometrically from the relationship of **h**, Ω and i, but it is more direct to do it analytically (see NASA CR-1005 for a geometric development).[3] From the chapter on orbit determination recall that

$$\cos i = \frac{\mathbf{h}\cdot\mathbf{K}}{h} \tag{9-35}$$

Differentiating both sides of Equation (9-35), we obtain

$$-\sin i \frac{di}{dt} = \frac{h\left(\dfrac{d\mathbf{h}}{dt}\cdot\mathbf{K}\right) - (\mathbf{h}\cdot\mathbf{K})\dfrac{dh}{dt}}{h^2} \tag{9-36}$$

$$= \frac{h\,(rF_s\mathbf{W} - rF_w\mathbf{S})\cdot\mathbf{K} - h\,\cos i\, rF_s}{h^2}$$

But **W** • **K** = cos i and **S** • **K** = sin i cos u, where u is the argument of latitude (angle from the ascending node to **r**), so

$$-\sin i\frac{di}{dt} = -\frac{rF_w\ \sin i \cos u}{na^2\sqrt{1-e^2}} \tag{9-37}$$

and

$$\frac{di}{dt} = \frac{rF_w\ \cos u}{na^2\sqrt{1-e^2}} \tag{9-38}$$

As was expected, changes in i result only from a perturbation along **W**.

In the derivation of dΩ/dt we note from the orbit determination chapter that

$$\cos\Omega = \frac{\mathbf{I}\cdot(\mathbf{K}\times\mathbf{h})}{|\mathbf{K}\times\mathbf{h}|} \tag{9-39}$$

Differentiating both sides we get

$$-\sin\Omega\frac{d\Omega}{dt}$$

$$= \frac{\mathbf{I}\cdot\left(\mathbf{K}\times\dfrac{d\mathbf{h}}{dt}\right)|\mathbf{K}\times\mathbf{h}| - \mathbf{I}\cdot(\mathbf{K}\times\mathbf{h})\dfrac{d}{dt}|\mathbf{K}\times\mathbf{h}|}{|\mathbf{K}\times\mathbf{h}|^2}$$

$$= \Big\{\ \mathbf{I}\cdot[\mathbf{K}\times(rF_S\mathbf{W} - rF_W\mathbf{S})]h\sin i$$

$$-h\cos\Omega\sin i\left(\frac{dh}{dt}\sin i + h\cos i\frac{di}{dt}\right)\Bigg\}\frac{1}{h^2\sin^2 i}$$

But $\mathbf{I} \cdot \mathbf{K} \times \mathbf{W} = \cos\Omega \sin i$ and $\mathbf{I} \cdot \mathbf{K} \times \mathbf{S} = \mathbf{I} \times \mathbf{K} \cdot \mathbf{S} = -\mathbf{J} \cdot \mathbf{S} = \sin\Omega \sin u - \cos\Omega \cos u \cos i$ (see Chapter 2 for the coordinate transformation for the PQW system and replace ω with u using spherical trigonometry). Substituting for di/dt and dh/dt from Equations (9-37) and (9-31) and canceling terms, we get

$$\frac{d\Omega}{dt} = \frac{rF_W \sin u}{h \sin i} \tag{9-40}$$

$$\boxed{\frac{d\Omega}{dt} = \frac{rF_W \sin u}{na^2\sqrt{1-e^2}\sin i}} \tag{9-41}$$

In the following derivation of $d\omega/dt$ an important difference from the previous derivation becomes apparent. Up to now "constants" such as a, e and i have been differentiated and the variation has obviously been due to the perturbations. Now, however, the state vectors of position and velocity will appear in the expression. Since we are considering only the time rate of change resulting from the perturbative force, not the changes resulting from the two-body reference motion, **r** will *not change to first order as a result of the perturbations* ($d\mathbf{r}/dt = 0$) but the derivative is only with respect to the perturbation forces, so

$$\frac{d\mathbf{v}}{dt} = \frac{\mathbf{F}_p}{m} = \mathbf{a}_p$$

Without this explanatory note, some of the following would be rather mysterious. This procedure is necessary to show the perturbation from the two-body reference orbit.

Consider the expression from orbit determination for finding u ($u = \omega + \nu$):

$$\frac{(\mathbf{K}\times\mathbf{h})\cdot\mathbf{r}}{|\mathbf{K}\times\mathbf{h}|} = r\cos(\omega+\nu) \tag{9-42}$$

Differentiating gives

$$\frac{|\mathbf{K}\times\mathbf{h}|\left(\mathbf{K}\times\frac{d\mathbf{h}}{dt}\cdot\mathbf{r}\right) - (\mathbf{K}\times\mathbf{h}\cdot\mathbf{r})\frac{d}{dt}|\mathbf{K}\times\mathbf{h}|}{|\mathbf{K}\times\mathbf{h}|^2} = -r\sin(\omega+\nu)\left(\frac{d\omega}{dt}+\frac{d\nu}{dt}\right)$$

which can be arranged to

$$\frac{d\omega}{dt} = \frac{-|\mathbf{K}\times\mathbf{h}|\left(\mathbf{K}\times\frac{d\mathbf{h}}{dt}\cdot\mathbf{r}\right) + (\mathbf{K}\times\mathbf{h}\cdot\mathbf{r})\frac{d}{dt}|\mathbf{K}\times\mathbf{h}| - \frac{d\nu}{dt}|\mathbf{K}\times\mathbf{h}|^2 r\sin(\omega+\nu)}{|\mathbf{K}\times\mathbf{h}|^2 r\sin(\omega+\nu)}$$

$$= \left\{\frac{1}{h^2 \sin^2 i\, r \sin u}\right\}\{-h \sin i[\mathbf{K} \times (rF_S\mathbf{W} - rF_W\mathbf{S}) \cdot \mathbf{r}]$$

$$+ (\mathbf{K} \times \mathbf{h} \cdot \mathbf{r})\left[\frac{dh}{dt}\sin i + h \cos i \frac{di}{dt}\right] - \frac{d\nu}{dt} h^2 r \sin^2 i \sin u\} \tag{9-43}$$

where

$$\mathbf{K} \times \mathbf{W} \cdot \mathbf{r} = r \sin i \cos u$$

$$\mathbf{K} \times \mathbf{S} \cdot \mathbf{r} = r\mathbf{S} \times \mathbf{R} \cdot \mathbf{K} = -r\mathbf{W} \cdot \mathbf{K} = -r \cos i$$

$$\mathbf{K} \times \mathbf{h} \cdot \mathbf{r} = rh \sin i \cos u$$

and dh/dt and di/dt are given by Equations (9-31) and (9-37). Now an expression for $d\nu/dt$ must be found. The true anomaly is affected by the perturbation resulting from movement of periapsis and also the node. From the conic equation,

$$r(1 + e \cos \nu) = \frac{h^2}{\mu} \tag{9-44}$$

Differentiating the terms that are affected by the perturbations gives

$$r\left(\frac{de}{dt}\cos \nu - e \sin \nu \frac{d\nu}{dt}\right) = \frac{2h}{\mu}\frac{dh}{dt} \tag{9-45}$$

$$\text{or } re \sin \nu \frac{d\nu}{dt} = r \cos \nu \frac{de}{dt} - \frac{2h}{\mu}\frac{dh}{dt} \tag{9-46}$$

At this point Equation (9-46) could be solved for $d\nu/dt$ and put in Equation (9-43) and a correct expression would result. However, experience has shown that this is algebraically rather complex and the resulting expression is not as simple as will result from another formulation of $d\nu/dt$. If we differentiate the identity

$$\mu re \sin \nu = h\, \mathbf{r} \cdot \mathbf{v} \tag{9-47}$$

$$\text{we get } \mu r \frac{de}{dt}\sin \nu + \mu re \cos \nu \frac{d\nu}{dt} = \frac{dh}{dt}\mathbf{r} \cdot \mathbf{v} + h\mathbf{r} \cdot \dot{\mathbf{v}} \tag{9-48}$$

Now multiply Equation (9-46) by $\mu \sin \nu$ and Equation (9-48) by $\cos \nu$ and add to get

$$\frac{d\nu}{dt} = \frac{1}{reh}\left[p \cos \nu\, \mathbf{r} \cdot \dot{\mathbf{v}} - (p + r)\sin \nu \frac{dh}{dt}\right] \tag{9-49}$$

Now Equation (9-43) becomes

$$\frac{d\omega}{dt} = \left[\frac{1}{h^2 r \sin^2 i \sin u}\right]\{-h \sin i[r^2 F_S \sin i \cos u + r^2 F_W \cos i]$$

$$+ rh \sin i \cos u[rF_S \sin i + rF_W \cos i \cos u]$$

$$-\frac{1}{reh}(p \cos \nu\, rF_R - (p + r)\sin \nu\, rF_S)h^2 r \sin^2 i \sin u\} \tag{9-50}$$

Writing $\left(\frac{d\omega}{dt}\right)_R$, $\left(\frac{d\omega}{dt}\right)_S$ and $\left(\frac{d\omega}{dt}\right)_W$ as the variations due to the three components of perturbation forces, after algebraic reduction, we obtain

$$\left(\frac{d\omega}{dt}\right)_R = -\frac{\sqrt{1-e^2}\cos\nu}{nae}F_R \tag{9-51}$$

$$\left(\frac{d\omega}{dt}\right)_S = \frac{p}{eh}\left[\sin\nu\left(1+\frac{1}{1+e\cos\nu}\right)\right]]F_S \tag{9-52}$$

$$\left(\frac{d\omega}{dt}\right)_W = -\frac{r\cot i\sin u}{na^2\sqrt{1-e^2}}F_W \tag{9-53}$$

$$\text{where}\quad \frac{d\omega}{dt} = \left(\frac{d\omega}{dt}\right)_R + \left(\frac{d\omega}{dt}\right)_S + \left(\frac{d\omega}{dt}\right)_W \tag{9-54}$$

As an aside, note that it can be shown that the change in ω resulting from the in-plane perturbations is directly related to the change in the true anomaly resulting from the perturbation,

$$\left(\frac{d\omega}{dt}\right)_{R,S} = -\frac{d\nu}{dt} \tag{9-55}$$

The derivation of dM_0/dt follows from the differentiation of the mean anomaly at epoch,

$$M_0 = E - e\sin E - n\left(t-t_0\right) \tag{9-56}$$

$$\text{So}\quad \frac{dM_0}{dt} = \frac{dE}{dt} - \frac{de}{dt}\sin E - e\cos E\frac{dE}{dt} - \frac{dn}{dt}\left(t-t_0\right)$$

$\frac{de}{dt}$ is known, and $\frac{dn}{dt} = -\frac{3\mu}{2na^4}\frac{da}{dt}$, $\frac{dE}{dt}$, $\sin E$ and $\cos E$ can be obtained from the expressions

$$\cos\nu = \frac{\cos E - e}{1 - e\cos E} \tag{9-57}$$

$$\sin\nu = \frac{\sqrt{1-e^2}\sin E}{1-e\cos E} \tag{9-58}$$

The result is (for elliptical orbits only)

$$\boxed{\begin{aligned}\frac{dM_0}{dt} &= -\frac{1}{na}\left(\frac{2r}{a}-\frac{1-e^2}{e}\cos\nu\right)F_R \\ &\quad -\frac{\left(1-e^2\right)}{nae}\left(1+\frac{r}{a\left(1-e^2\right)}\right)\sin\nu\, F_S - t\frac{dn}{dt}\end{aligned}} \qquad (9\text{-}59)$$

To use the variation of parameters formulation the procedure is as follows:

a. At $t = t_0$, calculate the six orbital elements.

b. Compute the perturbation force and transform it at $t = t_0$ to the **RSW** system.

c. Compute the six rates of change of the elements (right side of equations).

d. Numerically integrate the equations one step.

e. The integration has been for the perturbation of the elements, so the changes in elements must be added to the old values at each step.

f. From the new values of the orbital elements, calculate a position and velocity.

g. Go to step b and repeat until the final time is reached.

There are some limitations on this set of parameters. Note that some of them break down when $e = 1$ or $e = 0$. Special formulations using other parameter sets have been developed to handle near-circular and parabolic orbits. Note also that the dM_0/dt was derived for elliptic orbits only.

9.4.2 Variation of Parameters in the Universal Variable Formulation

Since there are many restrictions on the use of the perturbation equations developed in the previous sections, equations relating to the universal variables are developed here. There are no restrictions on the results of this development.

The parameters to be varied in this case are the six components of $\mathbf{r_0}$ and $\mathbf{v_0}$ (which certainly describe the orbit as well as the orbital elements). Thus, we would like to find the time derivative or $\mathbf{r_0}$ and $\mathbf{v_0}$ resulting from the perturbation forces. In Chapter 4 we described the position at any time as a function of $\mathbf{r_0}$, $\mathbf{v_0}$, f, $\dot{f}$, g and $\dot{g}$, where the f and g expressions were in terms of the universal variable, x. We can express $\mathbf{r_0}$ and $\mathbf{v_0}$ as

$$\begin{aligned}\mathbf{r}_0 &= F\mathbf{r} + G\mathbf{v} \\ \mathbf{v}_0 &= \dot{F}\mathbf{r} + \dot{G}\mathbf{v}\end{aligned} \qquad (9\text{-}60)$$

F, G, $\dot{F}$ and $\dot{G}$ are the same as f, g, $\dot{f}$ and $\dot{g}$ except t is replaced by $-t$ and x is replaced by $-x$ and $\mathbf{r}$ and $\mathbf{r_0}$ are interchanged. Thus,

$$F = 1 - \frac{x^2}{r} C(z)$$

$$G = -\left[\left(t - t_0\right) - \frac{x^3}{\sqrt{\mu}} S(z)\right] \tag{9-61}$$

$$\dot{F} = \frac{\sqrt{\mu}x}{rr_0}(1 - zS(z)) \tag{9-62}$$

$$\dot{G} = 1 - \frac{x^2}{r_0} C(z)$$

Note that

$$z = \frac{x^2}{a} = \alpha x^2 \tag{9-63}$$

where

$$\alpha = \frac{1}{a} = \frac{2}{r} - \frac{v^2}{\mu} \tag{9-64}$$

α is used to allow treatment of parabolic orbits. In fact, in the calculation of the derivatives of Equation (9-60), $\frac{d(\alpha \mathbf{r}_0)}{dt}$ and $\frac{d(\alpha \mathbf{v}_0)}{dt}$ will be computed. In the differentiation we consider $\mathbf{r}$ to be constant and the acceleration ($\dot{\mathbf{v}}$) to result only from the perturbing forces and do not consider changes resulting from the two-body reference motion. In this development we see from Equation (9-64) that

$$\frac{d\alpha}{dt} = -\frac{2}{\mu} \mathbf{v} \bullet \dot{\mathbf{v}} \tag{9-65}$$

Differentiating Equations (9-60) gives

$$\begin{aligned} \frac{d\left(\alpha \mathbf{r}_0\right)}{dt} &= \frac{d\left(\alpha F\right)}{dt}\mathbf{r} + \frac{d\left(\alpha G\right)}{dt}\mathbf{v} + \alpha G\dot{\mathbf{v}} \\ \frac{d\left(\alpha \mathbf{v}_0\right)}{dt} &= \frac{d\left(\alpha \dot{F}\right)}{dt}\mathbf{r} + \frac{d\left(\alpha \dot{G}\right)}{dt}\mathbf{v} + \alpha G\dot{\mathbf{v}} \end{aligned} \tag{9-66}$$

The derivatives in Equation (9-66) can be found from Equations (9-62). After algebraic reduction, we have

$$\frac{d(\alpha F)}{dt} = \frac{1}{\mu}\left(\frac{x r_0 \dot{F}}{\sqrt{\mu}} - 2\right)\mathbf{v}\cdot\dot{\mathbf{v}} - \frac{r_0 \dot{F}}{\sqrt{\mu}}\left(\alpha \frac{dx}{dt}\right) \tag{9-67}$$

$$\frac{d(\alpha G)}{dt} = \frac{-1}{\mu}\left[\frac{x r(1-F)}{\sqrt{\mu}} - \left(3t - 3t_0 + G\right)\right]\mathbf{v}\cdot\dot{\mathbf{v}} + \frac{r(1-F)}{\sqrt{\mu}}\left(\alpha \frac{dx}{dt}\right)$$

$$\frac{d(\alpha \dot{F})}{dt} = \frac{1}{r r_0 \sqrt{\mu}}\left[\alpha\sqrt{\mu}\,\left(G + t - t_0\right) + \alpha \times r\,(1-F) - 2x\right]\mathbf{v}\cdot\dot{\mathbf{v}}$$

$$+ \frac{\sqrt{\mu}}{r r_0}\,\left[1 - \alpha r\,(1-F)\right]\,\alpha \frac{dx}{dt} - \frac{\dot{F}}{r_0}\left(\alpha \frac{dr_0}{dt}\right) \tag{9-68}$$

$$\frac{d(\alpha \dot{G})}{dt} = \frac{1}{\mu}\left(\frac{x r \dot{F}}{\sqrt{\mu}} - 2\right)\mathbf{v}\cdot\dot{\mathbf{v}} - \frac{r\dot{F}}{\sqrt{\mu}}\left(\alpha \frac{dx}{dt}\right) + \frac{r(1-F)}{r_0^2}\left(\alpha \frac{dr_0}{dt}\right)$$

It is still necessary to find $\frac{dx}{dt}$ and $\frac{dr_0}{dt}$ by interchanging the roles of $\mathbf{r_0}$, $\mathbf{v_0}$ and $\mathbf{r}$, $\mathbf{v}$ as before, Equation (4-39), which is Kepler's Equation, becomes

$$\sqrt{\mu}\,\left(t - t_0\right) = \left(1 - r\alpha\right)\,x^3 S\,(z) + rx - \frac{\mathbf{r}\cdot\mathbf{v}}{\sqrt{\mu}}\,x^2 C(z) \tag{9-69}$$

Also Equation (4-40) becomes

$$r_0 = x^2\,C(z) - \frac{\mathbf{r}\cdot\mathbf{v}}{\sqrt{\mu}}\,x\,(1 - zS(z)) + r\,(1 - zC\,(z)) \tag{9-70}$$

Differentiating Equation (9-70), we obtain

$$\alpha \frac{dr_0}{dt} = \frac{1}{\mu}\,[2r\,(1-F) + \alpha x\sqrt{\mu}\,\left(t - t_0\right) + \alpha \mathbf{r}\cdot\mathbf{v}\left(G + t - t_0\right)$$

$$- \alpha x^2\,r]\,\mathbf{v}\cdot\mathbf{v} - \frac{\alpha r r_0 \dot{F}}{\mu}\mathbf{r}\cdot\dot{\mathbf{v}} + \left(x - \alpha\sqrt{\mu}\left(t - t_0\right) - \frac{\mathbf{r}\cdot\mathbf{v}}{\sqrt{\mu}}\right)\left(\alpha \frac{dx}{dt}\right) \tag{9-71}$$

Differentiating Equation (9-69) then gives

$$\alpha \frac{dx}{dt} = \frac{1}{\mu r_0}\,\left[x\left(r + r_0\right) - 2\sqrt{\mu}\left(t - t_0\right)\right.$$

$$\left. - \sqrt{\mu}(1 + r\alpha)\left(G + t - t_0\right)\right]\,\mathbf{v}\cdot\mathbf{v} + \frac{\alpha r(1-F)}{r_0\sqrt{\mu}}\mathbf{r}\cdot\mathbf{v} \tag{9-72}$$

Note that $\dot{\mathbf{v}}$ includes only the perturbative accelerations. Now the perturbed position and velocity can be calculated by numerical integration of Equations (9-66). The corresponding $\mathbf{x}$ is determined from Kepler's Equation (4-39), and then position and velocity are obtained from Equations (4-45) and (4-46). Note that Equation (9-65) also must be integrated to obtain α.

9.4.3 Example to Introduce General Perturbations

The variation of parameters method of the previous section is classified as a special perturbation technique only because the final integration is accomplished numerically. If these equations could be integrated analytically, the technique would be that of general perturbations. To enable an analytic treatment, the perturbation forces can be expressed in a power series and integrated analytically. Though this sounds simple enough, in practice the integration may be quite difficult. Also, a series solution method of integration is frequently used. In special perturbations, the result is an answer to one specific problem or set of initial conditions. In general perturbations, the result will cover many cases and will give a great deal more information on the perturbed orbit. This is especially true for long-duration calculations. To give the reader a beginning understanding of general perturbations, a very simple example will be treated here that uses the variation of parameters method and a series expansion of the perturbing force.

Consider the equation

$$\dot{x} - \alpha = \frac{\nu}{(1 + \mu x)^2} \tag{9-73}$$

where α is a constant and μ and ν are small numbers. Consider the term on the right to be a perturbation. In the absence of the perturbation, the solution would be

$$x = \alpha t + c \tag{9-74}$$

where c will be the parameter. The constant of integration, c, is analogous to the orbital elements in orbital mechanics. Since μ and ν are small, the solution will vary only slightly from Equation (9-74), so the variation of c to satisfy Equation (9-73) should be small. From Equation (9-74), allowing c to vary with time, we have

$$\dot{x} = \alpha + \dot{c}$$

Substituting $\dot{x}$ and x into Equation (9-73) and solving for $\dot{c}$, we find

$$\dot{c} = \frac{\nu}{[1 + \mu(\alpha t + c)]^2} \tag{9-75}$$

The right-hand side of Equation (9-75) is the perturbation and can be expanded as

$$\dot{c} = \nu\left[1 - 2\mu(\alpha t + c) + 3\mu^2(\alpha t + c)^2 - ...\right] \tag{9-76}$$

The lowest order approximation would be $\dot{c} = \nu$, but more practically we would consider

$$\dot{c} = \nu - 2\mu\nu(\alpha t + c)$$

$$\dot{c} + 2\mu\nu c - \nu = -2\mu\nu\alpha t \tag{9-77}$$

which can be easily solved for **c**:

$$c = \nu e^{-2\mu\nu t} - \alpha t + \frac{\alpha + \nu}{2\mu\nu} - \frac{\alpha}{2\mu\nu} e^{-2\mu\nu t} \quad \text{for } c(0) = \nu + \frac{1}{2\mu}$$

Then **x** will be given by Equation (9-74) correct to first order in μ. Though this example has no physical significance, it clearly demonstrates how the perturbation would be treated analytically using a series expansion, which is a common method of expression of the perturbation.

For the reader who wishes to pursue the general perturbation analysis, Brouwer and Clemence,[4] Moulton[5] and many papers discuss it in detail. Many other approaches could be treated in this chapter, such as the disturbing function, Lagrange's planetary variables or derivations from Hamiltonian mechanics, but to pursue the topic further here would be beyond the scope of this text.

9.5 COMMENTS ON INTEGRATION SCHEMES AND ERRORS

The topic of numerical integration in the context of special perturbation is both relevant and necessary. No matter how elegant the analytic foundations of a particular special perturbation method, the results after numerical integration can be meaningless or, at best, highly questionable, if the integration scheme is not accurate to a sufficient degree. Many results fall into this questionable category because the investigator has not understood the limitations of, or has not carefully selected, the integration method. Unfortunately, such understanding requires a combination of knowledge of numerical analysis and experience. The purpose of this section is not to analytically develop the various integration schemes but is to summarize several methods and help the reader understand what makes a particular integration scheme appropriate for the specific problem at hand.

9.5.1 Criteria for Choosing a Numerical Integration Method

The selection of an adequate integration scheme for a particular problem is too often limited to what is readily available for use rather than what is best. Computational efficiency with sufficient accuracy and algorithm robustness are the most important criteria. The problem to be solved should first be analyzed. The following questions may help to narrow the selection possibilities:

a. Is a wide range in the independent variables required (thus a large number of integration steps)?

b. Is the problem extremely sensitive to small errors?

c. Are the dependent variables changing rapidly (i.e., will a constant step size be satisfactory)?

Factors to be considered for any given method are speed, accuracy, storage and complexity. Desirable qualities may be in opposition to each other such that they all cannot be realized to the fullest extent. Some of the desirable qualities of a good integration scheme are the following

a. It should allow as large a step size as possible.

b. It should employ a variable step-size provision that is simple and fast.

c. It should be economical in terms of computing time.

d. The method should be stable such that errors do not exhibit exponential growth.

e. It should be as insensitive as possible to round-off errors.

f. It should have a maximum and minimum control on truncation error.

Obviously, all of these cannot be perfectly satisfied in any one method, but they are qualities to consider when evaluating an integration method.

9.5.2 Integration Errors

Before evaluating any particular method of numerical integration it is important to have some understanding of the kind of errors involved.

In any numerical integration method there are two primary kinds of errors that will always be present to some degree: round-off and truncation errors. Roundoff errors result from the fact that a computer can carry only a finite number of digits of any number. For instance, suppose that the machine can carry six digits and we wish to add $4.476276 + 3.388567 = 7.864843$. In the machine the numbers would be rounded off to six significant digits,

$$4.47628 + 3.38857 = 7.86485$$

The true answer rounded off would have a 4 in the sixth digit, but the rounding error has caused it to be a 5. It is clear that the occurrence of this round-off many times in the integration process can result in significant errors. Brouwer and Clemence[4] (p. 158) give a formula for the *probable* error after n steps as $0.1124n^{3/2}$ (in units of the last decimal) or $\log(0.1124n^{3/2})$ in number of decimal places. Baker (vol. 2, p. 270) gives the example of the error in 200 integration steps as

$$\log\left[0.1129(200)^{3/2}\right] \approx 2.5 \qquad (9\text{-}78)$$

If six places of accuracy are required then $6 + 2.5 \approx 9$ places must be carried in the calculations. The lesson here is that the fewer integration steps taken, the smaller the accumulated round-off error. The best inhibitor to the significant accumulation of round-off error is the use of double-precision arithmetic.

Truncation error is a result of an inexact solution of the differential equation. In fact, the numerical integration is actually an exact solution of the *difference* equation, which imperfectly represents the actual differential equation. Truncation error results from not using all of the series expression employed in the integration method. The larger the step size, the larger the truncation error, so the ideal here is to have small step sizes. Note that this is in opposition to the cure of round-off errors. Thus, errors are unavoidable in numerical integration and the object is to use a method that minimizes the sum of round-off and truncation errors. Round-off error is primarily a function of the machine (except for some resulting from poor programming technique) and truncation error is a function of the integration method.

9.6 NUMERICAL INTEGRATION METHODS

Numerical integration methods can be divided into single-step and multi-step categories. Multistep methods usually require a single-step method to start them at the beginning and after each step size change. Some representative single-step methods are Runge–Kutta, Gill, Euler–Cauchy and Bowie. Some of the multistep methods are Milne, Adams–Moulton, Gauss–Jackson (sum squared), Obrechkoff and Adams–Bashforth. Multi-step methods are often referred to as predictor–corrector methods. A few of the more common methods will be discussed here. In general, we will consider the integration of systems of first-order differential equations, though there are methods available to integrate some special second-order systems. Of course, any second-order system can be written as a system of first-order equations.

We will treat the first-order equation

$$\frac{dx}{dt} = f(t, x) \tag{9-79}$$

with initial conditions $x(t_0) = x_0$, where x and f could be vectors.

9.6.1 Runge–Kutta Method

There is actually a family of Runge–Kutta methods of varying orders. The technique approximates a Taylor series extrapolation of a function by several evaluations of the first derivatives at points within the interval of extrapolation. The order of a particular member of the family is the order of the highest power of the step size, h, in the equivalent Taylor series expansion. The formulas for the standard fourth-order method are

$$x_{n+1} = x_n + \frac{1}{6}(k_1 + 2k_2 + 2k_3 + k_4) \tag{9-80}$$

where

$$\begin{aligned} k_1 &= hf(t_n, x_n) \\ k_2 &= hf\left(t_n + \frac{h}{2}, x_n + \frac{k_1}{2}\right) \\ k_3 &= hf\left(t_n + \frac{h}{2}, x_n + \frac{k_2}{2}\right) \\ k_4 &= hf(t_n + h, x_n + k_3) \end{aligned} \tag{9-81}$$

and n is the increment number.

Another member of the Runge–Kutta family is the Runge–Kutta–Gill formula. It was developed for high-speed computers to use a minimum number of storage registers and to help control the growth of round-off errors. This formula is

$$x_{n+1} = x_n + \frac{1}{6}k_1 + \frac{1}{3}(1 - \sqrt{1/2})\, k_2 + \frac{1}{3}(1 + \sqrt{1/2})\, k_3 + \frac{1}{6}k_4 \quad (9\text{-}82)$$

where $k_1 = hf\,(t_n, x_n)$

$$\begin{aligned} k_2 &= hf\left(t_n + \frac{h}{2}, x_n + \frac{k_1}{2}\right) \\ k_3 &= hf\left(t_n + \frac{h}{2}, x_n + k_1\left[\sqrt{\frac{1}{2}} - \frac{1}{2}\right] + k_2\left[1 - \sqrt{\frac{1}{2}}\right]\right) \\ k_4 &= hf\left(t_n + h, x_n - \frac{k_2}{\sqrt{2}} + k_3\left[1 + \sqrt{\frac{1}{2}}\right]\right) \end{aligned} \quad (9\text{-}83)$$

The Runge–Kutta methods are stable and they do not require a starting procedure. They are relatively simple, easy to implement and have a relatively small truncation error, and the step size is easily changed. One of the disadvantages is that there is no simple way to determine the truncation error, so it is difficult to determine the proper step size. One obvious failure is that use is made only of the last calculated step. This idea leads to multistep or predictor–corrector methods.

9.6.2 Adams–Bashforth Multistep Method

Multistep methods take advantage of the history of the function being integrated. In general, they are faster than single-step methods, though at a cost of greater complexity and a requirement for a starting procedure at the beginning and after each step size change. Some multistep methods calculate a predicted value for $x_n + 1$ and then substitute $x_n + 1$ in the differential equation to get $\dot{x}_n + 1$, which is in turn used to calculate a corrected value of $x_n + 1$. These are naturally called "predictor–corrector" methods.

The Adams (sometimes called Adams–Bashforth) formula is

$$x_n + 1 = x_n + h \sum_{k=0}^{N} \alpha_k \nabla^k f_n \quad (9\text{-}84)$$

where N = the number of terms desired,
h = step size,
n = step number,

α = 1, 1/2, 5/12, 3/8, 251/720, 95/288,

k = 0, . . ., 5,

$\nabla^k f_n$ = backward difference operator

$= \nabla^{k-1} f_n - \nabla^{k-1} f_{n-1}$ and $\nabla^0 f_n = f_n$, and

$f_n = f(t_n, x_n)$.

The first few terms are

$$x_{n+1} = x_n + h\left(1 + \frac{1}{2}\nabla + \frac{5}{12}\nabla^2 + \frac{3}{8}\nabla^3 + \frac{251}{720}\nabla^4 + \frac{95}{288}\nabla^5\right)f_n \quad (9\text{-}85)$$

A method is available for halving or doubling the step size without a complete restart. Also, the local truncation error can be calculated (see references for detailed formulas).[3]

9.6.3 Adams–Moulton Formulas

The Adams–Bashforth method is only a multistep predictor scheme. In 1926 Moulton added a corrector formula to the method. The fourth-order formulas retaining third differences are given here. The predictor is

$$x_{n+1}^{(P)} = x_n + \frac{h}{24}(55f_n - 59f_{n-1} + 37f_{n-2} - 9f_{n-3}) + \frac{251}{720}h^5\frac{d^5 x(\xi)}{dt^5} \quad (9\text{-}86)$$

and the corrector is

$$x_{n+1}^{(C)} = x_n + \frac{h}{24}(9f_{n+1} + 19f_n - 5f_{n-1} + f_{n-2}) + \frac{19}{720}h^5\frac{d^5 x(\xi)}{dt^5} \quad (9\text{-}87)$$

where the last term in each of the equations is the truncation error term. In Equation (9-87), f_{n+1} is found from the predicted value:

$$f_n + 1 = f(t_{n+1} + 1, x_{n+1}^{(P)}) \quad (9\text{-}88)$$

In using the formulas, the error terms are not generally known, so the value used is the expression without those terms. For a discussion of the error terms see, for instance, Ralston's text on Numerical Analysis.[6] An estimate of the truncation error for the step from t_n to t_{n+1} is

$$\frac{19}{270}\left|x_{n+1}^{(C)} - x_{n+1}^{(P)}\right| \quad (9\text{-}89)$$

To use any predictor–corrector scheme the predicted value is computed first. Then the derivative corresponding to the predicted value is found and used to find the corrected value using the corrector formula. It is possible to iterate on the corrector formula until there is no significant change in x_{n+1} on successive iterations. The step size can be changed if the truncation error is larger than desired. This method, like other multistep methods, requires a single-step method to start to obtain f_n, f_{n-1}, etc. The Runge–Kutta method is suggested for a starter. The Adams–Moulton formulation is one of the more commonly used integration methods.

9.6.4 The Gauss–Jackson or Sum Squared (Σ^2) Method

The Gauss–Jackson method is one of the best, and most used, numerical integration methods for trajectory problems of the Cowell and Encke type. It is designed for the integration of systems of second-order equations and is faster than integrating two first-order equations. Its predictor alone is generally more accurate than the predictor and corrector of other methods, though it also includes a corrector. It exhibits especially good control of the accumulated round-off error effect. It is, though, more complex and difficult for the beginner to implement than the other methods. The general formula for solving the equation $\ddot{x} = f(t, x, \dot{x})$ is

$$x_n = h^2\left(\sum^2 \ddot{x}_n + \frac{1}{12}\ddot{x}_n - \frac{1}{240}\delta^2\ddot{x}_n + \frac{31}{60{,}480}\delta^4\ddot{x}_n - \frac{289}{36{,}288}\delta^6\ddot{x}_n + \ldots\right) \quad (9\text{-}90)$$

where

$$\ddot{x}_n = f(t_n, x_n, \dot{x}_n) = f(t_n) = f_n$$

and the central differences are defined as

$$\delta^0 f(t_n) = f(t_n)$$

$$\delta^1 f(t_n) = f\left(t_n + \frac{h}{2}\right) - f\left(t_n - \frac{h}{2}\right)$$

$$\delta^2 f(t_n) = f(t_n + h) + f(t_n - h) - 2f(t_n) = f_{n+1} + f_{n-1} - 2f_n$$

$$\delta^k f(t_n) = \delta^{k-1} f\left(t_n + \frac{h}{2}\right) - \delta^{k-1} f\left(t_n - \frac{h}{2}\right)$$

The definition of $\Sigma^2\ddot{x}_n$ can be found in several references (Baker, vol. 2, and NASA CR1005)[2,3] The procedure for calculating it will not be reiterated here since it becomes fairly involved and the reader can find it easily in the above-mentioned references.

9.6.5 Numerical Integration Summary

Experience has shown that the Gauss–Jackson method is clearly superior for orbital problems using the Cowell and Encke techniques. For normal integration of first-order equations such as occur in the variation of parameters technique, the Adams–Moulton or Adams (Adams–Bashforth) methods are preferred. The Runge–Kutta method is suggested for starting multistep methods. Round-off error is a function of the number of integration steps and is most effectively controlled by using double-precision arithmetic. Local truncation error can and should be calculated for multistep methods and should be used as a criterion for changing step size.

Table 9-1 gives a tabular comparison of a number of integration methods, including some that were not discussed in this chapter.

9.7 ANALYTIC FORMULATIONS OF PERTURBATIVE ACCELERATIONS

A few of the perturbation accelerations commonly used in practice will be presented in this section. Some of them become quite complex, but only the simpler forms will be treated. They will be stated with a minimal amount of discussion, but in sufficient detail such that they can be used in the methods of this chapter without extensive reference to more detailed works.

9.7.1 The Nonspherical Earth

The simplified gravitational potential of Earth, μ/r, is due to a spherically symmetric mass body and results in conic orbits. However, Earth is not a spherically symmetric body but is bulged at the equator, flattened at the poles and is generally asymmetric. If the potential function, ϕ, is known, the accelerations can be found from

$$\mathbf{a} = \nabla\phi = \frac{\delta\phi}{\delta x}\mathbf{I} + \frac{\delta\phi}{\delta y}\mathbf{J} + \frac{\delta\phi}{\delta z}\mathbf{K} \tag{9-91}$$

One such potential function, according to Vinti,[18] is

$$\phi = \frac{\mu}{r}\left[1 - \sum_{n=2}^{\infty} J_n \left(\frac{r_e}{r}\right)^n P_n \sin L\right] \tag{9-92}$$

where μ = gravitational parameter,
J_n = coefficients to be determined by experimental observation,
r_e = equatorial radius of Earth,
P_n = Legendre polynomials,
L = geocentric latitude, and
$\sin L = z/r$.

Table 9-1: Comparison of Integration Methods (from NASA SP-33, Vol. 1, Part 1).[7]

Method of Numerical Integration	Truncation Error	Ease of Changing Step Size	Speed	Stability	Round-off Error Accumulation
Single Step Methods					
Runge–Kutta	h^5	*	Slow	Stable	Satisfactory
Runge–Kutta–Gill	h^5	*	Slow	Stable	Satisfactory
Bowie	h^3	Trivial (step size varied by error control)	Fast	Stable	Satisfactory
Fourth-Order Multistep Predictor–Corrector					
Milne	h^5	Excellent	Very fast	Unstable	Poor
Adams–Moulton	h^5	Excellent	Very fast	Unconditionally stable	Satisfactory
Higher Order Multistep					
Adams Backward Difference	Arbitrary	Good	Very fast	Moderately stable	Satisfactory
Gauss–Jackson†	Arbitrary	Awkward and expensive	Fast	Stable	Excellent
Obrechkoff	h^7	Excellent	‡	Stable	Satisfactory
Special Second-Order Equation	$[\ddot{z} = f(t, z)]$				
Special Runge–Kutta	h^5	*	Slow	Stable	Satisfactory
Milne–Stormer	h^6	Excellent	Very fast	Moderately stable	Poor

*. R-K (single step) trivial to change steps, very difficult to determine proper size.

†. Gauss–Jackson is for second-order equations.

‡. Speed of Obrechkoff depends on complexity of the higher order derivatives required; it could be very fast.

The first seven terms of the expression are

$$\phi = \frac{\mu}{r}\left[1 - \frac{J_2}{2}\left(\frac{r_e}{r}\right)^2 \left(3 \sin^2 L - 1\right)\right.$$

$$- \frac{J_3}{2}\left(\frac{r_e}{r}\right)^3 \left(5 \sin^3 L - 3 \sin L\right)$$

$$- \frac{J_4}{8}\left(\frac{r_e}{r}\right)^4 \left(35 \sin^4 L - 30 \sin^2 L + 3\right)$$

$$- \frac{J_5}{8}\left(\frac{r_e}{r}\right)^5 \left(63 \sin^5 L - 70 \sin^3 L + 15 \sin L\right)$$

$$\left. - \frac{J_6}{16}\left(\frac{r_e}{r}\right)^6 \left(231 \sin^6 L - 315 \sin^4 L + 105 \sin^2 L - 5\right)\right] \quad (9\text{-}93)$$

Taking the partial derivative of ϕ gives

$$\ddot{x} = \frac{\delta\phi}{\delta x} = -\frac{\mu x}{r^3}\left[1 - J_2 \frac{3}{2}\left(\frac{r_e}{r}\right)^2 \left(5\frac{z^2}{r^2} - 1\right)\right.$$

$$+ J_3 \frac{5}{2}\left(\frac{r_e}{r}\right)^3 \left(3\frac{z}{r} - 7\frac{z^3}{r^3}\right) - J_4 \frac{5}{8}\left(\frac{r_e}{r}\right)^4 \left(3 - 42\frac{z^2}{r^2} + 63\frac{z^4}{r^4}\right)$$

$$- J_5 \frac{3}{8}\left(\frac{r_e}{r}\right)^5 \left(35\frac{z}{r} - 210\frac{z^3}{r^3} + 231\frac{z^5}{r^5}\right)$$

$$\left. + J_6 \frac{1}{16}\left(\frac{r_e}{r}\right)^6 \left(35 - 945\frac{z^2}{r^2} + 3465\frac{z^4}{r^4} - 3003\frac{z^6}{r^6}\right) + \ldots\right] \quad (9\text{-}94)$$

$$\ddot{y} = \frac{\delta\phi}{\delta y} = \frac{y}{x}\ddot{x} \quad (9\text{-}95)$$

$$\ddot{z}=\frac{\delta\phi}{\delta z}=-\frac{\mu z}{r^3}\left[1+J_2\frac{3}{2}\left(\frac{r_e}{r}\right)^2\left(3-5\frac{z^2}{r^2}\right)\right.$$
$$+J_3\frac{3}{2}\left(\frac{r_e}{r}\right)^3\left(10\frac{z}{r}-\frac{35}{3}\frac{z^3}{r^3}-\frac{r}{z}\right)$$
$$-J_4\frac{5}{8}\left(\frac{r_e}{r}\right)^4\left(15-70\frac{z^2}{r^2}+63\frac{z^4}{r^4}\right)$$
$$-J_5\frac{1}{8}\left(\frac{r_e}{r}\right)^5\left(315\frac{z}{r}-945\frac{z^3}{r^3}+693\frac{z^5}{r^5}-15\frac{r}{z}\right)$$
$$\left.+J_6\frac{1}{16}\left(\frac{r_e}{r}\right)^6\left(245-2205\frac{z^2}{r^2}+4851\frac{z^4}{r^4}-3003\frac{z^6}{r^6}\right)+\ldots\right] \quad (9\text{-}96)$$

Note that sin L is replaced by z/r. The first term in (x, y, z) is the two-body acceleration and the remaining terms are the perturbation accelerations resulting from Earth's nonsphericity. There have been various determinations of the J coefficients, which are slightly in variance, but a representative set of values is given here. The official Earth Gravitational Model EGM2008 has been publicly released by the U.S. National Geospatial-Intelligence Agency (NGA) EGM Development Team (www.earth-info.nga.ml) (also see Baker, vol. 1, p. 175).[8]

$$J_2 = (1082.63)\times 10^{-6}$$
$$J_3 = (-2.532)\times 10^{-6}$$
$$J_4 = (-1.62)\times 10^{-6}$$
$$J_5 = (-0.15 \pm 0.1)\times 10^{-6}$$
$$J_6 = (0.57 \pm 0.1)\times 10^{-6}$$
$$J_7 = (-0.44 \pm 0.1)\times 10^{-6}$$

It is obvious that the confidence factor diminishes beyond J_4. These equations include only the zonal harmonics—that is, those harmonics that are dependent only on that mass distribution, which is symmetric about the north–south axis of Earth (that is, they are not longitude dependent).

The even-numbered harmonics are symmetric about the equatorial plane and the odd-numbered harmonics are antisymmetric. There are also tesseral harmonics (dependent on *both* latitude and longitude) and sectorial harmonics (dependent on longitude *only*). A general expression to account for all three classes of harmonics in the potential function is given by EGM2008 - WGS 84 Version[17] but will not be presented here.

There are a few cautions to be pointed out in the use of the above formulation. The equations are formulated in the geocentric, equatorial coordinate system. Therefore, care must be taken to know Earth's current relationship to this inertial frame owing to rotation. The zonal harmonics are not longitude dependent, so the direction of the **X**-axis need not point to a fixed geographic point. However, if tesseral and sectorial harmonics were considered, then the frame would need to be fixed to the rotation of Earth. To use the acceleration equation in the perturbation methods it is necessary to transform the perturbation portion (not the $-\mu/r^3$ term) to the **RSW** system. To do this, simply use the **IJK** to **PQW** transformation of Chapter 2—substituting u (argument of latitude) for ω. Expressions for the Moon's triaxial ellipsoid potential can be found in Baker (vol. 1).[8]

9.7.2 Atmospheric Drag

The formulation of atmospheric drag equations is plagued with uncertainties of atmospheric fluctuations, frontal areas of the orbiting object (if not constant), the drag coefficient and other parameters. A fairly simple formulation will be given here (see NASA SP-33).[7] Drag, by definition, will be opposite to the velocity of the vehicle relative to the atmosphere. Thus, the perturbative acceleration is

$$\ddot{\mathbf{r}} = -\frac{1}{2} C_D \frac{A}{m} \rho v_a \dot{\mathbf{r}}_a \tag{9-97}$$

where C_D = the dimensionless drag coefficient associated with A,

A = the cross-sectional area of the vehicle perpendicular to the direction of motion,

m = vehicle mass,

$$\frac{C_D A}{m} = \frac{1}{\text{ballistic coefficient}},$$

ρ = atmospheric density at the vehicle's altitude,

$v_a = |\dot{r}_a|$ = speed of vehicle relative to the rotating atmosphere,

$$\dot{\mathbf{r}}_a = \begin{bmatrix} \dot{x} + \dot{\theta} y \\ \dot{y} - \dot{\theta} x \\ \dot{z} \end{bmatrix}$$

$$\dot{r} = \begin{bmatrix} \dot{x} \\ \dot{y} \\ \dot{z} \end{bmatrix}, \text{ the inertial velocity}$$

and

$\dot{\theta}$ = rate of rotation of the Earth

Once again **x**, **y**, **z** refer to the geocentric, equatorial coordinate system.

This formulation could be greatly complicated by the addition of an expression for theoretical density, altitude above an oblate Earth, etc. but the reader can find this in other documents. Equation (9-97) can give the reader a basic understanding of drag effects. Equations for lifting forces will not be presented here.

9.7.3 Sunlight Radiation Pressure

Radiation pressure, though small, can have considerable effect on large area/mass ratio satellites (such as Echo). In the geocentric, equatorial coordinate system the perturbative accelerations are

$$\begin{aligned} \ddot{x} &= f \cos A_{\odot} \\ \ddot{y} &= f \cos i_{\oplus} \sin A_{\odot} \\ \ddot{z} &= f \sin i_{\oplus} \sin A_{\odot} \end{aligned} \tag{9-98}$$

where f $= -4.5 \times 10^{-5}$ (A/m) cm/s^2 (multiply by 1 for zero reflection and 2 for full reflection),

A = cross section of vehicle exposed to the Sun (cm^2),

m = mass of vehicle (grams),

$A_{\odot}$ = mean right ascension of the Sun during computation, and

$i_{\oplus}$ = inclination of equator to ecliptic = 23.4349°.

9.7.4 Thrust

Thrust,**T**, can be handled quite directly by resolving the thrust vector into **x**, **y**, and **z** directions. Thus the perturbing acceleration is

$$\ddot{x} = \frac{T_x}{m}, \ddot{y} = \frac{T_y}{m}, \ddot{z} = \frac{T_z}{m} \tag{9-99}$$

Low-thrust problems can be treated using a perturbation approach such as the Encke method or variation of parameters, but high thrust should be treated using the Cowell technique since the thrust is no longer a small perturbation, but a major force.

Exercises

9.1 Verify the development of Equation (9-34), the variation of eccentricity.

9.2 In the variation of parameters using the universal variable it is asserted that one can develop f and g expressions to find $\mathbf{r}_0$ and $\mathbf{v}_0$ in terms of **r** and **v** by t, and x by (–t) and (–x). Prove this assertion.

9.3 Show how the potential function for the nonspherical Earth would be used in conjunction with Cowell's method. Write out the specific equations that would be used in a form suitable for programming.

9.4 Develop the equations of motion in spherical coordinates. See Equation (9-5).

9.5 Describe the process of rectification as used in the Encke method and variation of parameters method.

9.6 Develop a computer flow diagram for the Encke method including rectification of the reference orbit.

9.7 *Program the Cowell method including the perturbation resulting from the Moon.

9.8 *Program the Encke method for the above problem and compare results for (a) a near-Earth satellite of 200 km altitude and (b) a flight with an apogee of approximately 280,000 km.

9.9 *Verify at least one of the equations in Equations (9-67) and (9-68).

List of References

1. *Planetary Coordinates for the Years 1960–1980*. Her Majesty's Stationery Office, London.

2. Baker, Robert M. L. Jr. *Astrodynamics: Applications and Advanced Topics*. New York, NY, Academic Press, 1967.

3. Allione, M. S., Blackford, A. L., Mendez, J. C. and Whittouck, M. M. *Guidance, Flight Mechanics and Trajectory Optimization*. Vol. VI, "The N-Body Problem and Special Perturbation Techniques." NASA CR-1005, National Aeronautics and Space Administration, Washington, DC, Feb 1968.

4. Brouwer, D. and Clemence, G. M. *Methods of Celestial Mechanics*. New York, NY, Academic Press, 1961.

5. Moulton, F. R. *An Introduction to Celestial Mechanics*. New York, NY, Macmillan, 1914.

6. Ralston, A. and Wilf. *Mathematical Methods for Digital Computers*. New York, NY, John Wiley and Sons, Inc. 1960.

7. Townsend, G. E., Jr. *Orbital flight Handbook*. Vol. 1, Part 1. NASA SP-33, National Aeronautics and Space Administration, Washington, DC, 1963.

8. Baker, Robert M. L., Jr. and Makemson, Maud W. *An Introduction to Astrodynamics*. 2nd ed. New York, NY, Academic Press, 1967.

9. Battin, R. H. *Astronautical Guidance*. New York, NY, McGraw-Hill, 1964.

10. Herget, Paul. *The Computation of Orbits*. Published privately by the author, Ann Arbor, Michigan, 1948.

11. Plumber, M. C. *Dynamical Astronomy*. New York, NY, Cambridge University Press, 1918.

12. Smart, W. M. *Celestial Mechanics*. New York, NY, Longmans, 1953.

13. Hildebrand, F. B. *Introduction to Numerical Analysis*. New York, NY, McGraw-Hill, 1956.

14. Dubyago, A. D. *The Determination of Orbits*. New York, NY, Macmillan, 1961.

15. Escobal, P. R. *Methods of Orbit Determination*. New York, NY, John Wiley and Sons, Inc, 1965.

16. Escobal, P. R. *Methods of Astrodynamics*. New York, NY, John Wiley and Sons, Inc. 1968.

17. http://www.earth-info.nga.mil/GandG/wgs84/gravitymod/egm2008/egm08_wgs84.html, June 13, 2016.

18. J. P. Vinti. *Orbital and Celestial Mechanics*. Vol. 177, Reston, Virginia, American Institute of Aeronautics and Astronautics, Inc., 1998.

Chapter 10

Special Topics

"Toto, I've a feeling we're not in Kansas anymore."

-Dorothy in *The Wizard of Oz*

10.1 Historical Background

This chapter provides an overview of two topics that are of increasing importance to the space community. The first topic is the widespread use of general perturbation analytical models that are the primary means of maintaining the space object catalogs and the widely distributed two-line element (TLE) sets (or elset) that describe the space objects' orbit at a specific epoch using a specific definition of the orbital parameters. The second topic is the relative motion of satellites, driven by the increased interest in rendezvous and proximity operations (RPO). Both of those topics are presented and discussed in a manner designed to provide the reader with an overall understanding of the subject and to motivate further study of these two important applications of the fundamentals of astrodynamics.

10.2 General Perturbation Models

The previous chapter was a first step toward reality from the simplified theoretical foundation laid earlier in this text and developed several techniques useful for special perturbation models and propagating orbits using those models. The problem of developing general perturbation models was introduced in Section 9.4.3 and it was pointed out that the variation of parameters method used to formulate the equations would be general perturbations if they could be integrated analytically.

The purpose of this section is to provide a brief overview of the development of the most widely used general perturbation models and techniques, their limitations, and cautions about how they are used. Vallado[1] provides a detailed explanation of the development of general perturbation theory for astrodynamics. The fundamental problems of predicting the location of space objects at a particular point in time relate to how well the space objects' state vector are known, how accurately the perturbations to the two-body Kepler equations are known, how accurately the equations model the physical reality of forces such as gravity fields, drag forces, and variations in atmospheric density, and how accurately such mathematical expressions can be integrated (forward or backward) to maintain an acceptable degree of accuracy. With unlimited resources of computing time and numbers of observations, the prediction precision of any given space object can be made as small as the equations and observations allow for any point in time such that measurement errors and stochastic components of the perturbations become the limiting factors.

The dawn of the space age immediately brought about the problems of observing and cataloging an ever-increasing number of objects. Hoots[2] gives an excellent summary of the development of the analytical models and how their use evolved as computational capacity, observation fidelity and frequency and number of space objects increased. The U.S. Air Force and U.S. Navy each needed practical ways to track objects to meet their mission requirements. Each organization created a complete catalog of detectable space objects, with satellite tracking data forwarded continually to a central processing facility and updated orbital data distributed routinely to defense users.

Cataloged orbits have been represented by some type of *mean* orbital elements based on general perturbation techniques because of the calculational burden of using special perturbation techniques. Project SPACETRACK supported the development of original general perturbation models. In 1959, Brouwer developed a solution[4] for the motion of a near-Earth satellite under the influence of the zonal harmonics J_2, J_3, J_4 and J_5. This work was later published by Brouwer and Clemence.[5] Kozai[6] concurrently published another solution to the same problem. Both authors had similar ideas but they used different methods.

Hoots[2] points out that neither approach included a drag term for the satellites in their formulations. The concurrent atmospheric drag modeling efforts were extensive but the computational complexity precluded their use in the general perturbation models. Subsequent atmospheric models derived a density representation using power functions with integral exponents[7] that became potentially useful in general perturbation models. The importance of that work is that, when applied to an artificial satellite theory, it completely avoids series expansions that occur with exponential representations. This made possible the inclusion of drag in the Brouwer model in a more complete and compact manner. Most analytic orbit prediction models in the U.S. Space Surveillance Network today still have one of these two (Brouwer or Kozai) methods as their foundation.

A comprehensive model was developed by Lane[8] in 1965 with further improvements by Lane and Cranford[9] in 1969. A very important contribution to analytic satellite theory was made by Lyddane[10] in 1963. Lyddane showed that the Brouwer[3,4] solution based on Delaunay variables could be reformulated in terms of Poincaré variables to avoid the small divisors of eccentricity and the sine of inclination while maintaining the first-order character of the theory.

Several different analytic models, using general perturbations, have evolved and there are several methods in use for predicting space object position from a space catalog. The Navy extended the early work of Brouwer and Lyddane and developed what is known as the Position and Partials as functions of Time (PPT3) model. The Air Force demands of large numbers of objects in the catalog led to the development of the Simplified General Perturbation (SGP) model (Hilton and Kuhlman[11]). Computational limitations caused a different version of SGP to be developed that simplified the modeling of drag and higher order gravitational terms.

10.3 NORAD PROPAGATORS AND TWO-LINE ELEMENT SETS

The original NORAD orbit propagators are popular, relatively accurate and easy to use. The TLE sets required by these propagators are widely distributed on the Internet and maintained by AFSCN. Figure (10-1) shows the definition of the TLE format in the original 72-column "punch card" format.

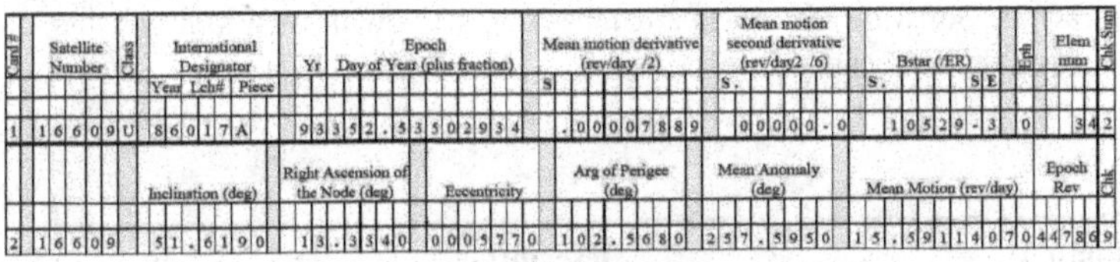

Fig. 10-1 Example Two-Line Element Set.

It is imperative to understand the relationship between the TLE sets and the general perturbation models. The first SGP model dates to 1964 and the first version specific to low-Earth orbit (SGP4) was developed in 1970. Other versions of SGP4 specific to deep space (SDP4) were also developed. The various versions of the models have been merged and SGP4 now refers to the AFSCN simplified general perturbations model. Observations are the basis for performing the orbit determination problem. The orbit determination method of solution uses the SGP4 perturbation model to generate TLE values. Those TLE values are distributed and any user can perform orbit predictions—*as long as the exact model is used to generate the TLEs*. Hoots[2] includes the equations for both the PPT3 and SGP4 models.

A large number of users had standardized on the originally adopted SGP method, so when the SGP4 update occurred many users did not want to update and requested that the TLEs support their continued use of SGP. To meet this need the TLEs have the SGP drag term (in rev/day) as well as the SGP4 drag term (in $1/r_{\oplus}$). The mean motion is the SGP Kozai compatible mean motion, to which SGP4 users must apply a correction prior to use, while SGP users can ingest it directly. This mode is indicated on the TLE first line by an ephemeris type of 0. It is possible, although very rare, to get a TLE with ephemeris type of 2, which indicates a "pure" SGP4 elset with Brouwer mean motion.

10.3.1 Two-Line Element Set Definition

The following is a typical TLE for the NOAA 14 spacecraft:

```
NOAA 14
1 23455U 94089A 16154.49791668-.00000011 00000-0 16871-4 0 9996
2 23455 98.7223 233.1912 0009980 153.3216 206.8472 14.14077396105175
```

The *mean* orbital elements contained in this data are Earth-centered-inertial (ECI) coordinates with respect to the true equator of date and the mean equinox of date. They do not include the effect of nutation. The following is a brief description of the data contained in each line of a TLE. Each item must appear in its column field in exactly the format specified. It is important to carefully note the units of each of the TLE variables.

Line 0

The two lines of TLE data are preceded by the name of the satellite of up to 24 characters. Software that reads a database of TLEs will often look for this name to find the correct data.

Line 1

Column	Description
01	line number of element data
03–07	satellite number
08	classification (u = unclassified)
10, 11	international designator (last two digits of launch year)
12–14	international designator (launch number of the year)
15–17	international designator (piece of the launch)
19, 20	epoch year (last two digits of year)
21–32	epoch (day of the year and fractional portion of the day)
34–43	first time derivative of the mean motion
45–52	second time derivative of mean motion (decimal point assumed)
54–61	bstar drag term (decimal point assumed)
63	ephemeris type
65–68	element number
69	checksum (modulo 10) (letters, blanks, periods, plus signs = 0; minus signs = 1)

Line 2

Column	Description
01	line number of element data
03–07	satellite number
09–16	orbital inclination (degrees)
18–25	right ascension of the ascending node (degrees)
27–33	orbital eccentricity (decimal point assumed)
35–42	argument of perigee (degrees)
44–51	mean anomaly (degrees)
53–63	mean motion (orbits per day)
64–68	revolution number at epoch (orbits)
69	checksum (modulo 10)
	All other columns are blank or fixed.

As of the time of the writing of this second edition, there are proposals being evaluated to redefine the TLE to a new three-line element (3LE) designation. Appendix D is the latest version and may be adopted as the standard.

10.3.2 Simplified General Perturbation (SGP4)

The original documentation for different SGP-related models is in Hoots,[12] published in 1980. At that time there were five separate models that represented space objects with different periods and different gravitational effects. Through the next two decades there were improvements to the algorithms and the deep-space and low-Earth orbit versions were merged. During this time there were various versions of SGP4 that were developed and distributed without consistent documentation and control. Hoots' 2004 paper documented the complete list of the latest (at that time) equations used in SGP4. In that summary of SGP4 equations it is clear that both Brouwer[4] and Kozai[6] should be reviewed to fully understand the nomenclature used. The SGP4 model is set in the Fundamental Katalog 5 (FK5) and the World Geodetic Survey 72 (WGS72) reference standards and refers to the Julian 2000 (J2000.0) epoch.

Vallado[13] addressed the problem of numerous code implementations of the SGP4 equations. Since changes to the original Spacetrack #3 code were not made publicly available, and the updated code that was used to generate the available TLEs was also not made publicly available, there has been uncertainty in the technical community about the correct usage of the available TLE data sets.

The TLE data sets used today are generally available from two sites for most users. Those two Internet locations are CelesTrak and AFSPC. A version of SGP4 software can be found at http://www.centerforspace.com/downloads/.

10.3.3 SGP4 Orbit Propagation Accuracy

There are many reasons for propagating the orbits of space objects and those different reasons have dramatically different requirements for absolute accuracy, timeliness, update frequency and forward projection of orbits for days or weeks. Common needs include the following:

a. rapidly searching a satellite's orbit to find visibility between the satellite and ground stations to determine times of visibility for communication antennas;
b. pointing communication antennas well enough in azimuth and elevation to keep the satellite in the antenna FOV during a communication pass;
c. performing satellite mission planning for days or weeks ahead by identifying approximate (within prediction errors) times for various payload functions;
d. using data from world-wide tracking locations to find specific satellites;
e. scanning multiple numbers of satellites for specific visibility conditions for specific sites;
f. performing continual scans of space object orbits to determine whether there are potential collisions; and
g. quickly comparing found space objects to the catalog to identify new objects such as missiles.

The approximately 22,000 objects in the space catalog each have a TLE based on a number of observations. The number of observations, and accuracy and priority of each observation, will vary and the frequency at which a space object's TLE is updated is not constant. However, each TLE in the catalog includes the epoch for that object and that is the time at which that particular TLE is most accurate, whether it is the time of the last measurement used to update the TLE, the center time of a batch of measurements used to update the TLE or the best fit to a series of measurements.

Owing to the limitations and approximations of analytic theory, the accuracy of predicting a space object's position when using a TLE, and the correct version of SGP4, is typically of the order of one kilometer at the time of epoch. For low-Earth orbit predictions the along-track error can grow by 10 to 20 kilometers per day while the cross-track error remains in the 10's of meters range (with approximately zero mean) for an extended time. The predominant source of the error is the approximation used in the modeling of the drag term.

The low accuracy of general perturbation techniques such as embodied in SGP4 is still very useful for many applications and the fact that TLE data sets are often the only information available on space objects at a particular epoch makes understanding their utility and limitations very important. The 2010 AIAA Astrodynamics Standard[14] is an excellent summary of recommended practices for constants, coordinate systems, force models, and propagation methods.

10.3.4 Converting a State Vector to a TLE

Analytical methods, such as SGP4, are based on *mean* variables because they describe the orbit in terms of secular variations and long-period terms. The osculating state vector measurements include short-term variations in the orbit and must be converted into mean quantities to be used. This section describes a numerical technique that can be used to estimate an SGP4-compatible TLE from *osculating* state vectors (position and velocity vectors). The computational steps of this algorithm are as follows:

(1) Compute the classical *osculating* orbital elements from the position $\mathbf{r}_{osc}$ and velocity $\mathbf{v}_{osc}$ (Section 2.4.2) at time t_0.

(2) Set the initial guess for the SGP4-compatible *mean* orbital elements (subscript sgp) to the osculating (subscript osc) values computed in step (1) as follows:

$n_{sgp} = n_{osc}$ = mean motion
$e_{sgp} = e_{osc}$ = orbital eccentricity
$i_{sgp} = i_{osc}$ = orbital inclination
$\omega_{sgp} = \omega_{osc}$ = argument of perigee
$\Omega_{sgp} = \Omega_{osc}$ = right ascension of the ascending node
$M_{sgp} = M_{osc}$ = mean anomaly

(3) The estimated orbital elements at epoch = t_0 create the first estimate of TLE_1. The SGP4 algorithm is used to convert TLE_1 back to osculating position and velocity vectors ($\mathbf{r}_{sgp}$ and $\mathbf{v}_{sgp}$) The SGP4 algorithm only does an initialization and calculation of the state vectors without propagating forward in time past t_0.

(4) The following set of vector nonlinear equations is then solved:

$$\begin{aligned} \mathbf{r}_{osc} - \mathbf{r}_{sgp} &= 0 \\ \mathbf{v}_{osc} - \mathbf{v}_{sgp} &= 0 \end{aligned} \qquad (10\text{-}1)$$

In each iteration in Equation (10-1), the values of $\mathbf{r}_{sgp}$ and $\mathbf{v}_{sgp}$ are updated with the SGP4 algorithm used in step (3).

This numerical method is designed to minimize the difference between the six components of the osculating measurement (state vector) input to the algorithm and the state vector (osculating estimate) computed by SGP4 during the iterative solution of the system of nonlinear equations. The first state vector estimate is relatively close to the original osculating measurement and the algorithm converges in a reasonable time. If there is only a single measurement to start with, then the values of the first time derivative of the mean motion, second time derivative of mean motion and the *bstar* drag term are not known. Because there is no forward propagation in the algorithm these SGP4 orbital elements are set to zero for all computations.

Lee[15] derives a similar approach to generating a TLE from a single measurement using a Taylor-series expansion of the functional expressions in the SGP4 equations. As expected, when propagating an orbit from a single osculating measurement, the forward error is a strong function of how well the drag term is estimated or derived from other sources.

10.3.5 Deriving Classical Orbital Elements from TLEs

Because of the large database of TLEs available it is often appropriate to use TLEs to define the classic orbital elements (COEs) for a satellite at a given epoch. The first assumption is that there is a TLE with an epoch close enough to the desired time to be useful. The COE can be integrated forward from the TLE epoch time to the desired time.[16] The three additional COEs not in the TLEs are calculated as shown in the following equations:

$$a_0 = \sqrt[3]{\frac{\mu_\oplus}{n_0^2}} \tag{10-2}$$

$$E_0 = M_0 + e_0 \sin E_0 \tag{10-3}$$

where the subscript "0" refers to the epoch time of the TLE. This is, of course, Kepler's equation from Chapter 1:

$$\nu_0 = \cos^{-1}\left[\frac{e_0 - \cos E_0}{e_0 \cos E_0 - 1}\right] \tag{10-4}$$

When working directly from a TLE it is important to check the units. The TLE value of mean motion, n_0, is in revolutions per day and must be converted to radians/s. 86,400 seconds can be used for the length of a solar day.

(Note that SGP4 is set in the Fundamental Katalog 5 (FK5) and World Geodetic Survey 72 (WGS72) reference standards, referenced to the Julian 2000 (J2000.0) epoch and the proper values of r_e and $\mu_\oplus$ must be used.)

The derivative quantities in the TLEs at epoch are used to move the COEs forward (or backward) to the desired epoch. Four of the COEs move slowly in time: n, e, Ω, and ω. A simple single-step integration is used for these elements since they do change slowly:

$$a_1 = a_0 + \dot{a}_0 \Delta t$$

$$\text{where } \dot{a} = -\frac{2a_0\dot{n}_0}{3n_0} \tag{10-5}$$

$$e_1 = e_0 + \dot{e}_0 \Delta t$$

$$\text{where } \dot{e} = -\frac{2(1 - e_0)\dot{n}_0}{3n_0} \tag{10-6}$$

The subscript 1 refers to the time of the desired epoch. Δt is $t_1 - t_0$.

The a_0 and e_0 derivative terms come from differentiating the equations for the radius at perigee and the mean motion ($r_p = a(1\text{-}e)$ and $n^2a^3 = \mu$) and are evaluated at t_0. The value of inclination in the TLE, i_0, is assumed to be constant. Then,

$$\Omega_1 = \Omega_0 + \dot{\Omega}_{J_2} \Delta t$$

where

$$\dot{\Omega}_{J_2} = \left[-\frac{3}{2} J_2 \left(\frac{r_\oplus}{a_0 \left(1 - e_0^2\right)} \right)^2 \cos i_0 \right] n_0 \tag{10-7}$$

$$= -2.064734896 \times 10^{14} a_0^{-3.5} \left(1 - e_0^2\right)^{-2} \cos i_0 \,(\text{deg/ day})$$

$$\omega_1 = \omega_0 + \dot{\omega}_{J_2} \Delta t$$

where

$$\dot{\omega}_{J_2} = \left[\frac{3}{4} J_2 \left(\frac{r_\oplus}{a_0 \left(1 - e_0^2\right)} \right)^2 \left(4 - 5 \sin^2 i_0\right) \right] n_0 \tag{10-8}$$

$$= 1.032367448 \times 10^{14} a_0^{-3.5} \left(1 - e_0^2\right)^{-2} \left(4 - 5 \sin^2 i_0\right) (\text{deg/ day})$$

The mean anomaly, M, moves rapidly but at a constant rate by definition. Therefore M_1 can be calculated from:

$$M_1 = M_0 + \dot{M}_0 \Delta t + \frac{\ddot{M}_0}{2} \Delta t^2 + \frac{\dddot{M}_0}{6} \Delta t^3$$

$$\text{where } \dot{M}_0 = n_0,\ \ddot{M}_0 = \dot{n}_0,\ \dddot{M}_0 = \ddot{n}_0 \tag{10-9}$$

$$\text{so } M_1 = M_0 + n_0 \Delta t + \frac{\dot{n}_0}{2} \Delta t^2 + \frac{\ddot{n}_0}{6} \Delta t^3$$

Given that all of the orbital elements are updated except for the true anomaly at epoch, n_0, Kepler's equation can be used with the new mean anomaly (M_1) to calculate the new eccentric anomaly (E_1) and then solve for n_1.

10.3.6 Converting Multiple GPS Measurements to TLEs

The increasing population of small satellites and number of ground stations outside of the traditional civil and military organizations have resulted in a need from many users for updated TLEs on a more regular and frequent basis than AFSCN can sometimes provide. Many small satellites have onboard GPS receivers that can provide multiple state vector measurements during a period of orbits or days. On-ground TLE-based propagators are used for such tasks as automated antenna pointing and the need for additional methods of deriving deriving TLEs that are SGP4 compatible has led to techniques that can determine TLE values from a sequence of GPS measurements.

Jochim[17] and Cho[18] each demonstrated methods of generating TLEs independent of AFSCN that can be used when AFSCN TLEs are not available with a near enough epoch. The computational limitations onboard a small satellite are another reason why a general perturbation method of orbit propagation is often preferred over a special perturbation method. While available TLEs can be uploaded to the satellite, the absence of a new enough TLE makes onboard processing of the GPS data attractive. The feasibility of the concept was illustrated based on real GPS navigation data from the TOPEX/Poseidon satellite and the MIR space station with an inherent data quality of 50–100 m. It was shown that 3 hours of data within a 4 day period are sufficient to keep the position error within 4 km and that is sufficient for the type of applications discussed above.

The final technique discussed is known as the *precise conversion of elements* and is described by Vallado.[1, 19] When numerous measurements are available this approach results in submeter errors relative to space catalog TLE values with the same epoch as the derived TLEs. The steps in this process are taken from Vallado and Crawford.[19] Definitions of the variables are as follows:

X: State of orbital model, either **XYZ** Cartesian position and velocity vectors or Keplerian orbital elements, etc.

A: Partial derivative matrix (Jacobian) relating the calculated "observation" value to the initial model state value. Typically we have the following when an analytical approach is used to compute **A**:

$$\mathbf{A} = \frac{\partial obs}{\partial \mathbf{X}_0} = \mathbf{H} \bullet \Phi = \frac{\partial obs}{\partial \mathbf{X}} \bullet \frac{\partial \mathbf{X}}{\partial \mathbf{X}_0} \tag{10-10}$$

W: Weighting matrix of observation accuracy; mathematically, this is the diagonal matrix of $1/\sigma^2$ but normally it's stored as the vector of the diagonal terms for efficient computation

b: Residual vector (obs – calculated, sometimes written as $\delta y = Y_{observed} - Y_{calculated}$).

$(\mathbf{A}^T\mathbf{W}\mathbf{A})^{-1}$: Covariance matrix.

The SGPR4 differential correction pseudo-algorithm to evaluate the measurements in Vallado and Crawford[19] is

- FOR i = 1 to the number of observations (N)
  - Propagate the nominal state $\mathbf{X_0}$ to the time of the observation using SGP4 (resulting in $\mathbf{X_1}$ in the TEME (true-equator, mean equinox) coordinate system)
  - Transform the TEME prediction $\mathbf{X_1}$ to the correct coordinate system and type of available observation data, $\mathbf{obs_1}$ (usually topocentric)
  - Find the **b** matrix as observed – nominal observations
  - Form the **A** matrix (select 1 of 2 approaches)
    - Finite (or central) differences
    - Analytical partials
  - H, Partials depending on observation type
  - Φ, Partials for state transition matrix.
  - Accumulate $\sum\mathbf{A^TWA}$ and $\sum\mathbf{A^TWB}$

END FOR
- Find $\mathbf{P} = (\mathbf{A^TWA})^{-1}$ using Gauss-Jordan elimination (LU decomposition and back-substitution)
- Solve $\delta\mathbf{x} = \mathbf{P} * (\mathbf{A^TWB})$
- Check RMS for convergence
- Update state $\mathbf{X_0} = \mathbf{X_0} + \delta\mathbf{x}$
- Repeat if not converged using updated state

On completion, the term $\mathbf{P} = (\mathbf{A^TWA})^{-1}$ is the covariance matrix.

If using singular value decomposition, replace the matrix inversion with the following:
- Use singular value decomposition to decompose the weighted Jacobian matrix $[\mathbf{S} \cdot \mathbf{A}]$ into **U**, **V** and **w**, where ($\mathbf{S} = \mathbf{A^TW}$)
- Inspect w_j and set any 'very small' w_j terms to zero rather than any $1/w_j$ term to near-infinite
- Solve $\delta\mathbf{x} = \mathbf{V} \cdot (\text{diag}[1/w_j]) \cdot \mathbf{U^T} \cdot \mathbf{S} \cdot \mathbf{B}$

Vallado and Crawford[19] also includes several examples of suggested methods of selecting state vector representations and matrix operations and cautions about numerical sensitivities in order to develop computer codes for these calculations.

10.4 RELATIVE MOTION OF SATELLITES

George William Hill was born in New York City in 1838 and, after his father moved the family to West Nyack, New York, grew up on a farm that was likely to have had severe drawbacks in the education of a young man. Teaching was frequently restricted to a few subjects on an elementary level but Hill had the good fortune to attend Rutgers College and come under the influence of Theodore Strong, who had translated Laplace's Mécanique Céleste into English. Hill quoted Strong as saying that "Euler is our great Master" and noted that Strong "scarcely had a book in his library published later than 1840." Poincaré said that to Strong Euler was "the god of mathematics" whose death marked the beginning of the decline of mathematics.[20]

Hill's knowledge of the techniques of the old masters strengthened his ingenuity in the creation of a new methodology: the extension of the Eulerian method using moving rectangular axes and the same first approximation. This device led to Hill's variational curve, the reference orbit in describing lunar motion. When Simon Newcomb became director of the American Ephemeris in 1877, he undertook the reconstruction of the theories and tables of lunar and planetary motion. Hill was induced to work on the theories of Jupiter and Saturn, known to be exceptionally difficult in the determination of their mutual perturbations. Because the Nautical Almanac office had meanwhile been transferred to Washington to be under the more immediate jurisdiction of the Navy Department, Hill resided there for a ten-year period beginning in 1882. His success with the Newcomb assignment represented one of the most important contributions to 19th-century mathematical astronomy. The calculation of the effects of the planets on the Moon's motion was a particular case of the famous three-body problem, which dates back to Newton (1686).[20]

Hill's "Researches in the Lunar Theory," was the first article in the first issue of the American Journal of Mathematics (1878) and, through its introduction of the periodic orbit, initiated a new approach to the study of three mutually attracting bodies. F. R. Moulton wrote in 1914 that no earlier work had approached it in practical application and no subsequent work had then surpassed it. The article became fundamental in the development of celestial mechanics.[20]

Hill's work[21] was based on describing the motion of a three-body system. He showed how a choice of coordinate systems and the use of rectangular coordinates leads to second-order differential equations. The equations describing lunar motion with respect to Earth, in the presence of a heliocentric gravitational field, became known as Hill's equations. There are several critical assumptions in the derivations:

- Earth and lunar orbits are circular.
- Only mean, not osculating, motions of the Sun and Moon are considered.
- Solar parallax is ignored.
- The inclination of the lunar orbit is ignored (relative to Earth's orbit with respect to the Sun).

In 1960 Clohessy and Wiltshire,[22] working for the Martin Company, developed control guidance laws for the autonomous rendezvous between satellites in orbit for the purpose of assembling multi-unit spacecraft. In order to solve for the guidance laws they presented the equations of relative motion for the rendezvous satellite in a coordinate frame of the target assembly. The differential equations describing the relative motion they presented were, in fact, Hill's equations applied to the three-body problem of two satellites in orbit around Earth, instead of the Moon and Earth in orbit around the Sun. The same equations used by the aerospace community are now commonly referred to as Clohessy–Wiltshire (CW) or Clohessy–Wiltshire–Hills (CWH) equations.

There are several textbooks that show the derivation of Hill's equations for satellite applications as they are very important in describing the relative motion of satellites for formation flying, rendezvous and proximity operations (RPO), satellite inspections, and other space operations that require controlled operations of two or more satellites in close proximity. The derivation here is based on the approach in Cloud et al.[16]

10.4.1 Motion Analysis Process

Chapter 3 developed equations for orbit changes to enable one satellite to rendezvous with another space object. For an observer on the ground the orbits of the two space objects could be described using COEs or state vectors in the geocentric coordinate system and the relative positions could be calculated from the difference of their position vectors. However, precision control of the relative position of the two satellites requires a local coordinate system that describes the motion of some satellite with respect to the other satellite and not with respect to Earth. We will describe one satellite as the *target* satellite (subscript t) and one as the *chase* satellite (subscript c).

The guidance and control aspects of the problem are beyond the scope of this book but the goal of analyzing the motion is to obtain equations that are effective for proximity operations. The coordinate system used includes a radial element, an in-track element and a cross-track element. The equations of motion are derived starting with two-body dynamics and simplifying assumptions include a circular orbit for the target and ignoring the higher order terms (J_2, J_4, etc.) and perturbing forces (atmospheric drag, solar pressure, etc.) on the satellites.

The coordinate system shown in Figure 10-2 is the target-fixed Radial, In-track, Cross-track (**RIC**) system. The **RIC** system is centered on the target satellite and the fundamental plane of the **RIC** system is the same as the orbit plane of the target satellite. The **C** vector is parallel to the angular momentum vector, **h**, of the target satellite orbit plane. The **R** vector is the principal direction of the **RIC** coordinate system and points in the same direction as the target satellite position vector, **r**. The **I** vector is defined by the right-hand rule for Cartesian coordinate systems and is

$$\mathbf{C} \times \mathbf{R} = \mathbf{I} \tag{10-11}$$

With the simplifying assumption of a circular orbit for the target satellite, the **I** vector is parallel to the satellite velocity vector, **v**. The **RIC** coordinate system rotates around the **C** vector at a rate equal to the rotation rate of the target satellite:

$$n_R = \sqrt{\mu_{\oplus} / a^3} \text{ radians/s} \tag{10-12}$$

The widely used **RSW** (or Gaussian) coordinate system is the same as the **RIC** system. Some organizations use the term LVLH (local vertical, local horizontal) for this coordinate system.

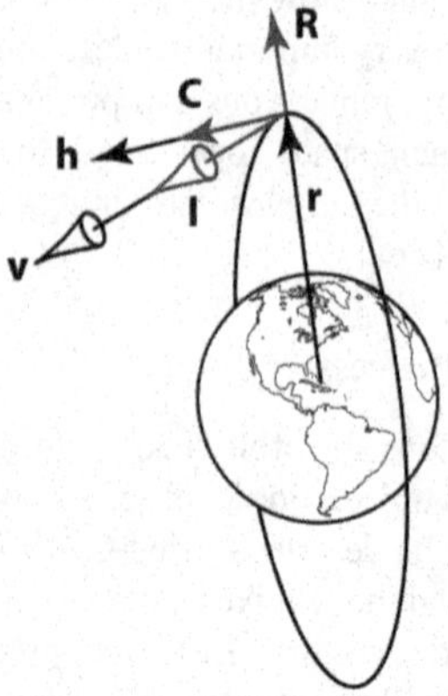

Fig. 10-2 RIC Coordinate System Definition.

Figure 10-3 shows the vector definitions used in the **RIC** coordinate system.

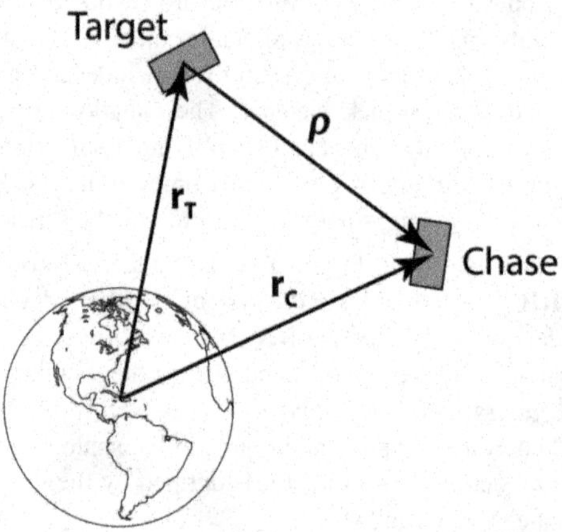

Fig. 10-3 Vector Definitions in the RIC Coordinate System.

The position vector of the target satellite, $\mathbf{r_T}$, is the vector from the center of Earth through the center of mass of the target satellite:

$$\mathbf{r}_T = r_t\mathbf{R} + 0\mathbf{I} + 0\mathbf{C} = \begin{bmatrix} r_T \\ 0 \\ 0 \end{bmatrix} \tag{10-13}$$

The magnitude is r_T and the vector is pointed along the **R** direction.

The vector, ρ, describing the position of the chase satellite with respect to the target satellite is

$$\rho = x\mathbf{R} + y\mathbf{I} + z\mathbf{C} = \begin{bmatrix} x \\ y \\ z \end{bmatrix} \tag{10-14}$$

Since x is the **R** component of relative position, a value of $x > 0$ indicates that the chase satellite is above (further away from Earth) the target satellite. The **I** component is y and $y > 0$ indicates that the chase satellite is in front of the target satellite. The **C** component, z, indicates whether the chase satellite is left ($z > 0$) or right ($z < 0$) when viewed from the target satellite direction of motion.

The position of the chase satellite is

$$\mathbf{r}_C = \mathbf{r}_T + \rho \tag{10-15}$$

10.4.2 Determining Relative Positions from COEs

To initiate a proximity operation as the chase satellite approaches the target satellite during the rendezvous phase of an operation it is important to determine the relative position of the chase and target satellites from ground-based observations or COEs. The chase satellite's COEs are a_C, e_C, i_C, Ω_C, ω_C and ν_C. In the ECI frame of reference the chase satellite position is

$$\mathbf{r}_C^{IJK} = r_i\mathbf{I} + r_j\mathbf{J} + r_k\mathbf{K} = \begin{bmatrix} r_i \\ r_j \\ r_k \end{bmatrix} \tag{10-16}$$

The target satellite's COEs are a_T, e_T, i_T, Ω_T, ω_T and ν_T. In the **RIC** frame of reference the target satellite position is:

$$\mathbf{r}_T^{RIC} = r_T\mathbf{R} + 0\mathbf{I} + 0\mathbf{C} = \begin{bmatrix} r_T \\ 0 \\ 0 \end{bmatrix} \tag{10-17}$$

where

$$r_T = \frac{a_T(1 - e_T^2)}{1 + e_T \cos \nu_T} \tag{10-18}$$

Determining the relative position from the COEs requires calculating the vector difference between $\mathbf{r}_T$ and $\mathbf{r}_C$ based on recognizing that the two vectors must be in the same coordinate frame. The chase satellite position is determined after two directional cosine matrices take the **IJK** position to the **PQW** position and then to the **RIC** position:

$$\mathbf{r}_C^{RIC} = \mathbf{C}_{PQW_T}^{RIC}\,\mathbf{C}_{IJK}^{PQW_T}\,\mathbf{r}_C^{IJK} \tag{10-19}$$

Section 2.6.5 derived the transformation from perifocal (**PQW**) to the geocentric-equatorial inertial (**IJK**) coordinate systems. That directional cosine matrix is orthonormal:

$$\left[\mathbf{C}_{IJK}^{PQW_T}\right]^{-1} = \left[\mathbf{C}_{IJK}^{PQW_T}\right]^{T} \tag{10-20}$$

From Equation (2-43) the transformed elements of the direction cosine matrix are now (in column form):

$$\begin{aligned}
C_{11} &= \cos\Omega\cos\omega - \sin\Omega\sin\omega\cos i\\
C_{21} &= -\cos\Omega\sin\omega - \sin\Omega\cos\omega\cos i\\
C_{31} &= \sin\Omega\sin i\\
C_{12} &= \sin\Omega\cos\omega + \cos\Omega\sin\omega\cos i\\
C_{22} &= -\sin\Omega\sin\omega + \cos\Omega\cos\omega\cos i\\
C_{32} &= -\cos\Omega\sin i\\
C_{13} &= \sin\omega\sin i\\
C_{23} &= \cos\omega\sin i\\
C_{33} &= \cos i
\end{aligned} \tag{10-21}$$

The directional cosine matrix to go from $\mathbf{PQW_T}$ to **RIC** is the rotation about the third axis:

$$\mathbf{C}_{PQW_T}^{RIC} = \begin{bmatrix} \cos\nu_T & \sin\nu_T & 0\\ -\sin\nu_T & \cos\nu_T & 0\\ 0 & 0 & 1 \end{bmatrix} \tag{10-22}$$

The position vector of the chase satellite in the target satellite reference frame is

$$\rho = \begin{bmatrix} x\\ y\\ z \end{bmatrix} = \mathbf{r}_C^{RIC} - \begin{bmatrix} r_T\\ 0\\ 0 \end{bmatrix} \tag{10-23}$$

10.4.3 Relative Equations of Motion

Developing the relative equations of motion between the chase satellite and the target satellite requires starting with the two-body equations of motion from Chapter 1. As in the first chapter there is the assumption of no external forces on the satellites except for Earth's gravitational field. All of the equations are developed in the target's **RIC** coordinate system. The "I" subscript in the equations indicates that the accelerations are calculated in an inertial reference frame. The two satellites' accelerations are:

$$\frac{d^2\mathbf{r}_T}{dt^2}\Big|_I = -\frac{\mu}{r_T^3}\mathbf{r}_T$$
$$\frac{d^2\mathbf{r}_C}{dt^2}\Big|_I = -\frac{\mu}{r_C^3}\mathbf{r}_C \qquad (10\text{-}24)$$

The vector difference is:

$$\frac{d^2\boldsymbol{\rho}}{dt^2}\Big|_I = \frac{d^2\mathbf{r}_T}{dt^2}\Big|_I - \frac{d^2\mathbf{r}_C}{dt^2}\Big|_I = -\frac{\mu}{r_C^3}\mathbf{r}_C + \frac{\mu}{r_T^3}\mathbf{r}_T$$
$$= -\frac{\mu}{r_T^3}\frac{r_T^3}{r_C^3}\mathbf{r}_C + \frac{\mu}{r_T^3}\mathbf{r}_T \qquad (10\text{-}25)$$

which can be arranged to get

$$r_C^{-3} = \left|\mathbf{r}_T + \boldsymbol{\rho}\right|^{-3} = r_T^{-3}\left(1 + \frac{2(\mathbf{r}_T \bullet \boldsymbol{\rho})}{r_T^2} + \frac{\rho^2}{r_T^2}\right)^{-3/2} \qquad (10\text{-}26)$$

Since

$$\left(\frac{\rho^2}{r_T^2}\right) \approx 0 \Rightarrow \frac{r_T^3}{r_C^3} \approx \left(1 + \frac{2(\mathbf{r}_T \bullet \boldsymbol{\rho})}{r_T^2}\right)^{-3/2} \qquad (10\text{-}27)$$

and

$$\left(1 + \frac{2(\mathbf{r}_T \bullet \boldsymbol{\rho})}{r_T^2}\right)^{-3/2} = 1 - \left(\frac{3(\mathbf{r}_T \bullet \boldsymbol{\rho})}{r_T^2}\right) + \text{higher order terms} \qquad (10\text{-}28)$$

we then obtain

$$\frac{r_T^3}{r_C^3} \approx 1 - \frac{3(\mathbf{r}_T \bullet \boldsymbol{\rho})}{r_T^2} \qquad (10\text{-}29)$$

Substituting and discarding higher order terms that are much less than 1 leads to:

$$\boxed{\frac{d^2\boldsymbol{\rho}}{dt^2}\Big|_I = -\frac{\mu_\oplus}{r_T^3}\begin{bmatrix} -2x \\ y \\ z \end{bmatrix}} \qquad (10\text{-}30)$$

Equation (10-30) shows the changes in relative positions between the target and chase satellites as a function of time—*as observed from inertial space*. This does not describe how relative position and velocity change in the **RIC** coordinate frame, which is a rotating reference frame.

Section 2.7.4 introduced the concept of derivatives in a moving reference frame and Equation (2-57) derived the general form of the transformation equation for any operator. The R subscript is for the rotating **RIC** frame and the I subscript is for the inertial frame. For velocity, we have:

$$\frac{d\mathbf{r}_T}{dt}\bigg|_I = \frac{d\mathbf{r}_T}{dt}\bigg|_R + \boldsymbol{\omega}_{R/I} \times \mathbf{r} \tag{10-31}$$

and, for acceleration, we have:

$$\frac{d^2\mathbf{r}_T}{dt^2}\bigg|_I = \frac{d^2\mathbf{r}_T}{dt^2}\bigg|_R + 2\boldsymbol{\omega}_{R/I} \times \frac{d\mathbf{r}_T}{dt}\bigg|_R + \frac{d\boldsymbol{\omega}_{R/I}}{dt} \times \mathbf{r} + \boldsymbol{\omega}_{R/I} \times \left(\boldsymbol{\omega}_{R/I} \times \mathbf{r}\right) \tag{10-32}$$

10.4.4 Clohessy–Wiltshire–Hills Equations

Because of the assumption of circular orbits the position, velocity, and acceleration vectors in a rotating **RIC** coordinate frame are:

$$\mathbf{r} = \begin{bmatrix} r \\ 0 \\ 0 \end{bmatrix}, \ \dot{\mathbf{r}} = \frac{d\mathbf{r}}{dt}\bigg|_R = \begin{bmatrix} 0 \\ 0 \\ 0 \end{bmatrix}, \ \ddot{\mathbf{r}} = \frac{d^2\mathbf{r}}{dt^2}\bigg|_R = \begin{bmatrix} 0 \\ 0 \\ 0 \end{bmatrix} \tag{10-33}$$

The rotation of the **RIC** coordinate frame is about the **C** (cross-track) axis at a constant angular velocity from Equation (10-12):

$$\omega_{R/I} = \begin{bmatrix} 0 \\ 0 \\ n_R \end{bmatrix}, \ \frac{d\omega_{R/I}}{dt} = \begin{bmatrix} 0 \\ 0 \\ 2n_R \end{bmatrix} \tag{10-34}$$

Substituting into Equations (10-31) and (10-32) for velocity and acceleration gives:

$$\frac{d\mathbf{r}}{dt}\bigg|_I = \begin{bmatrix} 0 \\ n_R r \\ 0 \end{bmatrix} \quad (\text{tangential velocity}) \tag{10-35}$$

$$\frac{d^2\mathbf{r}}{dt^2}\bigg|_I = \begin{bmatrix} -n_R^2 r \\ 0 \\ 0 \end{bmatrix} \quad (\text{centripetal acceleration}) \tag{10-36}$$

Equations (10-35) and (10-36) show the inertial velocity and acceleration in the inertial frame of reference.

The next step is to determine the acceleration of the chase satellite in the frame of reference of the target satellite by using Equation (2-59) again:

$$\dot{\boldsymbol{\rho}} = \frac{d^2\boldsymbol{\rho}}{dt^2}\bigg|_I - 2\boldsymbol{\omega} \times \dot{\boldsymbol{\rho}} - \dot{\boldsymbol{\omega}} \times \boldsymbol{\rho} - \boldsymbol{\omega} \times (\boldsymbol{\omega} \times \boldsymbol{\rho}) \tag{10-37}$$

Recognizing the following identities:

$$\boldsymbol{\rho} = \begin{bmatrix} x \\ y \\ z \end{bmatrix}, \quad \dot{\boldsymbol{\rho}} = \frac{d\boldsymbol{\rho}}{dt}\bigg|_R = \begin{bmatrix} \dot{x} \\ \dot{y} \\ \dot{z} \end{bmatrix}, \quad \ddot{\boldsymbol{\rho}} = \frac{d^2\boldsymbol{\rho}}{dt^2}\bigg|_R = \begin{bmatrix} \ddot{x} \\ \ddot{y} \\ \ddot{z} \end{bmatrix} \tag{10-38}$$

and

$$\boldsymbol{\omega} = \begin{bmatrix} 0 \\ 0 \\ n_T \end{bmatrix}, \quad \dot{\boldsymbol{\omega}} = \begin{bmatrix} 0 \\ 0 \\ 0 \end{bmatrix}, \quad 2\boldsymbol{\omega} \times \dot{\boldsymbol{\rho}} = \begin{bmatrix} -2n_T\dot{y} \\ 2n_T\dot{x} \\ 0 \end{bmatrix} \tag{10-39}$$

we then obtain

$$\boldsymbol{\omega} \times (\boldsymbol{\omega} \times \boldsymbol{\rho}) = \begin{bmatrix} -n_T^2 x \\ -n_T^2 \\ 0 \end{bmatrix}, \quad \frac{d^2\boldsymbol{\rho}}{dt^2}\bigg|_I = -n_T^2 \begin{bmatrix} -2x \\ y \\ z \end{bmatrix} \tag{10-40}$$

Substituting Equations (10-38) and (10-39) into Equation (10-37) leads to the three homogeneous, second-order differential CWH equations:

$$\begin{aligned} \ddot{x} - 3n_T^2 x - 2n_T\dot{y} &= 0 \\ \ddot{y} \qquad + 2n_T\dot{x} &= 0 \\ \ddot{z} + n_T^2 z \qquad &= 0 \end{aligned} \tag{10-41}$$

These equations describe the motion of the chase satellite relative to the target satellite, and are the basis for any guidance and control algorithms developed when controlling the proximity motion of the chase satellite. The cross-coupling between the **x** and **y** components of the position vector describes oscillatory motion in the plane that is along the track or "up" and "down." The **z** component is the cross-track motion and that is decoupled from the in-plane motion.

The limiting assumptions here are a circular orbit and no external forces except (two-body equation) gravity. Several authors and RPO applications have extended the CWH equations to include effects such as J_2 and J_4 perturbations, atmospheric drag, and eccentric orbits.

Exercises

10.1 XMM-Newton, also known as the High Throughput X-ray Spectroscopy Mission and the X-ray Multi-Mirror Mission, is an X-ray space observatory launched by the European Space Agency in December 1999 on an Ariane-5 rocket. The following TLE were published for that satellite.

XMM-NEWTON

1 25989U 99066A 16190.67148956 -.00000663 00000-0 00000+0 0 9991

2 25989 67.7194 21.9134 8222525 95.0698 359.9797 0.50152544 19164

Determine all of the variables and constants in the TLEs.

10.2 An unknown satellite was tracked at 1149 UTC on January 14, 2016, and the following state vector were generated in ECI coordinates:

$\mathbf{r} = -6132.0\mathbf{I} - 3380.0\mathbf{J} + 2472.0\mathbf{K}$ km
$\mathbf{v} = -3.369\mathbf{I} + 6.628\mathbf{J} + 2.433\mathbf{K}$ km/s

Determine the TLEs for this satellite.

10.3 This set of TLEs for an Iridium satellite was sent to your office:

IRIDIUM 7 [+]

1 24793U 97020B 16189.29473743 .00000203 00000-0 65255-4 0 9999

2 24793 86.3981 103.1602 0002265 84.1634 275.9820 14.34216459 3764

Determine the COEs for this satellite.

10.4 For the previous problem, propagate the COEs forward 17 minutes and calculate the new position and velocity vectors in ECI coordinates.

List of References

1. Vallado, D. A. *Fundamentals of Astrodynamics and Applications.* Microcosm Press, El Segundo, CA, 2004.

2. Hoots, F. R., Schumacher, P. W., and Glover, R. A. *History of Analytical Orbit Modeling in the U.S. Space Surveillance System.* Journal of Guidance, Control, and Dynamics, Vol. 27, No. 2, March–April 2004, pp. 174–185.

3. Brouwer, D. *Solution of the Problem of Artificial Satellite Theory Without Drag.* U.S. Air Force Cambridge Research Center, Geophysics Research Directorate, AFCRC-TN-59-638, Bedford, MA, Oct. 1959.

4. Brouwer, D. *Solution of the Problem of Artificial Satellite Theory Without Drag.* Astronomical Journal, Vol. 64, No. 1274, 1959, pp. 378–397.

5. Brouwer, D. and Clemence, G. M. *Methods of Celestial Mechanics.* New York, NY, Academic Press, 1961.

6. Kozai, Y. *The Motion of a Close Earth Satellite*. Astronomical Journal, Vol. 64, No. 1274, 1959, pp. 367–377.

7. Lane, M. H., Fitzpatrick, P. M., and Murphy, J. J. *On the Representation of Air Density in Satellite Deceleration Equations by Power Functions with Integral Exponents*. Deputy for Aerospace, Air Proving Ground Center, APGC-TDR-62-15, Eglin AFB, FL, March 1962.

8. Lane, M. H. *The Development of an Artificial Satellite Theory Using Power-Law Atmospheric Density Representation.* AIAA Paper 65-35, Jan. 1965.

9. Lane, M. H., and Cranford, K. H. *An Improved Analytical Drag Theory for the Artificial Satellite Problem.* AIAA Paper 69-925, Aug. 1969.

10. Lyddane, R. H. *Small Eccentricities or Inclinations in the Brouwer Theory of the Artificial Satellite.* Astronomical Journal, Vol. 68, No. 8, 1963, pp 555–558.

11. Hilton, C. G. *The SPADATS Mathematical Model.* Rep. ESD-TDR- 63-427, Aeronutronic Publ. U-2202, 5 Aug. 1963.

12. Hoots, F. R., and Roehrich, R. L. *Models for Propagation of the NORAD Element Sets.* Project SPACETRACK, Rep. 3, U.S. Air Force Aerospace Defense Command, Colorado Springs, CO, Dec. 1980.

13. Vallado, D. A., Crawford, P., and Hujsak, R. *Revisiting Spacetrack Report #3.* AIAA 2006-6753, 2006.

14. *Astrodynamics - Propagation Specifications, Technical Definitions, and Recommended Practices.* ANSI/AIAA S-131-2010, AIAA, Reston, VA, 2010.

15. Lee, B. S. *NORAD TLE Conversion from Osculating Orbital Elements.* J. Astron. and Space Sciences, Vol. 19, No. 4, 2002, pp. 395–402.

16. Cloud, D. J., et al. *Advanced Orbital Mechanics Study Guide Volume 2.* Advanced Space Operations School, Jan. 2012.

17. Jochim, E. F., et al. *GPS Based Onboard and Onground Operations for Small Satellites.* Acta Astronautica, Vol. 39, No. 9-12, 1996, pp. 917–922.

18. Cho, C. H., et al. *NORAD TLE Type Orbit determination of LEO Satellites Using GPS Navigation Solutions.* J. Astron. Space Sc., Vol. 19, No. 3, 2002, pp. 197–206.

19. Vallado, D. A. and Crawford, P. *SGP4 Orbit Determination.* AIAA-2008-6770, AIAA, Reston, VA, 2008.

20. Hill, G.W. *Complete Dictionary of Scientific Biography.* 2008. Encyclopedia.com. 20 May. 2016 <http://www.encyclopedia.com>.

21. Hill, G. W. *Researches in the Lunar Theory.* American Journal of Mathematics, Vol. 1, No. 1, 1878, pp. 5–26.

22. Clohessy, W. H. and Wiltshire, R. S. *Terminal Guidance System for Satellite Rendezvous.* Journal of Aerospace Sciences, Vol. 27, No. 9, Sept. 1960, pp. 653–658, 674.

Appendix A

Astrodynamic Constants

Table A-1: Geocentric Astrodynamic Constants.

	Canonical Units	English Units	Metric Units
Mean Equatorial Radius, $r_{\oplus}$	1 $DU_{\oplus}$	3443.918 nmi	6378.136 km
Time unit	1 $TU_{\oplus}$	13.44686457 solar min	806.8118744 solar s
Speed Unit	1 $DU_{\oplus}/TU_{\oplus}$	25936.24764 ft/s	7.90536828 km/s
Gravitational Parameter, $\mu_{\oplus}$	1 $DU_{\oplus}^3/TU_{\oplus}^2$	$1.407646882 \times 10^{16}$ ft^3/s^2	3.986004418e5 × 10^5 km^3/s^2
Angular Rotation, $\omega_{\oplus}$	0.0588336565 radians/$TU_{\oplus}$	0.2506844773 deg/min	$7.29211586 \times 10^{-5}$ radians/s

Table A-2: Heliocentric Astrodynamic Constants.

	Canonical Units	English Units	Metric Units
Mean Distance, Earth to Sun	1 AU	4.9081250×10^{11} ft	1.4959965×10^{8} km
Time Unit	1 $TU_{\odot}$	58.132821 days	5.0226757×10^{6} s
Speed Unit	1 $AU/TU_{\odot}$	9.7719329×10^{4} ft/s	29.784852 km/s
Gravitational Parameter, $\mu_{\odot}$	1 $AU^3/TU_{\odot}^2$	4.6868016×10^{21} ft^3/s^2	1.3271544×10^{11} km^3/s^2

Table A-3: Miscellaneous Constants and Conversions .*

Constant	Conversion	Units
second	$1.239446309 \times 10^{-3}$	$TU_{\oplus}$
degree/s	14.08152366	radians/$TU_{\oplus}$
radian/$TU_{\oplus}$	0.0710150424	degree/s
foot	$4.77881892 \times 10^{-8}$	$DU_{\oplus}$
statute mile	$2.52321639 \times 10^{-4}$	$DU_{\oplus}$
nautical mile	$2.903665564 \times 10^{-4}$	$DU_{\oplus}$
foot/s	$3.85560785 \times 10^{-5}$	$DU_{\oplus}/TU_{\oplus}$
km/s	0.1264963205	$DU_{\oplus}/TU_{\oplus}$
nautical mile/s	0.2342709	$DU_{\oplus}/TU_{\oplus}$

Table A-3: Miscellaneous Constants and Conversions (Continued).*

Constant	Conversion	Units
$2\pi/\sqrt{\mu_\oplus}$	$5.295817457 \times 10^{-8}$ $9.952004586 \times 10^{-3}$	s^3/ft^2 s^3/km^2
π	3.14159265359	
degree	0.0174532925199	radians
radian	57.2957795131	degrees
foot	0.3048 (exact)	meters
statute mile	1609.344 (exact)	meters
nautical mile	1852 (exact)	meters

*. Note: EGM-96/WGS-85.

Appendix B

Vector Review

Vector analysis is a branch of mathematics that saves time in the derivation of relationships involving multi-dimensional quantities. In three-dimensional space each vector equation represents three scalar equations. Thus, vector analysis is a powerful tool that makes many derivations easier and much shorter than otherwise would be possible.

In this section the fundamental vector operations will be discussed. A good understanding of these operations will be of significant help to the student in the use of this text.

B.1 DEFINITIONS

Vector. A vector is a quantity having both magnitude and direction. Examples are displacement, velocity, force and acceleration. The symbol used in the text to represent a vector quantity will be a bold-faced letter.

Scalar. A scalar is a quantity having only magnitude. Examples are mass, length of a vector, speed, temperature, time, or any real signed number. The symbol used in the text to represent a scalar quantity will be any signed number not bold-faced (e.g., 3 or A). An exception to this symbology is sometimes desirable when, for example, the scalar magnitude of **a** will be shown as $|\mathbf{a}|$.

Unit Vector. A unit vector is a vector having unit (1) magnitude. If **a** is a vector, not a null or zero vector, then

$$\frac{\mathbf{a}}{|\mathbf{a}|} = \mathbf{I}_a \tag{B-1}$$

where $\mathbf{I}_a$ is the symbol used to denote the unit vector in the direction of **a**. Certain sets of vectors are reserved for unit vectors in rectangular systems such as (**I**, **J**, **K**) and (**P**, **Q**, **W**).

Equality of Vectors. Two vectors **a** and **b** are said to be equal if and only if they are parallel and have the same sense of direction and the same magnitude, regardless of the position of their origins.

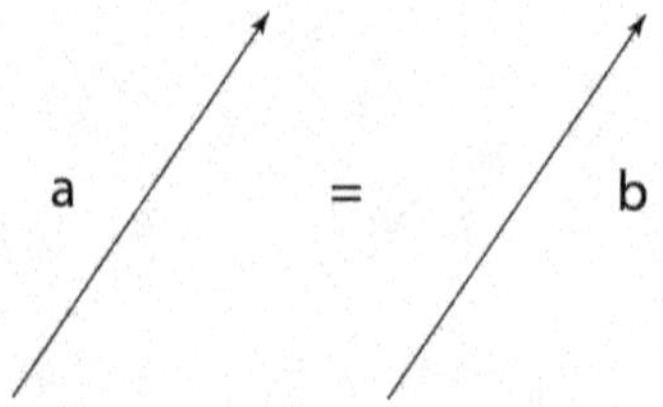

Fig. B-1 Equality of Vectors.

Coplanarand Collinear Vectors. Vectors that are parallel to the same straight line are said to be collinear. Two collinear vectors differ only by a scalar factor. Vectors that are parallel to the same plane are said to be coplanar (not necessarily parallel vectors).

If **a**, **b**, and **c** are coplanar vectors, and **a** and **b** are not collinear, it is possible to express **c** in terms of **a** and **b**.

The vector **c** may be resolved into components $\mathbf{c}_1$ and $\mathbf{c}_2$ parallel to **a** and **b**, respectively, so that $\mathbf{c} = \mathbf{c}_1 + \mathbf{c}_2$.

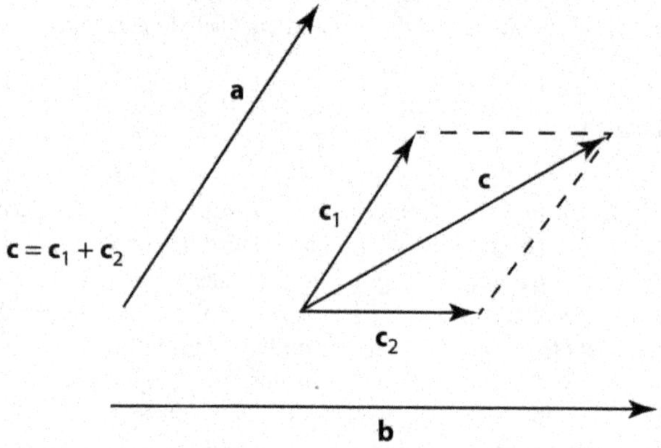

Fig. B-2 Coplanar Vectors.

Since $\mathbf{c}_1$ and **a** are collinearvectors, they differ only by a scalar factor x, such that

$$\mathbf{c}_1 = x\mathbf{a} \tag{B-2}$$

Similarly, since $\mathbf{c}_2$ and **b** are collinear we may write

$$\mathbf{c}_2 = y\mathbf{b} \tag{B-3}$$

Then

$$\mathbf{c} = \mathbf{c}_1 + \mathbf{c}_2 = x\mathbf{a} + y\mathbf{b} \tag{B-4}$$

B.2 VECTOR OPERATIONS

Addition of Vectors. The sum or result of two vectors **a** and **b** is a vector **c** obtained by constructing a triangle with **a** and **b** forming two sides of the triangle and **b** adjoined to **a**. The resultant **c** starts at the origin of **a** and ends at the terminus of **b**.

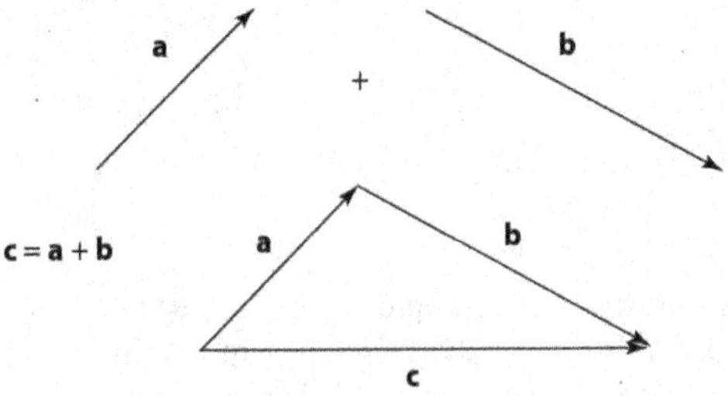

Fig. B-3 Vector Addition.

An obvious extension of this idea is the difference of two vectors. Suppose we wish to determine the resultant of **a** – **b**, which is the same as **a** + (–**b**).

Scalar or Dot Product. The "dot" product of two vectors **a** and **b**, denoted by **a•b**, is a scalar quantity defined by

$$\mathbf{a}\bullet\mathbf{b} = |\mathbf{a}|\ |\mathbf{b}|\cos\theta \qquad \text{(B-5)}$$

where θ is the angle between the two vectors.

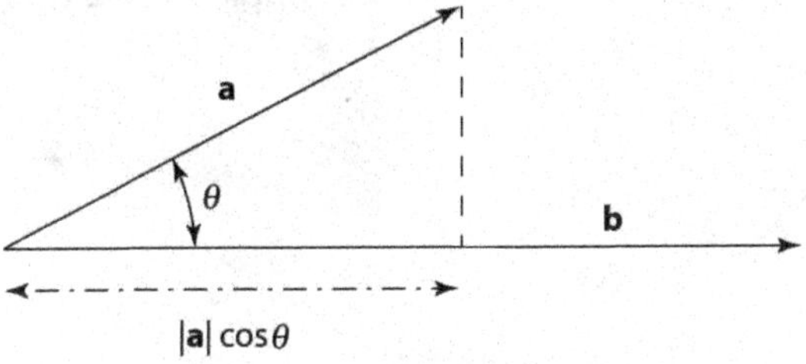

Fig. B-4 First Part of Dot Product.

From the definition of the dot product the following laws are derived:

(a) $\mathbf{a}\cdot\mathbf{b} = \mathbf{b}\cdot\mathbf{a}$ (Commutative law)

(b) $\mathbf{a}\cdot(\mathbf{b}+\mathbf{c}) = \mathbf{a}\cdot\mathbf{b} + \mathbf{a}\cdot\mathbf{c}$ (Distributive law) (B-6)

(c) $m(\mathbf{a}\cdot\mathbf{b}) = (m\mathbf{a})\cdot\mathbf{b} = \mathbf{a}\cdot(m\mathbf{b}) = (\mathbf{a}\cdot\mathbf{b})m$ (Associative law)

(d) $\mathbf{I}\cdot\mathbf{I} = \mathbf{J}\cdot\mathbf{J} = \mathbf{K}\cdot\mathbf{K} = 1$

$\mathbf{I}\cdot\mathbf{J} = \mathbf{J}\cdot\mathbf{K} = \mathbf{K}\cdot\mathbf{I} = 0$

where **I**, **J** and **K** are unit orthogonal vectors.

(e) If $\mathbf{a} = a_1\mathbf{I} + a_2\mathbf{J} + a_3\mathbf{K}$ (B-7)

$\mathbf{b} = b_1\mathbf{I} + b_2\mathbf{J} + b_3\mathbf{K}$

then

$\mathbf{a} \cdot \mathbf{b} = (a_1\mathbf{I} + a_2\mathbf{J} + a_3\mathbf{K}) \cdot (b_1\mathbf{I} + b_2\mathbf{J} + b_3\mathbf{K})$

$\mathbf{a} \cdot \mathbf{b} = a_1 b_1 + a_2 b_2 + a_3 b_3$

(f) $\mathbf{a} \cdot \mathbf{a} = a^2$ (B-8)

Differentiating both sides of the equation above yields

(g) $\mathbf{a} \cdot \dot{\mathbf{a}} = a\dot{a}$ (B-9)

Vector or Cross Product. The cross product of two vectors **a** and **b**, denoted by **a** × **b**, is a *vector* **c** such that the magnitude of **c** is the product of the magnitudes of **a** and **b** and the sine of the angle between them. The direction of the vector **c** is perpendicular to both **a** and **b** such that **a**, **b**, and **c** form a right-handed system. That is, if we say "**a** crossed with **b**" we mean that the resulting vector **c** will be in the direction of the extended thumb when the fingers of the right hand are closed from **a** to **b** through the smallest angle possible.

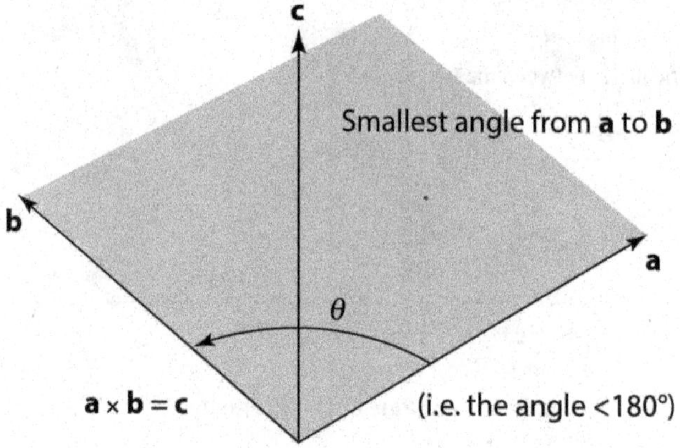

Fig. B-5 Vector or Cross Product.

$|\mathbf{c}| = |\mathbf{a}|\,|\mathbf{b}| \sin\theta$ (B-10)

If **a** and **b** are parallel, then $\mathbf{a} \times \mathbf{b} = 0$. In particular,

$\mathbf{a} \times \mathbf{a} = 0$ (B-11)

Figure B-6 shows the application of the right-hand rule to the problem of reversing the order of vector cross products.

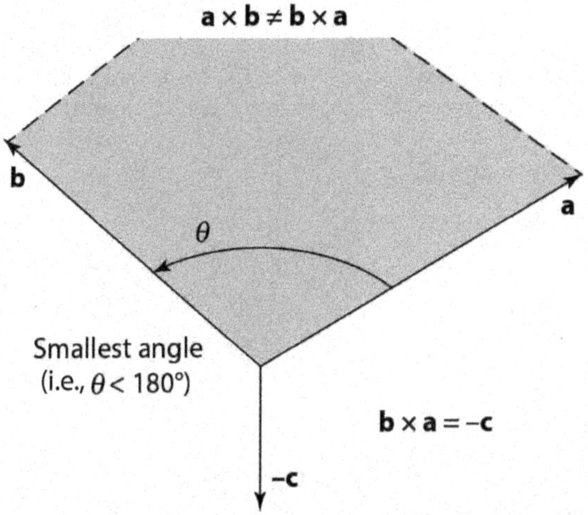

Fig. B-6 Cross-Product Direction Follows Right-Hand Rule.

In fact since $\mathbf{a} \times \mathbf{b} = -(\mathbf{b} \times \mathbf{a})$ the magnitudes are identical and the only difference due to the order in which the cross products are taken is the direction of the resultant vector **c**.

Other laws valid for the cross product are

(a) $\mathbf{a} \times (\mathbf{b} + \mathbf{c}) = \mathbf{a} \times \mathbf{b} + \mathbf{a} \times \mathbf{c}$ (Distributive law)

(b) $m(\mathbf{a} \times \mathbf{b}) = (m\mathbf{a}) \times \mathbf{b} = \mathbf{a} \times (m\mathbf{b}) = (\mathbf{a} \times \mathbf{b})(m)$ (Associative law)

(c) $\mathbf{I} \times \mathbf{I} = \mathbf{J} \times \mathbf{J} = \mathbf{K} \times \mathbf{K} = 0$

$\mathbf{I} \times \mathbf{J} = \mathbf{K}$

$\mathbf{J} \times \mathbf{K} = \mathbf{I}$

$\mathbf{K} \times \mathbf{I} = \mathbf{J}$

where **I**, **J** and **K** are unit orthogonal vectors.

(d) If $\mathbf{a} = a_1\mathbf{I} + a_2\mathbf{J} + a_3\mathbf{K}$

$\mathbf{b} = b_1\mathbf{I} + b_2\mathbf{J} + b_3\mathbf{K}$

then

$$\begin{aligned} \mathbf{a} \times \mathbf{b} &= (a_1\mathbf{I} + a_2\mathbf{J} + a_3\mathbf{K}) \times (b_1\mathbf{I} + b_2\mathbf{J} + b_3\mathbf{K}) \\ &= (a_2b_3 - a_3b_2)\mathbf{I} + (a_3b_1 - a_1b_3)\mathbf{J} + (a_1b_2 - a_2b_1)\mathbf{K} \end{aligned} \tag{B-12}$$

This result may be recognized as the expansion of the determinant

$$\mathbf{a} \times \mathbf{b} = \begin{bmatrix} \mathbf{I} & \mathbf{J} & \mathbf{K} \\ a_1 & a_2 & a_3 \\ b_1 & b_2 & b_3 \end{bmatrix} \tag{B-13}$$

Scalar Triple Product. Consider the scalar quantity

$$\mathbf{a} \cdot (\mathbf{b} \times \mathbf{c}) \tag{B-14}$$

If we let θ be the angle between **b** and **c** and α be the angle between the resultant of ($\mathbf{b} \times \mathbf{c}$) and the vector **a** then, from the definition of dot and cross products,

$$\mathbf{a} \cdot (\mathbf{b} \times \mathbf{c}) = a(bc \sin\theta) \cos \alpha \tag{B-15}$$

Thus if

$$\begin{aligned} \mathbf{a} &= a_1\mathbf{I} + a_2\mathbf{J} + a_3\mathbf{K} \\ \mathbf{b} &= b_1\mathbf{I} + b_2\mathbf{J} + b_3\mathbf{K} \\ \mathbf{c} &= c_1\mathbf{I} + c_2\mathbf{J} + c_3\mathbf{K} \end{aligned} \tag{B-16}$$

then

$$\begin{aligned} \mathbf{a}\cdot(\mathbf{b}\times\mathbf{c}) &= (a_1\mathbf{I} + a_2\mathbf{J} + a_3\mathbf{K})\cdot\begin{bmatrix} \mathbf{I} & \mathbf{J} & \mathbf{K} \\ b_1 & b_2 & b_3 \\ c_1 & c_2 & c_3 \end{bmatrix} \\ &= a_1(b_2c_3 - b_3c_2) + a_2(b_3c_1 - b_1c_3) + a_3(b_1c_2 - b_2c_1) \end{aligned} \tag{B-17}$$

But this is simply the expansion of the determinant

$$\mathbf{a}\cdot(\mathbf{b}\times\mathbf{c}) = \begin{bmatrix} a_1 & a_2 & a_3 \\ b_1 & b_2 & b_3 \\ c_1 & c_2 & c_3 \end{bmatrix} \tag{B-18}$$

You may easily verify that

$$\begin{bmatrix} a_1 & a_2 & a_3 \\ b_1 & b_2 & b_3 \\ c_1 & c_2 & c_3 \end{bmatrix} = \begin{bmatrix} c_1 & c_2 & c_3 \\ a_1 & a_2 & a_3 \\ b_1 & b_2 & b_3 \end{bmatrix} = \begin{bmatrix} b_1 & b_2 & b_3 \\ c_1 & c_2 & c_3 \\ a_1 & a_2 & a_3 \end{bmatrix} \tag{B-19}$$

or

$$\mathbf{a} \cdot (\mathbf{b} \times \mathbf{c}) = \mathbf{c} \cdot (\mathbf{a} \times \mathbf{b}) = \mathbf{b} \cdot (\mathbf{c} \times \mathbf{a}) \tag{B-20}$$

Vector Triple Product. The vector triple product, denoted by $\mathbf{a} \times (\mathbf{b} \times \mathbf{c})$, is a *vector* perpendicular to both ($\mathbf{b} \times \mathbf{c}$) and **a** such that it lies in the plane of **b** and **c**. The properties associated with this quantity are:

(a) $\mathbf{a} \times (\mathbf{b} \times \mathbf{c}) = (\mathbf{a} \cdot \mathbf{c})\mathbf{b} - (\mathbf{a} \cdot \mathbf{b})\mathbf{c}$ (B-21)

(b) $(\mathbf{a} \times \mathbf{b}) \times \mathbf{c} = (\mathbf{a} \cdot \mathbf{c})\mathbf{b} - (\mathbf{b} \cdot \mathbf{c})\mathbf{a}$ (B-22)

(c) $\mathbf{a} \times (\mathbf{b} \times \mathbf{c}) \neq (\mathbf{a} \times \mathbf{b}) \times \mathbf{c}$ (B-23)

B.3 VELOCITY

Velocity is the rate of change of position and is a directed quantity. Consider the case of a point P moving along the space curve shown in Figure B-7. The *position* of **P** is denoted by

$$\overline{O}\,\overline{P} = \mathbf{r} = x(t)\mathbf{I} + y(t)\mathbf{J} + z(t)\mathbf{K} \tag{B-24}$$

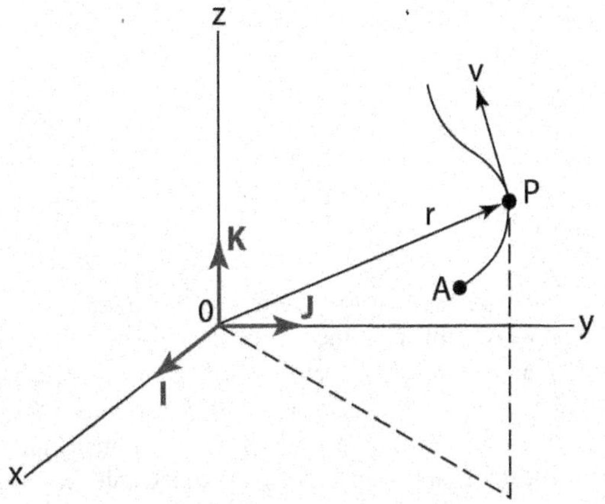

Fig. B-7 Velocity Vector.

Then the velocity relative to the (**I, J, K**) coordinate system is:

$$\mathbf{v} = \frac{d\mathbf{r}}{dt} = \frac{dx}{dt}\mathbf{I} + \frac{dy}{dt}\mathbf{J} + \frac{dz}{dt}\mathbf{K} \tag{B-25}$$

and the components of the particle velocity are

$$v_x = \frac{dx}{dt},\ v_y = \frac{dy}{dt},\ v_z = \frac{dz}{dt} \tag{B-26}$$

If we now let s be the arc length measured from some fixed point A on the curve to the point P, then the coordinates of P are functions of s:

$$\mathbf{r} = x(s)\mathbf{I} + y(s)\mathbf{J} + z(s)\mathbf{K} \tag{B-27}$$

and

$$\frac{d\mathbf{r}}{ds} = \frac{dx}{ds}\mathbf{I} + \frac{dy}{ds}\mathbf{J} + \frac{dz}{ds}\mathbf{K} \tag{B-28}$$

If we now take the dot product of **dr**/dt with itself, we obtain

$$\frac{d\mathbf{r}}{ds}\cdot\frac{d\mathbf{r}}{ds} = \left(\frac{dx}{ds}\right)^2 + \left(\frac{dy}{ds}\right)^2 + \left(\frac{dz}{ds}\right)^2 \tag{B-29}$$

But

$$dx^2 + dy^2 + dz^2 = ds^2 \tag{B-30}$$

Thus

$$\frac{d\mathbf{r}}{ds} \cdot \frac{d\mathbf{r}}{ds} = \frac{ds^2}{ds^2} = 1 \tag{B-31}$$

which, from the definition of a dot product, means that

$$\frac{d\mathbf{r}}{ds} = \text{unit vector} \triangleq \mathbf{T} \tag{B-32}$$

is tangent to the curve at point P. Thus we may write

$$\mathbf{v} = \frac{d\mathbf{r}}{dt} = \frac{d\mathbf{r}}{ds}\frac{ds}{dt} = \frac{ds}{dt}\mathbf{T} \tag{B-33}$$

which more simply stated means that the velocity is the rate of change of arc length traveled and is always tangent to the path of the particle.

This result may be applied to a unit vector (which has constant magnitude). The instantaneous velocity of a unit vector is the rate of change of arc length traveled and is tangent to the path. In this special case then, the velocity associated with a unit vector is always perpendicular to the unit vector and has a magnitude of ω, the instantaneous angular vector of the unit vector.

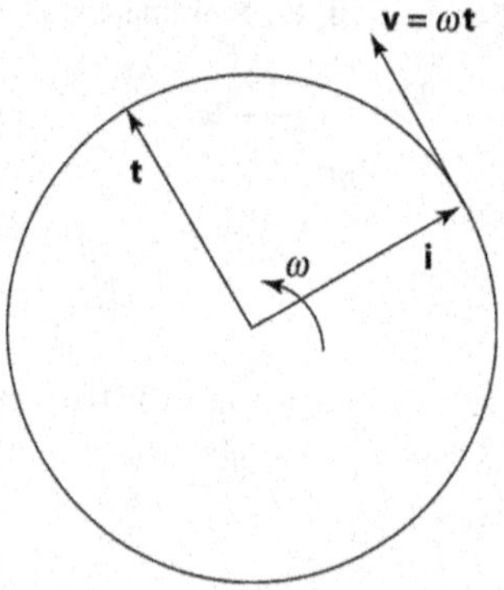

Fig. B-8 Velocity Vector Defined with the Angular Rate and Perpendicular Unit.

Appendix C

Gauss Problem

Simply, the Gauss target problem can be defined as finding the orbit between two known positions $\mathbf{r_1}$ and $\mathbf{r_2}$ given the time of flight (TOF) between these two positions. The situation is shown on the following page. Since the target satellite will move during the transfer, the first step is to calculate the new position and velocity vectors. This process can be accomplished using universal variables, but as long as the analysis uses only two-body dynamics, it's much easier to find the orbit elements, move the appropriate elements through time, and recalculate the position and velocity vectors.

The orbit transfer between the interceptor and the target, after it has moved through the TOF, remains and is generally classified as the Gauss problem. In general, there are two solutions to this problem, the long and the short way illustrated below, so the flight direction must also be specified. Chapter 5 gives an excellent description of solutions to the Gauss problem. The method that uses universal variables is especially useful for most applications since the orbit transfers that will be examined may be of any orbit type. Universal variables allow solutions with just one set of equations.

The new orbit is now determined since an ($\mathbf{r}$, $\mathbf{v}$) pair is now defined. The difficult part in this process is finding a new value for z at every step. Luckily, the z vs. t curve is continuous and is a monotonically increasing function (Figure C-2). This means that there is always one unique solution for every time t. Section 5.3.1 suggests a Newton iteration scheme for finding the next value for z. Unfortunately, this method does not always work since it can result in a very negative value for z, which then makes y negative, causing the above algorithm to stop. Another way to pick a new z value is to use a bisection technique. If done correctly, this method will always supply a solution, although it may take a few more iterations. In this case, the object is to find the value of z that corresponds to the given TOF. This can be accomplished by first bounding the correct value of z, picking a trial value of z that is halfway between these bounds, readjusting the bounds based on the results of this trial value, and continuing the process until a value of z is reached that results in a TOF sufficiently close to the given time.

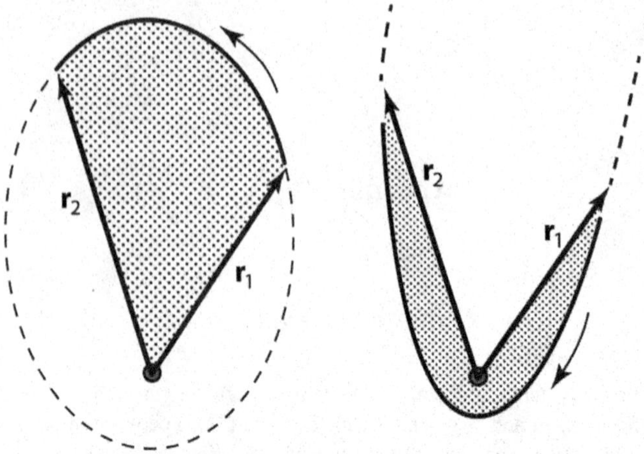

Figure C-1 Gauss Problem Geometry.

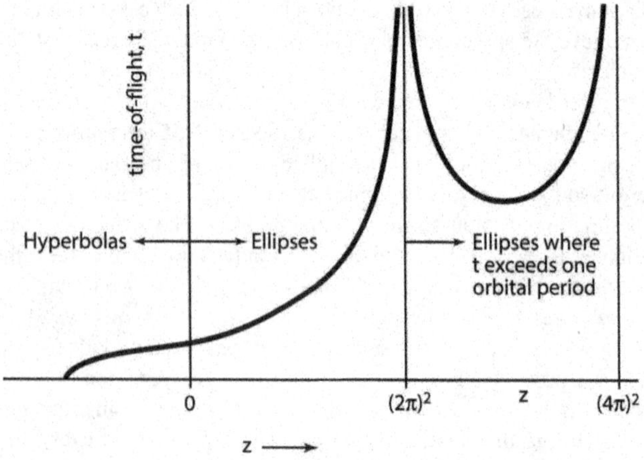

Figure C-2 Z vs. Time.

Given: r_{int} = interceptor position vector at initial time,

v_{int} = interceptor velocity vector at initial time,

r_{tgt} = target position vector at initial time,

v_{tgt} = target velocity vector at initial time,

TOF = time of flight to intercept.

Find: Δv_1 = change in velocity required to intercept in TOF.

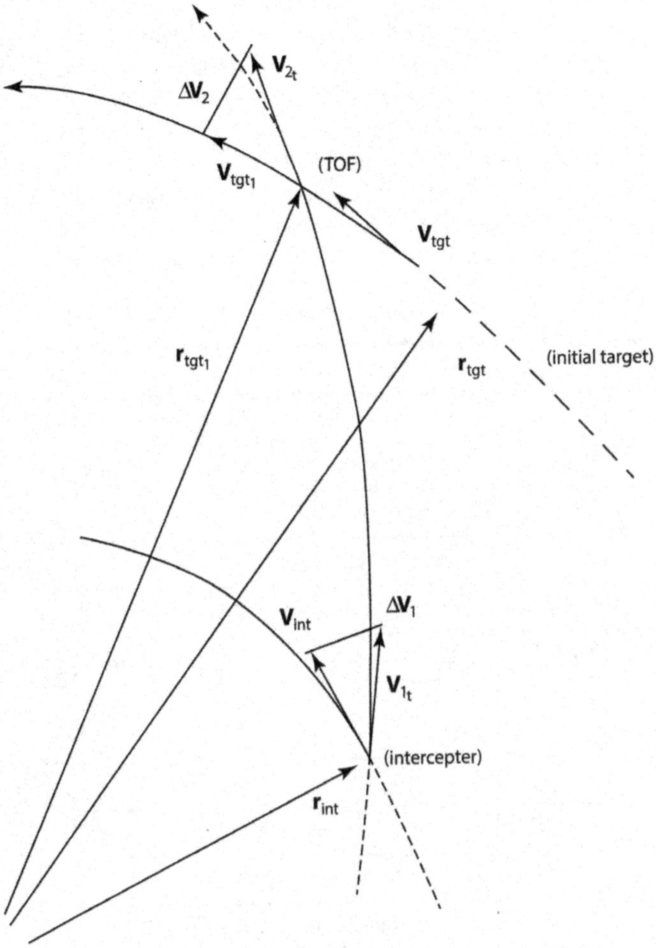

Figure C-3 Geometry for Target Problem.

This project is to develop a computer subroutine (or procedure) that will take inputs of the positions, velocities, direction of motion and time of flight and calculate the required change in velocity to satisfy those conditions. In a computer language such as MATLAB the subroutine could be

$\Delta V = \text{Gauss}\ (r_1, r_2, \text{dm}, \text{time}, v_1, v_2)$

This procedure solves the Gauss problem of orbit determination and returns the velocity vectors at each of two given position vectors. The solution uses universal variables for calculations and a bisection technique for updating **z**. This method is slower than the Newton iteration discussed in Chapter 5, but it does not suffer problems with negative **z** values and is valid for ellipses less than one revolution, parabolas and hyperbolas. Also note the selection of a small error bound since the algorithm is very sensitive to changes in this variable. A value of 0.001 will converge in say 10 iterations instead of 25 iterations with a value of 0.00001, and the accuracy will differ in the third to fourth decimal place. It is recommended to keep the higher accuracy for cases like the example, BMW pg. 230, Problem 5.10. (Refer to Figure 5-2 for ranges of **z**.) Table C-1 lists some suggested variable definitions for use in the project computer program.

Table C-1: Computer Program Variables.

		Variable	Range
Inputs:	R_1	r_{int}, IJK position vector of interceptor	km
	R_2	r_{tgt1}, IJK position vector of target after time	km
	DM	direction of motion	'L', 'S'
	Time	t_0 time between R_1 and R_2	s
Outputs:	V_1	v_{1tran}, IJK velocity of transfer orbit	km/s
	V_2	v_{2tran}, IJK velocity vector of transfer orbit	km/s
Local Variables:	VarA	variable of the iteration; not the semi-major axis!	
	Y	Y	
	F, G	f, g, f and g expressions	
	GDOT	$\dot{g}$, derivative of g expression	
	Xold	x_o, old value of universal variable x	
	Zold	z_o, old value of z	
	Znew	z_n, new value of z	
	Cnew	C, C(z) function	
	Snew	S, S(z) function	
	TimeNew	T, new time	s
References:			
Sec. 5.3.1		(Uses a Newton iteration)	

ALGORITHM Outline for GAUSS $(\bar{r}_1, \bar{r}_2, \text{dm}, t_0, \bar{v}_1, \bar{v}_2)$

$$\text{Cos}\,\Delta\nu = \frac{r_1 \cdot r_2}{r_1\ \ r_2}$$

dm = + (Short Way) or (Long Way)

VarA = $\text{dm}\sqrt{r_1 r_2 (1 + \cos(\Delta\nu))}$. If VarA = 0.0, the orbit is not possible

Guess $Z_0 = 0.0$, therefore C = 1/2 and S = 1/6

Set bounds: Upper = $4\pi^2$ and Lower = -4π

BEGIN LOOP:

$$y_n = r_1 + r_2 - \frac{\text{VarA}(1 - Z_0 S)}{\sqrt{C}}$$

Check if VarA > 0.0 and y < 0.0, then re-adjust lower bound of Z until y > 0.0

$$x_0 = \sqrt{\frac{y_n}{C}}$$

$$t = \frac{x_0^3 S + \text{VarA}\sqrt{y_n}}{\sqrt{\mu}}$$

If $t \le t_0$, reset lower bound = Z_0

else

If $t > t_0$, reset upper bound = Z_0

$$Z_n = \frac{\text{upper} + \text{lower}}{2}$$

Calculate C and S:

If $Z_n \le 0.0$

$$C = \frac{1}{2!} - \frac{Z_n}{4!} + \frac{Z_n^2}{6!} - \frac{Z_n^3}{8!} + \frac{Z_n^4}{10!} - \frac{Z_n^5}{12!} \qquad S = \frac{1}{3!} - \frac{Z_n}{5!} + \frac{Z_n^2}{7!} - \frac{Z_n^3}{9!} + \frac{Z_n^4}{11!} - \frac{Z_n^5}{13!}$$

If $Z_n \le 0.0$

$$C = \frac{1 - \cos(\sqrt{Z_n})}{Z_n} \qquad S = \frac{\sqrt{Z_n} - \sin(\sqrt{Z_n})}{\sqrt{Z_n^3}}$$

$Z_o = Z_n$

Check if the first guess is too close

UNTIL $|t - t_o| < 0.000001$

END LOOP

Evaluate f and g coefficients:

$$f = 1 - \frac{y_n}{r_1}, \qquad g = \text{VarA}\sqrt{\frac{y_n}{\mu}}$$

$$v_1 = \frac{r_2 - f r_1}{g}$$

$$\dot{g} = 1 - \frac{y_n}{r_2}$$

$$v_2 = \frac{\dot{g} r_2 - r_1}{g}$$

Appendix D

Proposed Three-Line Element Set Definition

Table D-1: Example of Current Two-line (TLE) and Proposed Three-Line (3LE) Element Sets.

Current TLE Example	ISS (ZARYA) 1 25544U 98067A 04236.56031392 .00020137 00000-0 16538-3 0 9993 2 25544 51.6335 344.7760 0007976 126.2523 325.9359 15.70406856328903 --
Reference line	12345678901234567890123456789012345678901234567890123456789012345678901234567890123456789 1 2 3 4 5 6 7
New TLE	ISS (ZARYA) 1 000025544U 19980067A___ 2004236.56031392 .00020137 .00000-0 .16538-3 0XXX3 2 000025544U 51.6335 344.7760 .0007976 126.2523 325.9359 15.70406856 _32890 3
Notional third line of 3LE	3 000025544SFVEY __999 AAAAA BBBBBBBBBB CDDEFGGGG HHHHHHHH IIII J KKKKKKK 3

Table D-2: Position and Description Key for 3LE .

Current Columns	Current Example	New Columns	New Example	Description
Line 1				
1	1	1	1	Line Number
3–7	25544	3–11	000025544	Satellite Catalog Number
8	U	12	U	Elset Classification – Original; Elset Line Classification – New
10–17	98067A (with 2 blank spaces following)	14–25	19980067A (with 3 blank spaces following)	International Designator (Year, # launch of year, piece)
19–32	04236.56031392	27–42	2004236.56031392	Element Set Epoch (UTC)
34–43	(leading blank space for sign) .00020137	44–53	(leading blank space for sign) .00020137	1st Derivative of the Mean Motion with Respect to Time
45–52	(leading blank space for sign) 00000-0	55–63	(leading blank space for sign) .00000-0	2nd Derivative of the Mean Motion with Respect to Time (decimal point assumed in original), or Solar Radiation Pressure (Cr*A/m or "AGOM")

Table D-2: Position and Description Key for 3LE (Continued).

Current Columns	Current Example	New Columns	New Example	Description
54–61	(leading blank space for sign) 16538-3	65–73	(leading blank space for sign) .16538-3	B* Drag Term (decimal point assumed in original)
63	0	75	0	Element Set Type
65–68	999		[MOVED TO THIRD LINE FOR OPSEC REASONS]	Element Number
		76	[SINGLE LETTER – TBD]	Object Type
		77	[SINGLE LETTER – TBD]	Extrapolation Theory Used to Generate Element Set, Indicator for Whether Columns 55–63 Represent AGOM or 2nd Derivative of Mean Motion
		78	[SINGLE LETTER – TBD]	RCS Size Bin
69	3	79	3	Checksum

Table D-2: Position and Description Key for 3LE (Continued).

Current Columns	Current Example	New Columns	New Example	Description
Line 2				
1	2	1	2	Line Number
3–7	25544	3–11	000025544	Satellite Catalog Number
		12	U	Elset Line Classification
9–16	(leading blank space) 51.6335	14–21	(leading blank space) 51.6335	Orbit Inclination (degrees)
18–25	344.7760	23–30	344.7760	Right Ascension of Ascending Node (degrees)
27–33	0007976	32–39	.0007976	Eccentricity (decimal point assumed in original)
35–42	126.2523	41–48	126.2523	Argument of Perigee (degrees)
44–51	325.9359	50–57	325.9359	Mean Anomaly (degrees)
53–63	15.70406856	59–69	15.70406856	Mean Motion (revolutions/day)
64–68	32890	71–76	(new leading blank space) 32890	Revolution Number at Epoch
69	3	79	3	Checksum

Table D-3: Position and Description Key for Third Line of 3LE .

Columns	Notional Content	Description
Line 3		
1	2	Line Number
3–11	000025544	Satellite Catalog Number
12	S	Elset Line Classification
13–16	FVEY	Elset Line Distribution Marking
18–22	#####	Elset Number
24–28	#.#+/–E	Last Calculated Error Growth Rate (in km/day)
30–39	+/–ddd.ddddd	Time of Last Observation (used) Offset Relative to Epoch (in +/– fractional days)
41	[Single Letter – TBD]	Object Status (table lookup)
42, 43	[Two Letters – TBD]	Primary Payload Mission Code (table lookup)
44	[Single Letter – TBD]	Primary Mission Status (table lookup)
45	[Single Letter – TBD]	Maneuverability Flag (Y/N/U; could develop table at later date)
46–49	[Three- or Four-Letter Code – TBD]	Payload Owner/Operator; country code for single entities, consortium designator for non-single entities (table lookup)
51–58	#.###+/–E_	Median Radar Cross Section (in meters squared, followed by character designating radar frequency type used)
60–63	##.#	Median Visual Magnitude of Object Scaled to 36,000 km

Table D-3: Position and Description Key for Third Line of 3LE (Continued).

Columns	Notional Content	Description
65	[SINGLE LETTER – TBD]	Tumble Rate (binned, table lookup)
67–73	#.###+/–E	Drag Coefficient
74–78		[Left Blank for Future Use]
79	3	Checksum

INDEX

F

G

H

I

U

V

Z